# Heath Mathematics

Walter E. Rucker • Clyde A. Dilley • David W. Lowry

D. C. Heath and Company
Lexington, Massachusetts Toronto

**Illustrations:** Leo Abbett, Maryann Cocca, Chris Czernota, James R. Hamilton, Len Reno, Andrew Shiff, Maxie Chambliss, Lynn Titleman, George Ulrich, Joe Veno/Gwen Goldstein

**Photography:** Jonathan Barkan: 53; Bevilacqua/Sygma: 242; Fredrik D. Bodin: 59, 190; Andrew Brilliant/Carol Palmer: 9, 28, 114 (top), 171, 240, 301, 317, 337; © Rene Burri/Magnum: 1; Tim Carlson/Stock, Boston: 10, 11; E.R. Degginger/ Slidemakers: 133; Focus on Sports: 254, 255; F.P.G./Thomas Zimmerman: 227; Jonathan Goell: 93, 152, 153, 300; Grant Heilman/Grant Heilman Photography: 156, 157; I.B.M./Sanders H. Millens: 89; Paul Johnson: 32, 73, 114 (bottom), 115, 237, 253, 279; Talbot D. Lovering: 97, 285; Tom Magno: 6, 19, 88, 94, 98, 112, 113, 163, 208, 225, 247, 258, 268, 269, 290, 298, 305, 308, 325; Mike Malyszko: 91; The Metropolitan Museum of Art, purchase Gebhard Gourgand Foundation Gift, 1973: 205; Morse Photography: Cover, Title page; Julie O'Neil: 124, 125, 186, 187, 194, 201, 220, 221, 337, 341; Schwinn Bicycle Company: 127; Ozzi Smith/Shostal Associates: 51; Deidra Delano Stead: 34, 70, 196, 197, 203, 251, 281, 292, 321; Jerry Wachter/Focus on Sports: 184; Cary Wolinsky/Stock, Boston: 23



Published simultaneously in Canada.

Printed in the United States of America.

International Standard Book Number: 0-669-07417-9

1

**Place Value** 1–22

Thousands • Comparing numbers • Rounding to the nearest ten or hundred • Rounding to the nearest thousand or ten thousand • Millions • Billions • Roman numerals

**Chapter Checkup • Chapter Project • Chapter Review • Chapter Challenge • Major Checkup**

2

**Addition and Subtraction** 23–58

Addition facts • More than two addends • Adding 2-digit numbers • Adding 3-digit numbers • Adding larger numbers • Adding 3 or more numbers • Subtraction facts • Subtracting 2-digit numbers • Subtracting 3-digit numbers • Zeros in subtraction • Subtracting larger numbers • Computers in banking • Skill Game • Problem solving—steps and too much information

**Chapter Checkup • Chapter Project • Chapter Review • Chapter Challenge • Major Checkup**

3

**Multiplication** 59–96

Multiplication facts • Multiples and common multiples • Multiplying 2- and 3-digit numbers • Multiplying larger numbers • Problem solving • Multiplying by 10 and 100 • Multiplying by multiples of 10 and 100 • Multiplying by a 2-digit number • Skill Game • Multiplying by a 3-digit number • Computers store information • Problem solving—two or more steps

**Chapter Checkup • Chapter Project • Chapter Review • Chapter Challenge • Major Checkup**

4

**Division** 97–132

Division facts • Factors and common factors • Problem solving—choose the operation • Dividing a 2-digit number • Division with regrouping • Dividing greater numbers • Zeros in division • Finding an average • Dividing by a 2-digit number • Addition, subtraction, multiplication, and division • Skill Game • Problem solving • Problem solving—not enough information

**Chapter Checkup • Chapter Project • Chapter Review • Chapter Challenge • Major Checkup**

5

**Geometry** 133–162

Ready or Not? • Points and segments • Rays, angles, and lines • Measuring angles • Kinds of angles • Parallel and perpendicular lines • Circles • Congruent figures • Line of symmetry • Triangles and quadrilaterals • Graphing number pairs • Problem solving—organizing and interpreting data

**Chapter Checkup • Chapter Project • Chapter Review • Chapter Challenge • Major Checkup**

6

**Fractions—Addition and Subtraction** 163–204

Ready or Not? • Using fractions • Equivalent fractions • Lowest terms • Skill Game • Comparing fractions • Adding • Adding, unlike denominators • Subtracting • Subtracting, unlike denominators • Problem solving • Whole numbers, mixed numbers, and fractions • Ratios • Probability • Problem solving—using graphs and maps

**Chapter Checkup • Chapter Project • Chapter Review • Chapter Challenge • Major Checkup**

7 **Fractions—Multiplication and Division** 205–226

Ready or Not? • Fraction of a number • Using mixed numbers • Multiplying fractions • Dividing fractions • Problem solving—reading a pictograph • Problem solving

**Chapter Checkup • Chapter Project • Chapter Review • Chapter Challenge • Major Checkup**

8 **Decimals—Addition and Subtraction** 227–250

Ready or Not? • Tenths and hundredths • Thousandths • Comparing and ordering decimals • Rounding to the nearest whole number • Adding decimals • Subtracting decimals • More about adding and subtracting decimals • Computer timing • Skill Game • Problem solving

**Chapter Checkup • Chapter Project • Chapter Review • Chapter Challenge • Major Checkup**

9 **Measurement** 251–284

Time • Millimeters and centimeters • Meters and kilometers • Perimeter • Area • Volume • Problem solving—drawing pictures • Liquid volume and weight—metric • Inch • Inch, foot, yard, and mile • Perimeter, area, and volume • Liquid volume and weight • Problem solving

**Chapter Checkup • Chapter Project • Chapter Review • Chapter Challenge • Major Checkup**

10 **Consumer Mathematics** 285–304

Making change • Computing prices • Computing discounts • Sales tax • Computer service • Sales

**Chapter Checkup • Chapter Project • Chapter Review • Chapter Challenge • Major Checkup**

11 **Mixed Numbers—Addition and Subtraction** 305–324

Ready or Not? • Fractions and mixed numbers • Adding and subtracting mixed numbers • More about adding mixed numbers • More about subtracting mixed numbers • Skill Game • Problem solving

**Chapter Checkup • Chapter Project • Chapter Review • Chapter Challenge • Major Checkup**

12 **Decimals—Multiplication and Division** 325–344

Ready or Not? • Multiplying decimals • Dividing a decimal • Estimating • Skill Game • Problem solving—more than one way to solve

**Chapter Checkup • Chapter Project • Chapter Review • Chapter Challenge • Major Checkup**

**Getting Ready for Computer Programming** 346–355

**Skill Test** 356–359

**Extra Practice** 360–378

**Extra Problem Solving** 379–389

**Glossary** 390–393

**Index** 394–396

# Place Value

# 1

# Thousands

Below are three ways to write about the number of blocks.

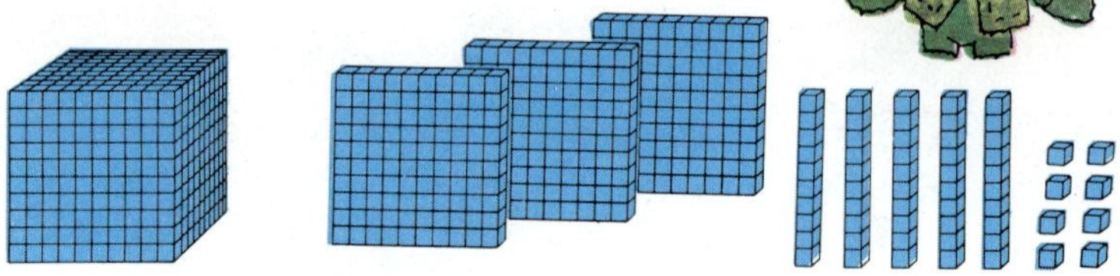

| Thousands | Hundreds | Tens | Ones |
|---|---|---|---|
| 1 | 3 | 5 | 8 |

**standard numeral** **1358**

The 1 stands for 1 thousand, or 1000.

The 3 stands for 3 hundreds, or 300.

The 5 stands for 5 tens, or 50.

The 8 stands for 8 ones, or 8.

**words** one thousand three hundred fifty-eight

## EXERCISES

**How many blocks? Give the standard numeral.**

**1.**

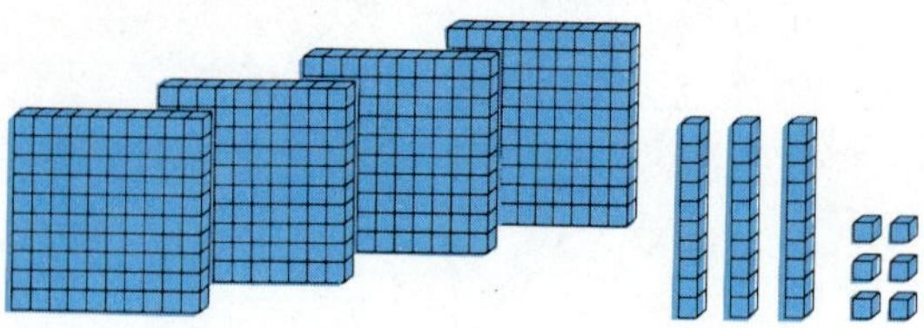

**2.**

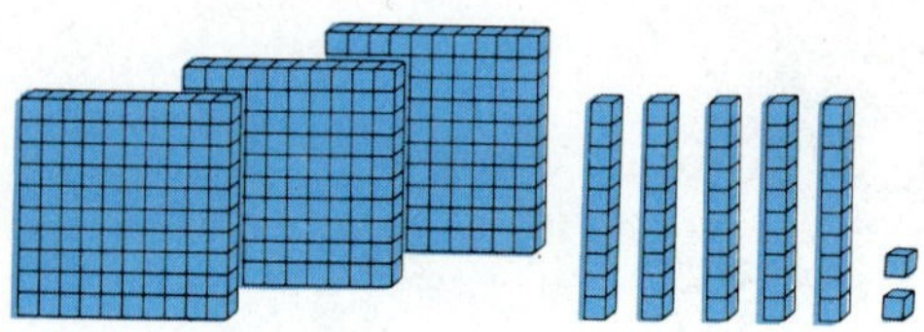

**3.**

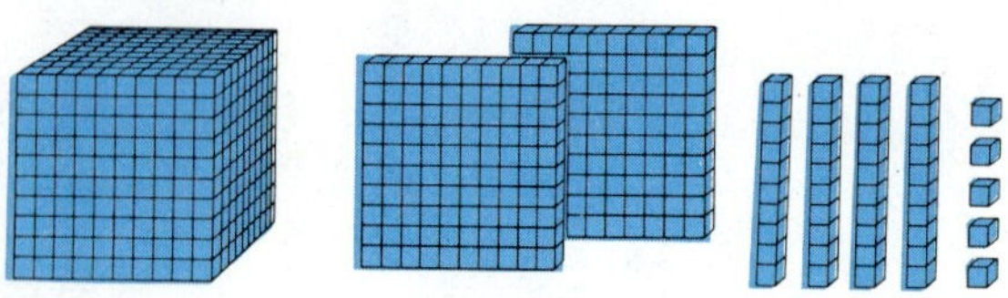

**4.**

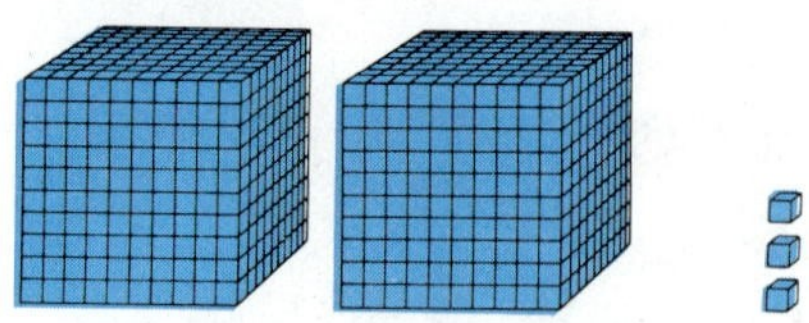

**What does the red digit stand for?**
**Give two answers.**

5. 8364

3 hundreds
300

6. 3821
7. 7564
8. 5266
9. 4278
10. 9037
11. 7428
12. 3569
13. 4386
14. 5271

**Build the standard numeral.**

15. 2 in the tens place
    8 in the ones place
    3 in the thousands place
    7 in the hundreds place

16. 5 in the hundreds place
    9 in the ones place
    0 in the tens place
    6 in the thousands place

17. 3 tens
    4 thousands
    9 hundreds
    6 ones

18. 5 hundreds
    2 ones
    7 thousands
    1 ten

**Write the standard numeral.**

19. four hundred sixty-three
20. nine hundred forty
21. one thousand five hundred
22. three thousand two hundred twelve
23. seven thousand two hundred sixty-one
24. four thousand two hundred ninety
25. eight thousand seventy
26. nine thousand six

We can also write **expanded numerals.**

3475 = 3000 + 400 + 70 + 5

**Write as expanded numerals.**

27. 586
28. 698
29. 746
30. 839
31. 2816
32. 3574
33. 9281
34. 6453

# More about thousands

| Thousands | | | | | |
|---|---|---|---|---|---|
| Hundreds | Tens | Ones | Hundreds | Tens | Ones |
| 2 | 8 | 1 | 4 | 3 | 6 |

We can use a comma to separate the thousands.

281,436

- 6 ones, or 6
- 3 tens, or 30
- 4 hundreds, or 400
- 1 thousand, or 1000
- 8 ten thousands, or 80,000
- 2 hundred thousands, or 200,000

Here are two more ways to write the number.

281 thousand, 436

two hundred eighty-one thousand, four hundred thirty-six

**EXERCISES**

**Which digit is in the** **357,168**

1. thousands place?
2. ones place?
3. tens place?
4. ten thousands place?
5. hundreds place?
6. hundred thousands place?

**What does the red digit stand for?**
**Give two answers.**

7. 382,719

8 ten thousands
80,000

8. 91,692

9. 528,314

10. 291,754

11. 362,846

12. 555,555

13. 444,444

**Write the standard numeral.**

14. 26 thousand, 219

15. 453 thousand, 506

16. forty-five thousand, two hundred seventy-four

17. two hundred fifty-six thousand, five hundred twenty

18. six hundred thirty thousand, three hundred six

19. two hundred thousand, two hundred

20. two hundred two thousand

21. two hundred thousand, two

**Write the next five numbers.**

22. 2017

23. 1998

24. 45,397

25. 217,358

26. 309,996

27. 799,998

**Write each number in expanded form.**

42356 = 40000 + 2000 + 300 + 50 + 6

28. 38,921

29. 57,496

30. 35,142

★ 31. 238,145

★ 32. 726,574

★ 33. 953,884

# Comparing numbers

Here is how to compare a guess to the correct number of marbles.

| | Guess | Number of marbles |
|---|---|---|
| 1. Start at the left. Compare the digits in the ten thousands place. They are the same. | **12,426** | **13,010** |

2. Compare the digits in the thousands place: $2 < 3$.

**12,426** $<$ **13,010**
is less than

**13,010** $>$ **12,426**
is greater than

Other examples.

**749,263** $<$ **749,402**

**593,468** $>$ **593,439**

**EXERCISES**

**Less than (<) or greater than (>)?**

1. 532 ● 534
2. 682 ● 79
3. 742 ● 651
4. 6905 ● 6991
5. 5934 ● 5926
6. 8342 ● 8432
7. 783 ● 2873
8. 6914 ● 6710
9. 3999 ● 4000
10. 56,984 ● 6884
11. 34,291 ● 35,384
12. 72,916 ● 75,388
13. 26,593 ● 26,504
14. 391,782 ● 390,700
15. 458,371 ● 453,871
16. 100,000 ● 99,999

| *River* | *Length in Kilometers* |
|---|---|
| Arkansas | 2333 |
| Mississippi | 3778 |
| Missouri | 3725 |
| Ohio | 1578 |
| Red | 1855 |
| Rio Grande | 3033 |
| Snake | 1670 |

**Solve.**

17. Which river is longest?
18. Which river is shortest?
19. Which river is the second longest?
20. How many of the rivers are longer than 2500 kilometers?
21. How many of the rivers are longer than the Rio Grande?
22. How many of the rivers are shorter than 3000 kilometers?

★ 23. Arrange the lengths in order from longest to shortest.

# Rounding a number to the nearest ten or hundred

Here is how to round numbers to the nearest ten or nearest hundred.

**EXAMPLE 1.** Round 4128 to the nearest ten.

4128 is between 4120 and 4130. It is nearer 4130. So, round up to 4130.

**EXAMPLE 2.** Round 6742 to the nearest hundred.

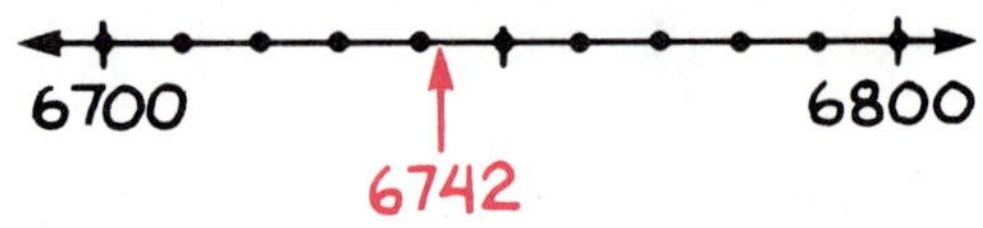


6742 is between 6700 and 6800. It is nearer 6700. So, round down to 6700.

| When a number is halfway between, round up. |
|---|

**EXAMPLE 3.** Round 1465 to the nearest ten.

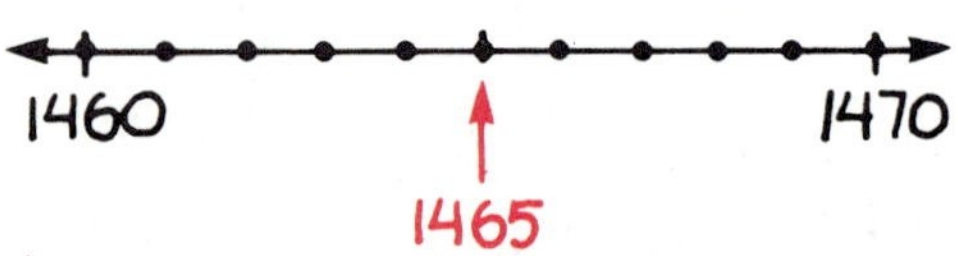


1465 is halfway between 1460 and 1470. So, round up to 1470.

**EXAMPLE 4.** Round 9250 to the nearest hundred.

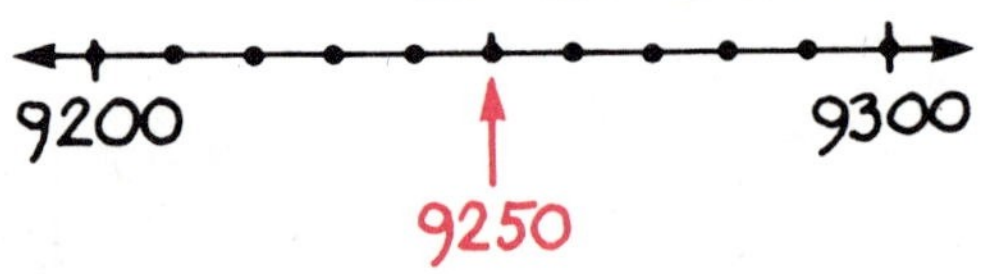


9250 is halfway between 9200 and 9300. So, round up to 9300.

**EXERCISES**

**Round to the nearest ten.**

**1.** 28 **2.** 37 **3.** 75 **4.** 86 **5.** 94

**6.** 109 **7.** 173 **8.** 255 **9.** 382 **10.** 416

**11.** 563 **12.** 608 **13.** 715 **14.** 851 **15.** 995

**16.** 1328 **17.** 2653 **18.** 4802 **19.** 3777 **20.** 5859

**Round to the nearest hundred.**

| | | | | |
|---|---|---|---|---|
| **21.** 418 | **22.** 529 | **23.** 375 | **24.** 650 | **25.** 738 |
| **26.** 936 | **27.** 853 | **28.** 719 | **29.** 591 | **30.** 950 |
| **31.** 1348 | **32.** 1685 | **33.** 1911 | **34.** 2250 | **35.** 2846 |
| **36.** 3873 | **37.** 4146 | **38.** 5220 | **39.** 6714 | **40.** 7992 |

Round $3.87 to the nearest dollar.

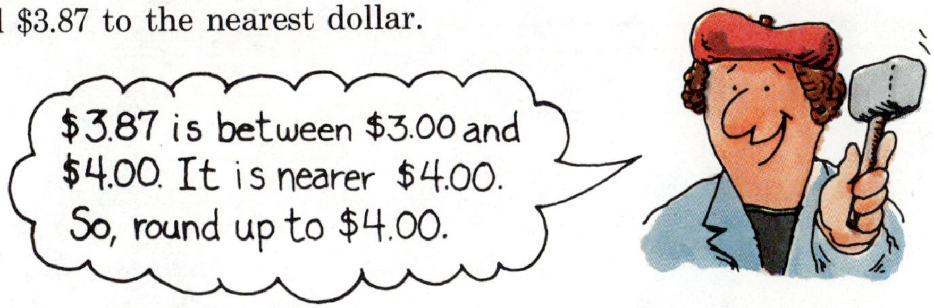


**Round to the nearest dollar.**

| | | | | |
|---|---|---|---|---|
| **41.** $4.16 | **42.** $2.75 | **43.** $8.37 | **44.** $5.81 | **45.** $7.93 |
| **46.** $16.39 | **47.** $27.09 | **48.** $41.53 | **49.** $87.16 | **50.** $9.71 |
| **51.** $30.42 | **52.** $49.68 | **53.** $12.50 | **54.** $.59 | **55.** $.32 |

## *Challenge!*

**Guess my number.**

# Rounding a number to the nearest thousand or ten thousand

The *exact* seating capacity of the Georgia Tech Stadium is

**59,245**

Here is how to round this number to the nearest thousand or ten thousand.

**EXAMPLE 1.** Round 59,245 to the nearest thousand.

59,245 is between 59,000 and 60,000. It is nearer 59,000. So, round down to 59,000.

**EXAMPLE 2.** Round 59,245 to the nearest ten thousand.

59,245 is between 50,000 and 60,000. It is nearer 60,000. So, round up to 60,000.

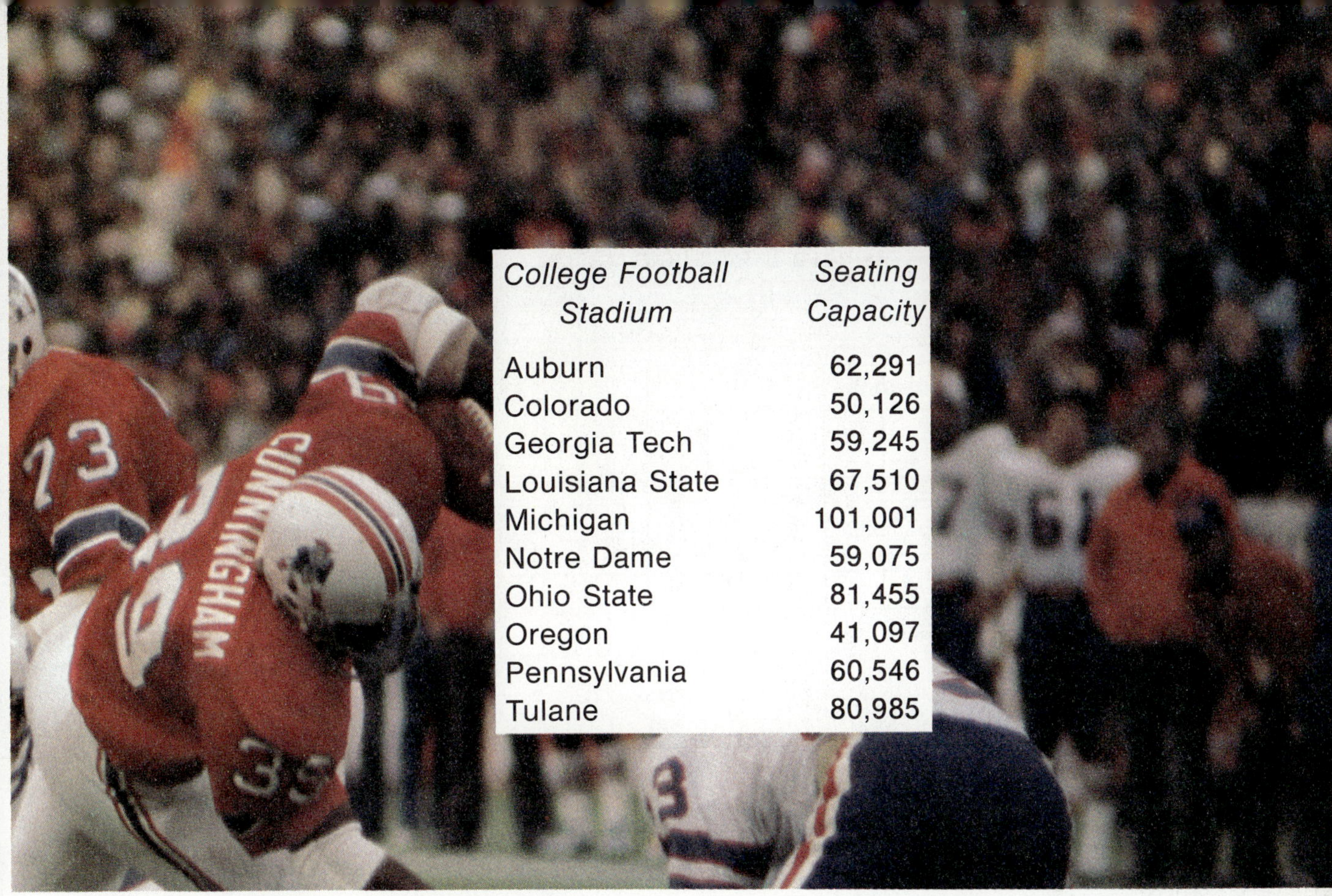


| College Football Stadium | Seating Capacity |
|---|---|
| Auburn | 62,291 |
| Colorado | 50,126 |
| Georgia Tech | 59,245 |
| Louisiana State | 67,510 |
| Michigan | 101,001 |
| Notre Dame | 59,075 |
| Ohio State | 81,455 |
| Oregon | 41,097 |
| Pennsylvania | 60,546 |
| Tulane | 80,985 |

## EXERCISES

1. Which stadium has the greatest seating capacity?
2. Which has the least?
3. How many stadiums can hold more than 75,000?
4. Round the seating capacity of each stadium to the nearest thousand.
5. Round the seating capacity of each stadium to the nearest ten thousand.
6. Arrange the seating capacities in order from greatest to least.
7. Our stadium seats more than Colorado but fewer than Georgia Tech. What college do I play for?

# Millions

A million is 1000 thousands.

On your tenth birthday you have lived about this many seconds:

| Millions | | | Thousands | | | | | |
|---|---|---|---|---|---|---|---|---|
| Hundreds | Tens | Ones | Hundreds | Tens | Ones | Hundreds | Tens | Ones |
| 3 | 1 | 5 | 5 | 3 | 2 | 8 | 0 | 0 |

We start at the right and use commas to set off groups of three digits, called periods. The commas separate the thousands and millions.

**315,532,800**

315 million, 532 thousand, 800

three hundred fifteen million, five hundred thirty-two thousand, eight hundred

**EXERCISES**

**Read aloud.**

1. 8,234,512
2. 5,067,825
3. 7,208,310
4. 52,259,186
5. 73,429,381
6. 60,716,429
7. 293,160,824
8. 301,298,634
9. 597,005,320
10. 400,290,067
11. 502,820,006
12. 840,312,968

**Give the digit that is in the** **463,572,100**

13. tens place.
14. hundred thousands place.
15. hundreds place.
16. ten millions place.
17. thousands place.
18. ten thousands place.
19. millions place.
20. hundred millions place.

**Write the standard numeral.**

21. eight million, two hundred forty-five thousand, nine hundred five
22. sixty-two million, seven hundred forty-three thousand, two hundred eighty-four
23. one hundred eighty-one million, five hundred eleven thousand, two hundred
24. six hundred nineteen million, three hundred sixty-two thousand, twenty-five

## *Challenge!*

**True or false?**

25. By your second birthday you had lived more than 60 million seconds.
26. By your sixth birthday you had lived fewer than 200 million seconds.
27. By now you have lived more than 250 million seconds.
28. At some time between your fourth and fifth birthdays you had lived exactly 150 million seconds.

| Your birthday | Your approximate age in seconds |
|---|---|
| First | 31,557,600 |
| Second | 63,115,200 |
| Third | 94,672,800 |
| Fourth | 126,230,400 |
| Fifth | 157,788,000 |
| Sixth | 189,345,600 |
| Seventh | 220,903,200 |
| Eighth | 252,460,800 |
| Ninth | 284,018,400 |
| Tenth | 315,576,000 |

# Billions

Would you believe . . .

The United States is less than 7 billion seconds old!

A billion is 1000 million.
A billion is a *very* large number.

| Billions | | | Millions | | | Thousands | | | | | |
|---|---|---|---|---|---|---|---|---|---|---|---|
| H | T | O | H | T | O | H | T | O | H | T | O |
| 3 | 8 | 2 | 5 | 1 | 6 | 3 | 0 | 5 | 4 | 1 | 7 |

**382,516,305,417**

382 billion, 516 million, 305 thousand, 417

Three hundred eighty-two billion, five hundred sixteen million,
three hundred five thousand, four hundred seventeen

**EXERCISES**

**Read aloud.**

1. 6,382,534
2. 29,782,568
3. 583,297,156
4. 9,348,291,765
5. 8,429,635,218
6. 6,382,491,708
7. 25,296,481,107
8. 98,364,273,081
9. 63,829,543,000
10. 237,642,812,438
11. 400,360,298,000
12. 370,400,000,000
13. 863,000,000,000
14. 890,000,000,000
15. 604,003,200,010
16. 502,060,300,020
17. 700,040,000,330
18. 200,002,020,200

**Tell which digit is in the** **538,264,527,019**

19. millions place.
20. billions place.
21. thousands place.
22. ten millions place.
23. hundred billions place.
24. hundred thousands place.
25. hundred millions place.
26. ten billions place.

**Write the standard numeral.**

27. 325 billion, 426 million, 843
28. 247 billion
29. 86 billion, 926 million
30. sixty-seven billion, three hundred million, one thousand, seventy-five
31. two hundred billion, five hundred twelve million, forty-two thousand, fifty-eight

**Round to the nearest million.**

32. 8,319,406
33. 34,952,164
34. 258,500,000

**Round to the nearest billion.**

35. 7,861,752,000
36. 59,261,493,280
37. 356,281,743,500

Sometimes commas are not written. Then you have to put them in yourself.

**Copy these and write in the commas.**

38. 4700000
39. 56307014
40. 5734168900
41. 89314632166
42. 483914256
43. 1748265443
44. 316284391052
45. 370800591325

# Roman numerals

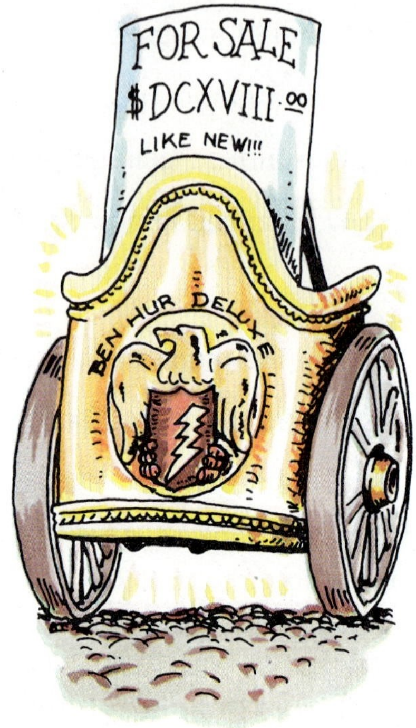


The Romans of long ago used letters to write numerals.

Here are the symbols they used for the numbers 1 through 10.

| 1 | 2 | 3 | 4 | 5 | 6 | 7 | 8 | 9 | 10 |
|---|---|---|---|---|---|---|---|---|---|
| I | II | III | IV | V | VI | VII | VIII | IX | X |

The Romans did not use a place-value system. They wrote the symbols side by side and usually added the values.

1+1+1     5+1+1

III = 3     VII = 7     XXV = 25

XV = 15     XVIII = 18     XXVI = 26

The Romans had a special rule.

> If a symbol for a smaller number is written before a symbol for a larger number, subtract.

IV = 4     IX = 9     XIV = 14     XXIX = 29

5-1     10-1

Here are some more symbols that were used by the Romans.

| 50 | 100 | 500 | 1000 |
|---|---|---|---|
| L | C | D | M |

XL = 40     XC = 90     DCXIV = 614     MCD = 1400

## EXERCISES

**Write standard numerals.**

| | | | | |
|---|---|---|---|---|
| **1.** III | **2.** V | **3.** VII | **4.** IX | **5.** XVI |
| **6.** XXXV | **7.** XXIII | **8.** XL | **9.** LX | **10.** XLV |
| **11.** LXVII | **12.** XC | **13.** CX | **14.** CLXV | **15.** CCL |
| **16.** DC | **17.** CD | **18.** DCL | **19.** CDLXXV | **20.** DCXLI |

**Write Roman numerals.**

| | | | | |
|---|---|---|---|---|
| **21.** 9 | **22.** 14 | **23.** 16 | **24.** 45 | **25.** 58 |
| **26.** 73 | **27.** 91 | **28.** 140 | **29.** 282 | **30.** 356 |
| **31.** 400 | **32.** 468 | **33.** 527 | **34.** 802 | **35.** 920 |

## *Challenge!*

**Write the dates of these famous inventions in standard numerals.**

**36.** Card-punch Computer

MDCCCLXX

**37.** Adding Machine

MDCXLII

**38.** Slide Rule

MDCXX

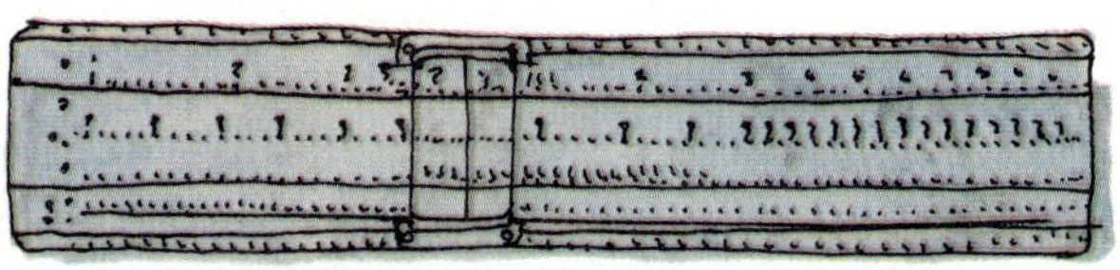

**39.** Personal Computer

MCMLXXVI

# CHAPTER CHECKUP

**What does the red digit stand for?**
**Give two answers.** [pages 2–5]

**1.** 8539 **2.** 71,642 **3.** 46,983

**4.** 273,416 **5.** 803,251 **6.** 669,382

< **or** >**?** [pages 6–7]

**7.** 382 ● 392 **8.** 425 ● 415

**9.** 8964 ● 900 **10.** 58,132 ● 59,423

**11.** 7803 ● 7964 **12.** 637,812 ● 637,714

[pages 8–11]

**Round**

**13.** 3756 to the nearest ten.

**14.** 28,165 to the nearest thousand.

**15.** 494,253 to the nearest hundred.

**16.** 665,000 to the nearest ten thousand.

[pages 12–13]

**In 463,198,027, which digit is in the**

**17.** millions place? **18.** ten thousands place?

**19.** thousands place? **20.** ten millions place?

**21.** hundred thousands place? **22.** hundred millions place?

**Write as standard numerals.** [pages 16–17]

**23.** XXXIV **24.** LXXXVII **25.** MDC **26.** MCDXCII

**Write as Roman numerals.** [pages 16–17]

**27.** 42 **28.** 96 **29.** 354 **30.** 1776

# CHAPTER PROJECT

1. Make these digit cards:

   3 4 2 5

2. How many 2-digit numbers can you build with these digits?

   4 2

3. How many 2-digit numbers can you build with these digits?

   5 4 2

4. How many 3-digit numbers can be built with these digits?

   5 4 2

5. How many 4-digit numbers can be built with these digits?

   3 5 2 4

# CHAPTER REVIEW

**3258**

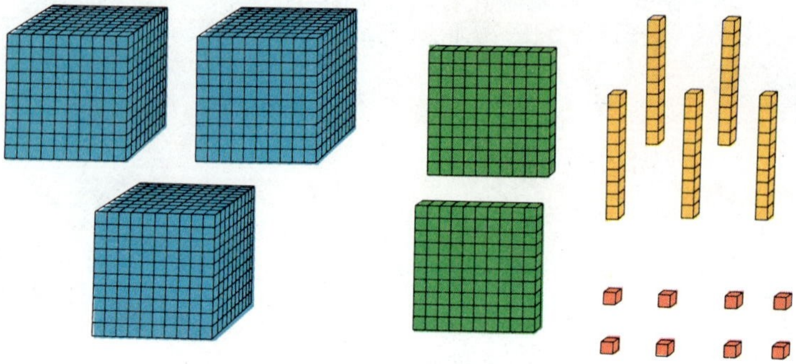

1. The 3 stands for 3 _?_.
2. The 2 stands for 2 _?_.
3. The 5 stands for 5 _?_.
4. The 8 stands for 8 _?_.
5. There are _?_ blocks.
6. The number of blocks (in words) is three thousand two hundred _?_.

**Less than (<) or greater than (>)?**

7. 3258 ● 3285
8. 6001 ● 5968

**Round**

9. 3284 to the nearest thousand.
10. 3284 to the nearest ten.

| Millions | | | Thousands | | | | | |
|---|---|---|---|---|---|---|---|---|
| Hundreds | Tens | Ones | Hundreds | Tens | Ones | Hundreds | Tens | Ones |
| 7 | 1 | 8 | 3 | 4 | 6 | 2 | 0 | 5 |

**Which digit is in the**

11. ten thousands place?
12. ten millions place?
13. hundred thousands place?
14. hundred millions place?

# CHAPTER CHALLENGE

More than 5000 years ago, the Egyptians developed a system for writing about numbers.

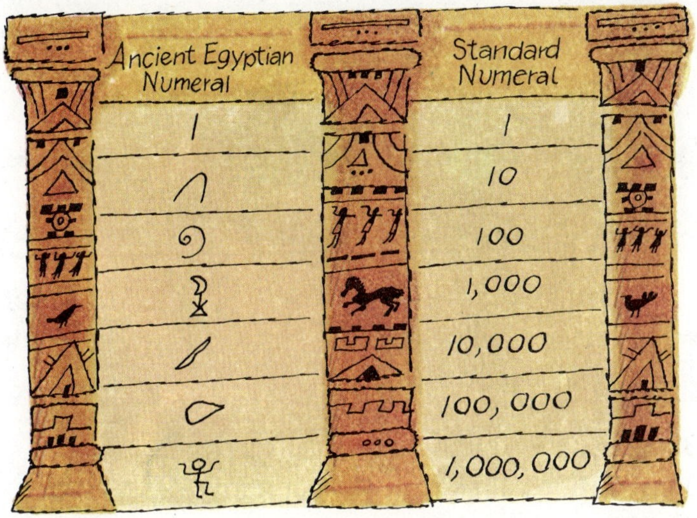


The Egyptians simply combined the symbols and added their values. Since they did not have a place-value system, they could write the symbols in any order.

**18 =** 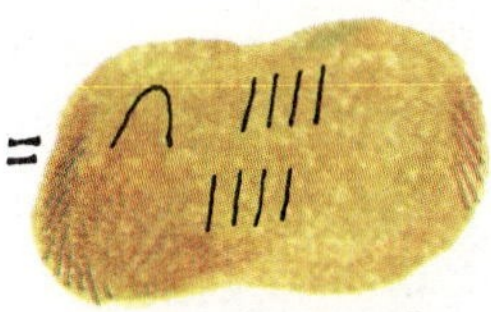 **123 =** 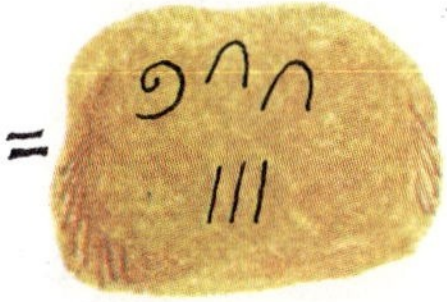 **4362 =** 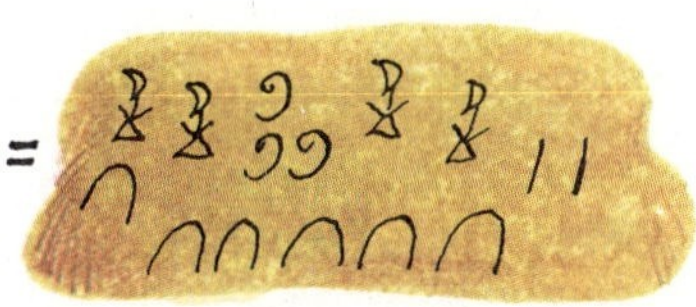

**Write, using Egyptian numerals.**

**1.** 6 **2.** 34 **3.** 201 **4.** 524

**5.** 1241 **6.** 24,352 **7.** 242,100 **8.** 1,240,000

**Give standard numerals.**

**9.**  **10.**  **11.**  **12.** 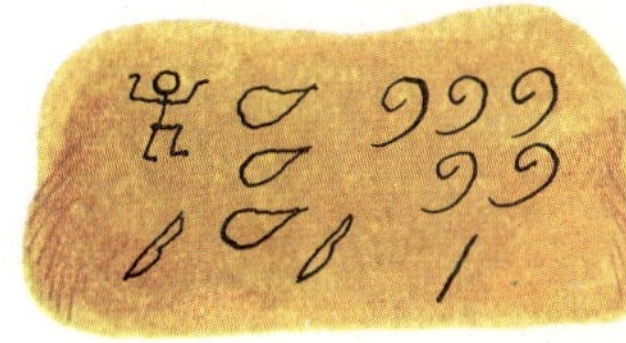

# MAJOR CHECKUP

STANDARDIZED FORMAT

Choose the correct letter.

1. In 9634 the 9 stands for
   a. 9 tens
   b. 9 thousands
   c. 9 hundreds
   d. none of these

2. In 8216 the 2 stands for
   a. 20
   b. 2000
   c. 200
   d. none of these

3. In 374,927 the digit in the ten thousands place is
   a. 2
   b. 7
   c. 4
   d. none of these

4. The number 100,000 greater than 638,125 is
   a. 639,125
   b. 638,225
   c. 637,125
   d. none of these

5. Which number is less than 82,156?
   a. 735,216
   b. 83,000
   c. 82,099
   d. none of these

6. Which number is greatest?
   a. 567,834
   b. 576,834
   c. 574,834
   d. 577,834

7. 38,265 rounded to the nearest thousand is
   a. 40,000
   b. 38,000
   c. 39,000
   d. none of these

8. 255,000 rounded to the nearest ten thousand is
   a. 260,000
   b. 250,000
   c. 255,000
   d. none of these

9. In 38,649,254 the 8 is in the
   a. thousands place
   b. hundred thousands place
   c. millions place
   d. none of these

10. The number 10 million greater than 694,831,725 is
    a. 695,831,725
    b. 684,831,725
    c. 704,831,725
    d. none of these

11. The Roman numeral for 46 is
    a. XLVI
    b. LXVI
    c. XLIV
    d. none of these

12. Stadium X has 64,753 seats. Stadium Y has 63,984 seats. Stadium Z has 64,735 seats. Which stadium has the most seats?
    a. X
    b. Y
    c. Z
    d. none of these

# Addition and Subtraction

# 2

# Basic addition facts

The numbers that you add are called addends. The answer is called the sum. It is important that you memorize all the basic addition facts.

$$3 + 5 = 8$$

addends (3, 5) sum (8)

**The Order Property**
Changing the order of the addends does not change the sum.

$$4 + 3 = 3 + 4$$
$$8 + 5 = 5 + 8$$

**The Adding 0 Property**
The sum of any number and 0 is that same number.

$$7 + 0 = 7$$
$$9 + 0 = 9$$

**EXERCISES**

**Give each sum.**

1. If 9 + 2 = 11, then 2 + 9 = ?.
2. If 8 + 5 = 13, then 5 + 8 = ?.
3. If 7 + 5 = 12, then 5 + 7 = ?.
4. If 9 + 6 = 15, then 6 + 9 = ?.
5. If 8 + 4 = 12, then 4 + 8 = ?.
6. 5 + 3 = ? and 3 + 5 = ?.
7. 7 + 3 = ? and 3 + 7 = ?.
8. 6 + 4 = ? and 4 + 6 = ?.

9. 6 + 0 = ?  10. 7 + 0 = ?  11. 5 + 0 = ?  12. 0 + 0 = ?

13. 9 + 0 = ?  14. 4 + 0 = ?  15. 8 + 0 = ?  16. 27 + 0 = ?

**Study the shortcut. Then use the shortcut to find each sum.**

**7 + 7 = 14**

**7 + 6 = ?**

**8 + 5 = 13**

**8 + 7 = ?**

17. If 8 + 2 = 10, then 8 + 3 = ?.

18. If 5 + 5 = 10, then 5 + 6 = ?.

19. If 8 + 2 = 10, then 8 + 4 = ?.

20. If 6 + 6 = 12, then 6 + 7 = ?.

21. If 8 + 2 = 10, then 8 + 5 = ?.

22. If 9 + 9 = 18, then 9 + 8 = ?.

23. If 7 + 3 = 10, then 7 + 4 = ?.

24. If 7 + 7 = 14, then 7 + 6 = ?.

25. If 7 + 3 = 10, then 7 + 6 = ?.

26. If 8 + 8 = 16, then 8 + 7 = ?.

27. If 5 + 5 = 10, then 5 + 7 = ?.

28. If 6 + 4 = 10, then 6 + 5 = ?.

**Find the missing addends.**

29. 8 + ? = 10  30. 8 + ? = 11  31. 8 + ? = 12

32. 7 + ? = 10  33. 7 + ? = 12  34. 7 + ? = 13

35. 9 + ? = 10  36. 9 + ? = 11  37. 9 + ? = 14

38. 9 + ? = 15  39. 9 + ? = 16  40. 9 + ? = 17

# Basic addition facts

WHAT'S YOUR SCORE?

**All the basic addition facts are listed below. See how quickly you can list the sums.**

| | | | | | | | |
|---|---|---|---|---|---|---|---|
| 1. 0 +3 | 2. 1 +0 | 3. 2 +5 | 4. 0 +8 | 5. 1 +5 | 6. 2 +8 | 7. 0 +2 | 8. 7 +5 |
| 9. 2 +1 | 10. 9 +2 | 11. 6 +3 | 12. 1 +7 | 13. 2 +4 | 14. 3 +8 | 15. 5 +8 | 16. 2 +6 |
| 17. 4 +8 | 18. 5 +4 | 19. 8 +2 | 20. 1 +4 | 21. 0 +0 | 22. 3 +4 | 23. 8 +7 | 24. 3 +7 |
| 25. 3 +2 | 26. 1 +8 | 27. 1 +6 | 28. 1 +9 | 29. 0 +4 | 30. 8 +6 | 31. 2 +9 | 32. 2 +3 |
| 33. 5 +7 | 34. 2 +7 | 35. 1 +2 | 36. 8 +4 | 37. 7 +9 | 38. 0 +5 | 39. 4 +5 | 40. 7 +4 |
| 41. 4 +0 | 42. 4 +3 | 43. 1 +3 | 44. 2 +0 | 45. 7 +8 | 46. 3 +3 | 47. 1 +1 | 48. 6 +2 |
| 49. 3 +0 | 50. 6 +0 | 51. 5 +3 | 52. 8 +5 | 53. 9 +7 | 54. 4 +7 | 55. 7 +1 | 56. 8 +3 |
| 57. 3 +6 | 58. 9 +8 | 59. 7 +3 | 60. 4 +9 | 61. 9 +1 | 62. 0 +9 | 63. 3 +9 | 64. 7 +2 |
| 65. 0 +7 | 66. 4 +6 | 67. 9 +6 | 68. 2 +2 | 69. 9 +0 | 70. 9 +5 | 71. 0 +1 | 72. 7 +7 |
| 73. 5 +2 | 74. 9 +9 | 75. 4 +2 | 76. 3 +5 | 77. 6 +6 | 78. 6 +7 | 79. 6 +8 | 80. 5 +6 |
| 81. 0 +6 | 82. 4 +1 | 83. 4 +4 | 84. 3 +1 | 85. 6 +5 | 86. 6 +4 | 87. 7 +0 | 88. 9 +3 |
| 89. 5 +0 | 90. 8 +0 | 91. 5 +9 | 92. 6 +1 | 93. 8 +8 | 94. 5 +5 | 95. 7 +6 | 96. 5 +1 |
| | | 97. 6 +9 | 98. 8 +1 | 99. 8 +9 | 100. 9 +4 | | |

**Check your answers.**

| | | | | | | | |
|---|---|---|---|---|---|---|---|
| **1.** 3 | **2.** 1 | **3.** 7 | **4.** 8 | **5.** 6 | **6.** 10 | **7.** 2 | **8.** 12 |
| **9.** 3 | **10.** 11 | **11.** 9 | **12.** 8 | **13.** 6 | **14.** 11 | **15.** 13 | **16.** 8 |
| **17.** 12 | **18.** 9 | **19.** 10 | **20.** 5 | **21.** 0 | **22.** 7 | **23.** 15 | **24.** 10 |
| **25.** 5 | **26.** 9 | **27.** 7 | **28.** 10 | **29.** 4 | **30.** 14 | **31.** 11 | **32.** 5 |
| **33.** 12 | **34.** 9 | **35.** 3 | **36.** 12 | **37.** 16 | **38.** 5 | **39.** 9 | **40.** 11 |
| **41.** 4 | **42.** 7 | **43.** 4 | **44.** 2 | **45.** 15 | **46.** 6 | **47.** 2 | **48.** 8 |
| **49.** 3 | **50.** 6 | **51.** 8 | **52.** 13 | **53.** 16 | **54.** 11 | **55.** 8 | **56.** 11 |
| **57.** 9 | **58.** 17 | **59.** 10 | **60.** 13 | **61.** 10 | **62.** 9 | **63.** 12 | **64.** 9 |
| **65.** 7 | **66.** 10 | **67.** 15 | **68.** 4 | **69.** 9 | **70.** 14 | **71.** 1 | **72.** 14 |
| **73.** 7 | **74.** 18 | **75.** 6 | **76.** 8 | **77.** 12 | **78.** 13 | **79.** 14 | **80.** 11 |
| **81.** 6 | **82.** 5 | **83.** 8 | **84.** 4 | **85.** 11 | **86.** 10 | **87.** 7 | **88.** 12 |
| **89.** 5 | **90.** 8 | **91.** 14 | **92.** 7 | **93.** 16 | **94.** 10 | **95.** 13 | **96.** 6 |
| | | **97.** 15 | **98.** 9 | **99.** 17 | **100.** 13 | | |

**101.** Your score is the number correct. What is your score?

**102.** Practice the facts that you missed. See if you can get a better (higher) score.

**103.** See how fast you can give the sums orally. Have a classmate keep a tally of your mistakes.

**104.** Use graph paper. Make a graph of your scores like the one on the computer screen.

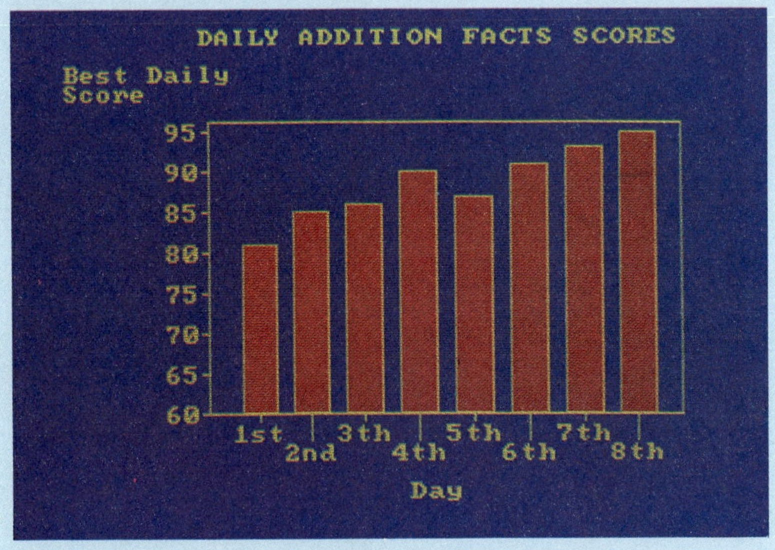

# More than two addends

**The Grouping Property**

Changing the grouping of the addends does not change the sum.

$$(3 + 2) + 4 = 3 + (2 + 4)$$

Remember to work inside the grouping symbols first.

Since you can change the order and the grouping of the addends without changing the sum, you can add numbers in any order.

Add these numbers in different orders.

Do you always get the same sum?

**EXERCISES**

**Give each sum.**

**1. a.** $(3 + 5) + 4$ **b.** $3 + (5 + 4)$

**2. a.** $(5 + 1) + 6$ **b.** $5 + (1 + 6)$

**3. a.** $(4 + 2) + 6$ **b.** $4 + (2 + 6)$

**4. a.** $(3 + 3) + 5$ **b.** $3 + (3 + 5)$

**5. a.** $(3 + 4) + 5$ **b.** $3 + (4 + 5)$

**6. a.** $(6 + 2) + 3$ **b.** $6 + (2 + 3)$

**Add.**

2 5 8

| 7. | 8. | 9. | 10. | 11. | 12. |
|---|---|---|---|---|---|
| 3 | 2 | 4 | 8 | 3 | 6 |
| 4 | 6 | 6 | 5 | 7 | 6 |
| +7 | +8 | +5 | +5 | +7 | +4 |

| 13. | 14. | 15. | 16. | 17. | 18. |
|---|---|---|---|---|---|
| 5 | 6 | 9 | 2 | 7 | 8 |
| 5 | 8 | 7 | 8 | 4 | 3 |
| +4 | +4 | +1 | +7 | +6 | +7 |

| 19. | 20. | 21. | 22. | 23. | 24. |
|---|---|---|---|---|---|
| 3 | 2 | 3 | 1 | 3 | 8 |
| 9 | 4 | 4 | 7 | 6 | 6 |
| 2 | 3 | 7 | 2 | 2 | 1 |
| +1 | +8 | +2 | +8 | +7 | +2 |

| 25. | 26. | 27. | 28. | 29. | 30. |
|---|---|---|---|---|---|
| 5 | 1 | 3 | 9 | 6 | 8 |
| 8 | 4 | 7 | 3 | 3 | 1 |
| 1 | 2 | 6 | 1 | 5 | 4 |
| 9 | 8 | 4 | 8 | 4 | 2 |
| +5 | +6 | +4 | +2 | +7 | +7 |

**Give each missing addend.**

**31.** 5 + ? = 8  **32.** 6 + ? = 10  **33.** 4 + ? = 9

**34.** 6 + ? = 14  **35.** 9 + ? = 15  **36.** 8 + ? = 16

**37.** ? + 4 = 13  **38.** ? + 8 = 13  **39.** ? + 7 = 15

## *Challenge!*

**In how many different orders can the numbers be added?**

**40.** 7 2  **41.** 8 1 4  **42.** 9 5 3 6

# Adding 2-digit numbers

You can find the total number of blocks by adding in columns.

**Step 1.** Add in ones column.

  42
+23
   5

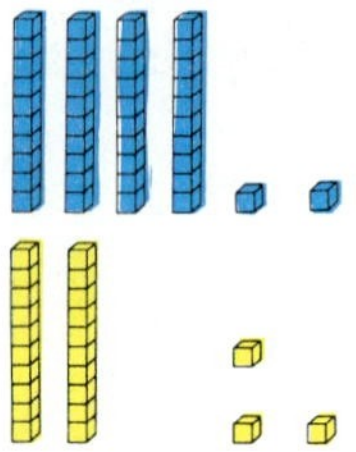

**Step 2.** Add in tens column.

  42
+23
 65

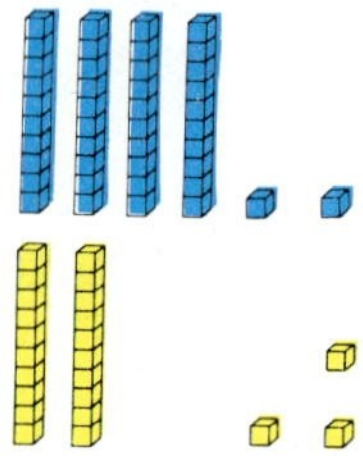

In this example, 10 ones are regrouped for 1 ten.

**Step 1.** Add ones.

  38
+25

13 ones

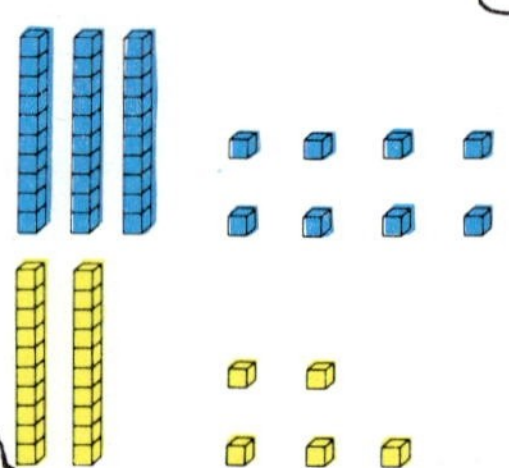

**Step 2.** Regroup.

13 ones = 1 ten + 3 ones

  1
  38
+25
   3

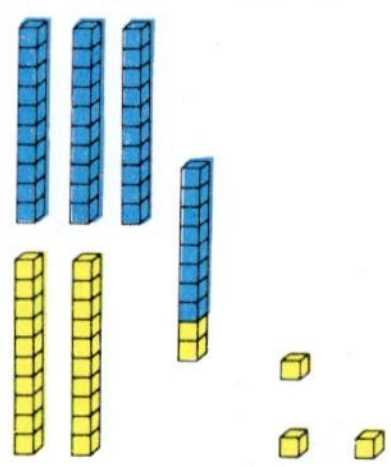

**Step 3.** Add tens.

  1
  38
+25
 63

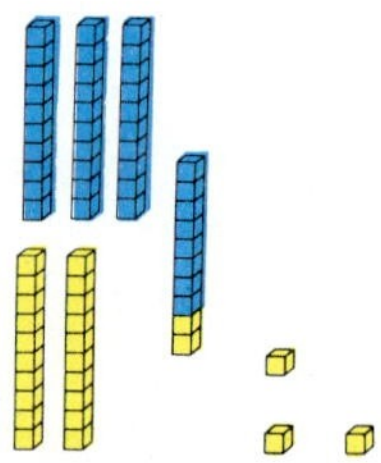

**EXERCISES**

**Add.**

| 1. 24 + 12 | 2. 43 + 22 | 3. 40 + 30 | 4. 50 + 35 | 5. 53 + 43 | 6. 30 + 19 |
|---|---|---|---|---|---|
| 7. 32 + 32 | 8. 54 + 13 | 9. 25 + 24 | 10. 51 + 35 | 11. 41 + 53 | 12. 44 + 52 |
| 13. 42 + 23 | 14. 23 + 42 | 15. 31 + 16 | 16. 52 + 15 | 17. 26 + 61 | 18. 45 + 43 |

19. 58 + 27

20. 39 + 25

21. 52 + 38

22. 27 + 67

23. 39 + 38

24. 49 + 23

25. 38 + 7

26. 56 + 38

27. 49 + 36

28. 72 + 18

29. 35 + 55

30. 27 + 19

31. 43 + 87

32. 58 + 46

33. 19 + 36

34. 43 + 8

35. 53 + 28

36. 28 + 53

37. 56 + 48

38. 43 + 88

39. 72 + 9

40. 56 + 88

41. 55 + 89

42. 75 + 49

43. 38¢ + 29¢ = 67¢

44. 15¢ + 29¢

45. 36¢ + 27¢

46. 53¢ + 38¢

47. $.26 + .87 = $1.13

48. $.46 + .29

49. $.36 + .98

50. $.48 + .76

**Solve.**

51.

What is the total price?

52.

What is the total price?

53. Sarah bought 2 sandwiches for 47¢ each. How much did she spend?

54. Jon had 82¢. He wanted to buy a 35¢ milk and a 49¢ hamburger. Did he have enough money?

## KEEPING SKILLS SHARP

**Round to the nearest hundred.**

1. 168
2. 541
3. 650
4. 1252
5. 1963

**Round to the nearest thousand.**

6. 3410
7. 2678
8. 5500
9. 78,360
10. 92,500

# Adding 3-digit numbers

First estimate the sum by rounding each addend to the nearest hundred and adding.

$$\begin{array}{r} 142 \\ +163 \\ \hline \end{array} \quad \text{rounds to} \quad \begin{array}{r} 100 \\ +200 \\ \hline 300 \end{array}$$

The sum should be about 300.

**Step 1.** Add ones.

$$\begin{array}{r} 142 \\ +163 \\ \hline 5 \end{array}$$

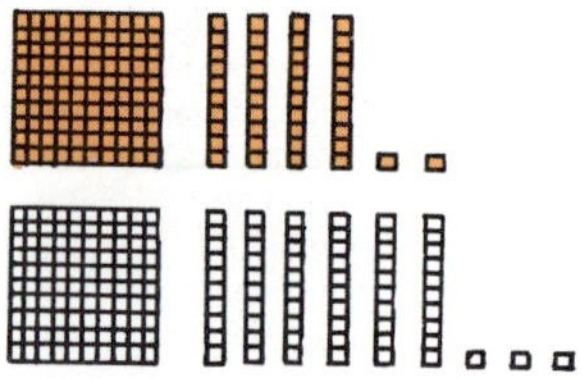

**Step 2.** Add tens.

$$\begin{array}{r} 142 \\ +163 \\ \hline 5 \end{array}$$

10 tens

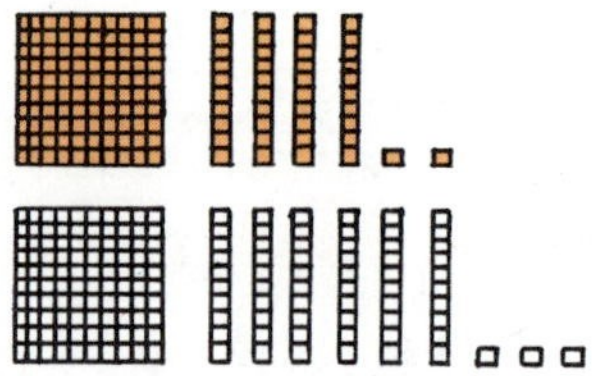

**Step 3.** Regroup.

$$\begin{array}{r} \overset{1}{1}42 \\ +163 \\ \hline 05 \end{array}$$

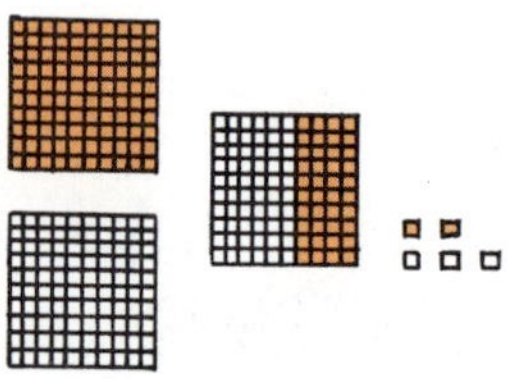

**Step 4.** Add hundreds.

$$\begin{array}{r} \overset{1}{1}42 \\ +163 \\ \hline 305 \end{array}$$

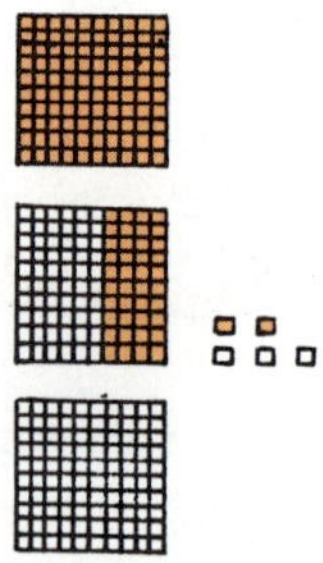

Is the sum near the estimate? If so, the addition was probably done correctly.

In this example, we regroup twice. First estimate the sum, and then study the steps.

**Step 1.** Add ones and regroup.

$$\begin{array}{r} 2\overset{1}{9}8 \\ +174 \\ \hline 2 \end{array}$$

**Step 2.** Add tens and regroup.

$$\begin{array}{r} \overset{1}{2}\overset{1}{9}8 \\ +174 \\ \hline 72 \end{array}$$

**Step 3.** Add hundreds.

$$\begin{array}{r} \overset{1}{2}\overset{1}{9}8 \\ +174 \\ \hline 472 \end{array}$$

## EXERCISES

**First estimate the sum. Then add.**

| | | | | |
|---|---|---|---|---|
| 1. 242 + 173 | 2. 439 + 270 | 3. 272 + 81 | 4. 521 + 390 | 5. 143 + 681 |
| 6. 359 + 476 | 7. 566 + 389 | 8. 475 + 365 | 9. 394 + 78 | 10. 293 + 529 |
| 11. 653 + 12 | 12. 553 + 486 | 13. 443 + 258 | 14. 345 + 482 | 15. 521 + 278 |
| 16. 976 + 847 | 17. 627 + 848 | 18. 902 + 641 | 19. 578 + 645 | 20. 438 + 717 |
| 21. 845 + 974 | 22. 478 + 396 | 23. 846 + 795 | 24. 296 + 385 | 25. 574 + 86 |

| | | | |
|---|---|---|---|
| 26. $3.58 + 2.74 = $6.32 | 27. $7.56 + 2.43 | 28. $6.75 + 2.83 | 29. $8.39 + 3.95 |
| 30. $2.95 + 2.95 | 31. $4.98 + 2.76 | 32. $7.85 + 5.29 | 33. $5.29 + .85 |

BOWLING SCORES

| Name | Game 1 | Game 2 | Game 3 |
|---|---|---|---|
| Bill | 114 | 98 | 117 |
| Joan | 125 | 101 | 95 |
| Lou | 100 | 106 | 115 |
| Susan | 114 | 96 | 118 |

**Solve.**

34. What was Bill's total score for his first two games?
35. What was Bill's total for all three games?
36. What was Joan's total score?
37. Who had the greater total, Lou or Susan?

***Challenge!***

**Give the missing digits.**

38. 526 + 38■ = 9■1

39. 4■6 + 39■ = 823

40. ■5■ + 3■7 = ■055

# Adding larger numbers

You add larger numbers the same way you add smaller numbers.

**Step 1.** Add ones and regroup.

$$\begin{array}{ccccc} & & & 1 & \\ & 7 & 5 & 8 & 4 \\ + & 2 & 9 & 0 & 8 \\ \hline & & & & 2 \end{array}$$

**Step 2.** Add tens.

$$\begin{array}{ccccc} & & & 1 & \\ & 7 & 5 & 8 & 4 \\ + & 2 & 9 & 0 & 8 \\ \hline & & & 9 & 2 \end{array}$$

**Step 3.** Add hundreds and regroup.

$$\begin{array}{ccccc} & 1 & & 1 & \\ & 7 & 5 & 8 & 4 \\ + & 2 & 9 & 0 & 8 \\ \hline & & 4 & 9 & 2 \end{array}$$

**Step 4.** Add thousands.

$$\begin{array}{ccccc} & 1 & & 1 & \\ & 7 & 5 & 8 & 4 \\ + & 2 & 9 & 0 & 8 \\ \hline 1 & 0, & 4 & 9 & 2 \end{array}$$

Other examples:

$$\begin{array}{cccccc} & 1 & 1 & 1 & & \\ & 5 & 9, & 6 & 7 & 8 \\ + & 2 & 4, & 3 & 8 & 0 \\ \hline & 8 & 4, & 0 & 5 & 8 \end{array} \qquad \begin{array}{ccccccc} & 1 & & 1 & 1 & 1 & \\ & 8 & 9 & 2, & 6 & 3 & 5 \\ +1 & 1 & 4 & 2, & 7 & 8 & 5 \\ \hline 1, & 0 & 3 & 5, & 4 & 2 & 0 \end{array}$$

## EXERCISES

**Give each sum.**

| | | | |
|---|---|---|---|
| **1.** 3829 <br> +1746 | **2.** 4973 <br> +9618 | **3.** 5348 <br> +645 | **4.** 8475 <br> +6397 |
| **5.** 74,235 <br> +9,847 | **6.** 69,184 <br> +5,986 | **7.** 35,792 <br> +34,596 | **8.** 48,635 <br> +48,635 |
| **9.** 59,483 <br> +72,965 | **10.** 71,358 <br> +42,848 | **11.** 49,658 <br> +27,498 | **12.** 89,653 <br> +17,846 |
| **13.** 593,482 <br> +28,567 | **14.** 783,619 <br> +7,478 | **15.** 395,826 <br> +438,599 | **16.** 742,856 <br> +174,698 |
| **17.** \$354.25 <br> +486.19 | **18.** \$736.28 <br> +540.75 | **19.** \$483.97 <br> +594.86 | **20.** \$562.88 <br> +947.05 |

**Solve.**

**21.**

What is the total price?

**22.**

What was the total attendance?

**23.** David put his baseball cards in 2 large boxes. He put 1396 cards in each box. How many cards did he have?

**24.** Susan had 296 photos in one book and 187 photos in another book. How many photos did she have?

**25.** The Clark family drove their car 10,483 kilometers one year and 19,768 kilometers the next. How far did they drive during the two years?

**26.** In 1959, the population of Oakville was 28,746. During the next 20 years the population increased by 19,658. What was the population then?

# Adding 3 or more numbers

| Step 1.<br>Add ones. | Step 2.<br>Regroup. | Step 3.<br>Add tens. |
|---|---|---|
| 58<br>26<br>34<br>+45<br>(23)<br>8+6=14<br>14+4=18<br>18+5=23 | 2<br>58<br>26<br>34<br>+45<br>3 | 2<br>58<br>26<br>34<br>+45<br>163 |

| Step 1.<br>Add ones and regroup. | Step 2.<br>Add tens and regroup. | Step 3.<br>Add hundreds and regroup. |
|---|---|---|
| 2<br>546<br>383<br>154<br>501<br>+957<br>1 | 2 2<br>546<br>383<br>154<br>501<br>+957<br>41 | 2 2<br>546<br>383<br>154<br>501<br>+957<br>2541 |

**EXERCISES**

**Give each sum.**

| | | | | |
|---|---|---|---|---|
| **1.** 56<br>23<br>+48 | **2.** 29<br>74<br>+63 | **3.** 801<br>746<br>+594 | **4.** 363<br>775<br>+457 | **5.** 955<br>145<br>+628 |
| **6.** 63<br>42<br>78<br>+91 | **7.** 45<br>45<br>45<br>+45 | **8.** 258<br>346<br>91<br>+759 | **9.** 7425<br>6813<br>4259<br>+647 | **10.** 24,752<br>76,874<br>80,257<br>+323 |
| **11.** 51<br>38<br>29<br>74<br>+51 | **12.** 59<br>61<br>58<br>62<br>+60 | **13.** 272<br>353<br>438<br>686<br>+147 | **14.** 3456<br>2851<br>1234<br>659<br>+154 | **15.** 76,325<br>81,804<br>97,235<br>936<br>+5,563 |

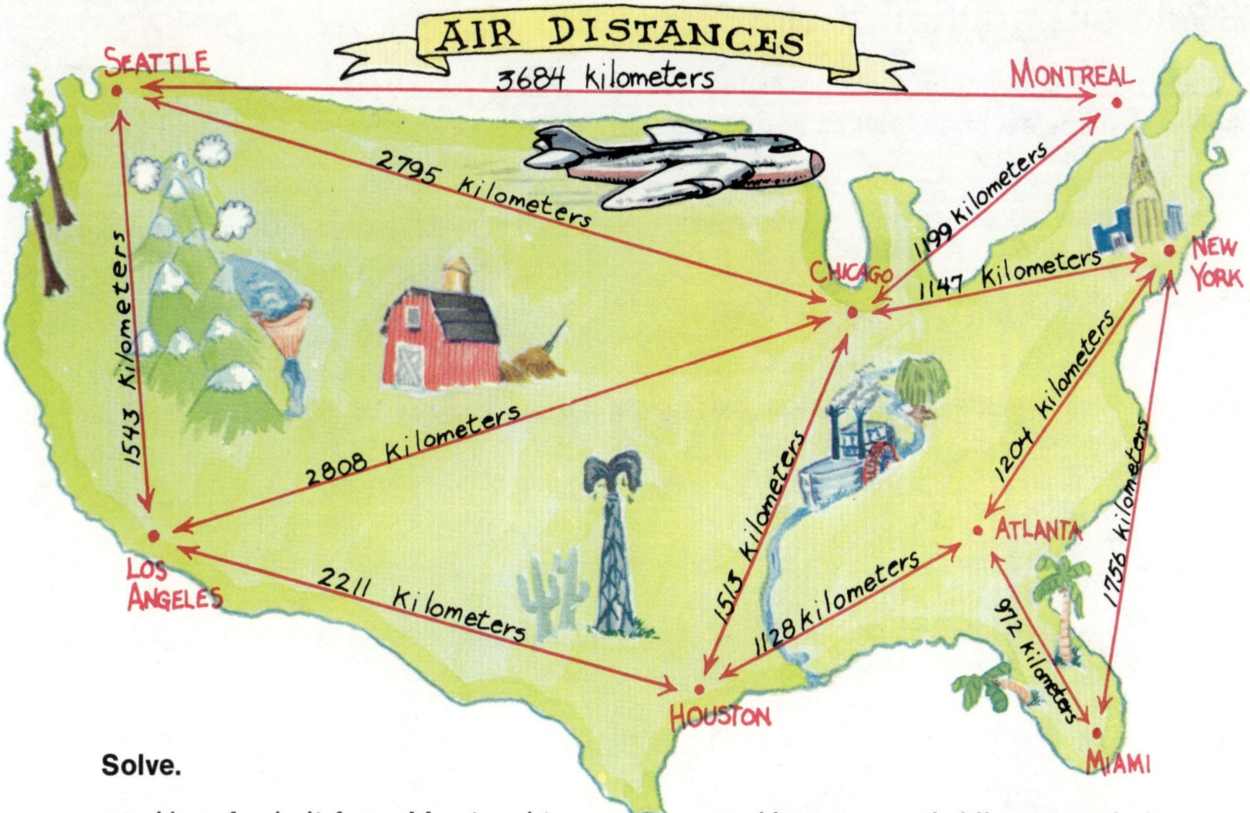


**Solve.**

16. How far is it from Montreal to Chicago by air?

17. How many air kilometers is it from New York to Miami?

18. A jet flying from New York to Miami stopped at Atlanta. How far did it fly?

19. How long is the Seattle–Los Angeles–Houston flight?

20. Which is longer, a New York–Atlanta flight or a New York–Chicago flight?

21. How many kilometers is a Montreal–Chicago–Los Angeles flight?

22. How long is a Los Angeles–Houston–Atlanta–New York flight?

23. Find the total distance of a New York–Chicago–Los Angeles–Seattle flight.

***Challenge!***

24. I started in Chicago, flew to 3 cities, and returned to Chicago. I traveled about 4900 kilometers. What cities did I visit?

# Basic subtraction facts

WHAT'S YOUR SCORE?

**All the basic subtraction facts are listed below.**
**See how many seconds it takes you to write the differences.**

| | | | | | | | |
|---|---|---|---|---|---|---|---|
| 1. 3 −3 | 2. 1 −0 | 3. 7 −5 | 4. 8 −8 | 5. 6 −5 | 6. 10 −8 | 7. 2 −2 | 8. 12 −5 |
| 9. 3 −1 | 10. 11 −2 | 11. 9 −3 | 12. 8 −7 | 13. 6 −4 | 14. 11 −8 | 15. 13 −8 | 16. 8 −6 |
| 17. 12 −8 | 18. 9 −4 | 19. 10 −2 | 20. 5 −4 | 21. 0 −0 | 22. 7 −4 | 23. 15 −7 | 24. 10 −7 |
| 25. 5 −2 | 26. 9 −8 | 27. 7 −6 | 28. 10 −9 | 29. 4 −4 | 30. 14 −6 | 31. 11 −9 | 32. 5 −3 |
| 33. 12 −7 | 34. 9 −7 | 35. 3 −2 | 36. 12 −4 | 37. 16 −9 | 38. 5 −5 | 39. 9 −5 | 40. 11 −4 |
| 41. 4 −0 | 42. 7 −3 | 43. 4 −3 | 44. 2 −0 | 45. 15 −8 | 46. 6 −3 | 47. 2 −1 | 48. 8 −2 |
| 49. 3 −0 | 50. 6 −0 | 51. 8 −3 | 52. 13 −5 | 53. 16 −7 | 54. 11 −7 | 55. 8 −1 | 56. 11 −3 |
| 57. 9 −6 | 58. 17 −8 | 59. 10 −3 | 60. 13 −9 | 61. 10 −1 | 62. 9 −9 | 63. 12 −9 | 64. 9 −2 |
| 65. 7 −0 | 66. 10 −6 | 67. 15 −6 | 68. 4 −2 | 69. 9 −0 | 70. 14 −5 | 71. 1 −0 | 72. 14 −7 |
| 73. 7 −2 | 74. 18 −9 | 75. 6 −2 | 76. 8 −5 | 77. 12 −6 | 78. 13 −7 | 79. 14 −8 | 80. 11 −6 |
| 81. 6 −6 | 82. 5 −1 | 83. 8 −4 | 84. 4 −1 | 85. 11 −5 | 86. 10 −4 | 87. 7 −0 | 88. 12 −3 |
| 89. 5 −0 | 90. 8 −0 | 91. 14 −9 | 92. 7 −1 | 93. 16 −8 | 94. 10 −5 | 95. 13 −6 | 96. 6 −1 |
| | | 97. 15 −9 | 98. 9 −1 | 99. 17 −9 | 100. 13 −4 | | |

Remember that subtraction is finding the missing addend. The answer is called the **difference.**

**Check your answers.**

| | | | | | | | |
|---|---|---|---|---|---|---|---|
| **1.** 0 | **2.** 1 | **3.** 2 | **4.** 0 | **5.** 1 | **6.** 2 | **7.** 0 | **8.** 7 |
| **9.** 2 | **10.** 9 | **11.** 6 | **12.** 1 | **13.** 2 | **14.** 3 | **15.** 5 | **16.** 2 |
| **17.** 4 | **18.** 5 | **19.** 8 | **20.** 1 | **21.** 0 | **22.** 3 | **23.** 8 | **24.** 3 |
| **25.** 3 | **26.** 1 | **27.** 1 | **28.** 1 | **29.** 0 | **30.** 8 | **31.** 2 | **32.** 2 |
| **33.** 5 | **34.** 2 | **35.** 1 | **36.** 8 | **37.** 7 | **38.** 0 | **39.** 4 | **40.** 7 |
| **41.** 4 | **42.** 4 | **43.** 1 | **44.** 2 | **45.** 7 | **46.** 3 | **47.** 1 | **48.** 6 |
| **49.** 3 | **50.** 6 | **51.** 5 | **52.** 8 | **53.** 9 | **54.** 4 | **55.** 7 | **56.** 8 |
| **57.** 3 | **58.** 9 | **59.** 7 | **60.** 4 | **61.** 9 | **62.** 0 | **63.** 3 | **64.** 7 |
| **65.** 7 | **66.** 4 | **67.** 9 | **68.** 2 | **69.** 9 | **70.** 9 | **71.** 1 | **72.** 7 |
| **73.** 5 | **74.** 9 | **75.** 4 | **76.** 3 | **77.** 6 | **78.** 6 | **79.** 6 | **80.** 5 |
| **81.** 0 | **82.** 4 | **83.** 4 | **84.** 3 | **85.** 6 | **86.** 6 | **87.** 7 | **88.** 9 |
| **89.** 5 | **90.** 8 | **91.** 5 | **92.** 6 | **93.** 8 | **94.** 5 | **95.** 7 | **96.** 5 |
| | | **97.** 6 | **98.** 8 | **99.** 8 | **100.** 9 | | |

**101.** To find your score, add the number of seconds and the number of mistakes.

**102.** Practice the facts that you missed. See if you can get a better (lower) score.

**103.** Work with a classmate. Practice your subtraction facts orally each day. Keep a record of your best daily score.

**104.** Use graph paper. Make a graph of your scores like the one on the computer screen.

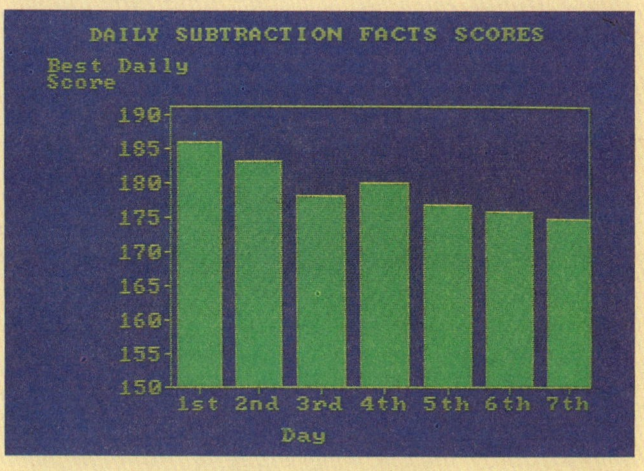

# Subtracting 2-digit numbers

You can find the number of blocks left by subtracting in columns.

**Step 1.** Subtract in ones column.

$$\begin{array}{r} 58 \\ -23 \\ \hline 5 \end{array}$$

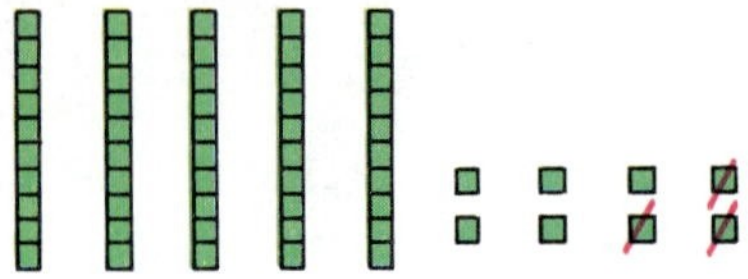

**Step 2.** Subtract in tens column.

$$\begin{array}{r} 58 \\ -23 \\ \hline 35 \end{array}$$

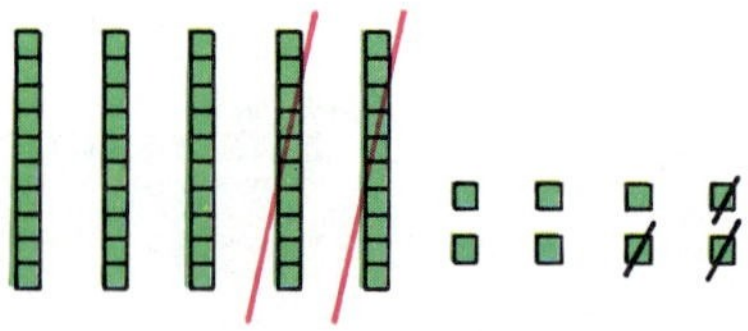

Addition and subtraction are related.

Since subtraction is finding a missing addend, we can check subtraction by addition.

| | |
|---|---|
| blocks left | 35 |
| blocks taken away | +23 |
| blocks in all | 58 |

It checks!

In this example, 1 ten is regrouped for 10 ones.

**Step 1.** Not enough ones.

$$\begin{array}{r} 62 \\ -48 \\ \hline \end{array}$$

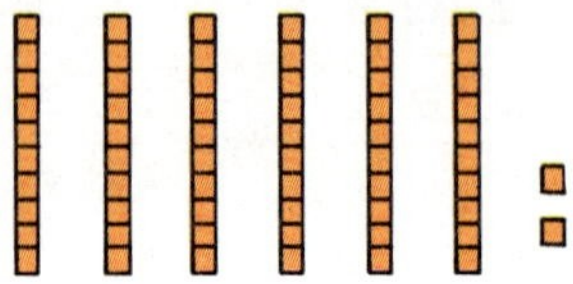

**Step 2.** Regroup.

$$\begin{array}{r} \overset{5}{\not{6}}\,{}^{1}2 \\ -48 \\ \hline \end{array}$$

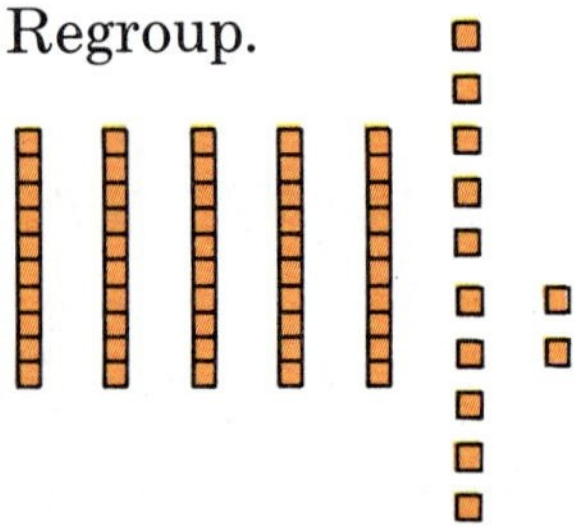

**Step 3.** Subtract.

$$\begin{array}{r} \overset{5}{\not{6}}\,{}^{1}2 \\ -48 \\ \hline 14 \end{array}$$

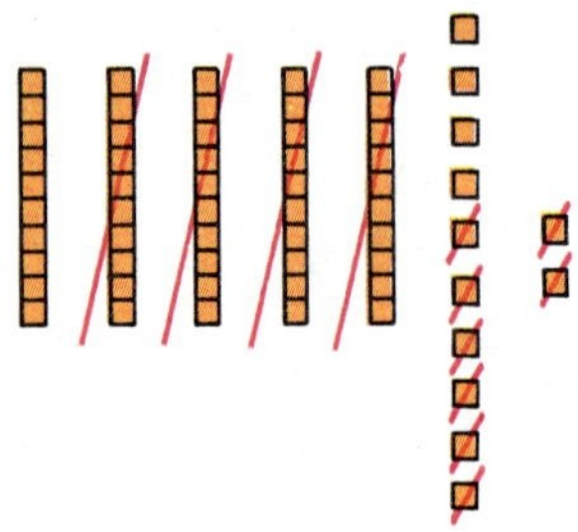

What two numbers would you add to check the subtraction?

**EXERCISES**

**First subtract. Then check by addition.**

1. $\begin{array}{r} 98 \\ -58 \\ \hline \end{array}$ 2. $\begin{array}{r} 75 \\ -53 \\ \hline \end{array}$ 3. $\begin{array}{r} 89 \\ -26 \\ \hline \end{array}$ 4. $\begin{array}{r} 77 \\ -36 \\ \hline \end{array}$ 5. $\begin{array}{r} 68 \\ -42 \\ \hline \end{array}$ 6. $\begin{array}{r} 88 \\ -35 \\ \hline \end{array}$

7. $\begin{array}{r} 86 \\ -35 \\ \hline \end{array}$ 8. $\begin{array}{r} 67 \\ -40 \\ \hline \end{array}$ 9. $\begin{array}{r} 97 \\ -27 \\ \hline \end{array}$ 10. $\begin{array}{r} 69 \\ -28 \\ \hline \end{array}$ 11. $\begin{array}{r} 96 \\ -35 \\ \hline \end{array}$ 12. $\begin{array}{r} 76 \\ -42 \\ \hline \end{array}$

**Give each difference.**

13. 52 − 18
14. 60 − 22
15. 90 − 46
16. 73 − 58
17. 80 − 39
18. 62 − 45
19. 92 − 35
20. 71 − 58
21. 82 − 46
22. 53 − 29
23. 63 − 48
24. 91 − 48
25. 72 − 22
26. 61 − 56
27. 45 − 28
28. 81 − 27
29. 64 − 9
30. 60 − 42
31. 93 − 43
32. 83 − 7
33. 85 − 43
34. 94 − 36
35. 84 − 35
36. 65 − 56
37. 52¢ − 26¢
38. 64¢ − 29¢
39. 78¢ − 39¢
40. 58¢ − 25¢
41. 59¢ − 38¢
42. 70¢ − 56¢

**Solve.**

43. 

a. What is the cost of 1 adult ticket and 1 child's ticket?

b. What is the cost of 2 adult tickets?

c. Is $5 enough to buy 3 children's tickets?

44. 

a. How much more does the large drink cost?

b. How much do 2 large drinks cost?

c. Does a large drink cost more than 2 small drinks?

## Challenge!

45. The feature movie begins at 7:30. It lasts 1 hour 42 minutes. What time does the feature movie end?

# Subtracting 3-digit numbers

**EXAMPLE.**
$$\begin{array}{r} 365 \\ -182 \\ \hline \end{array}$$

**Step 1.** Subtract ones.

$$\begin{array}{r} 365 \\ -182 \\ \hline 3 \end{array}$$

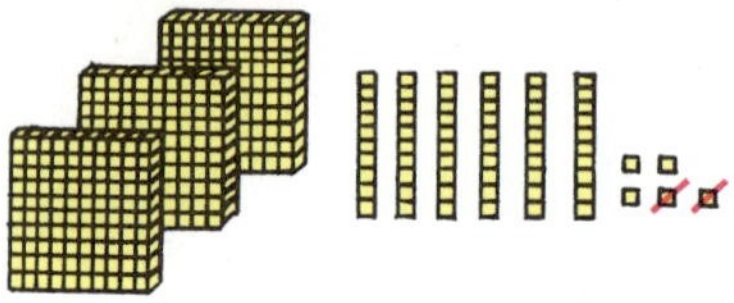

**Step 2.** Regroup.

$$\begin{array}{r} \overset{2}{\not 3}\,{}^{1}6\,5 \\ -1\,8\,2 \\ \hline 3 \end{array}$$

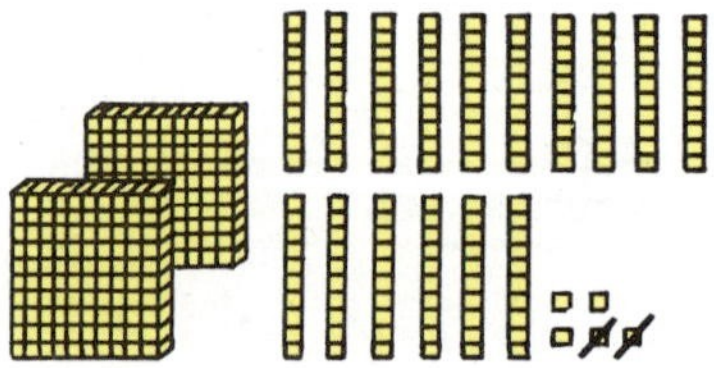

**Step 3.** Subtract tens.

$$\begin{array}{r} \overset{2}{\not 3}\,{}^{1}6\,5 \\ -1\,8\,2 \\ \hline 8\,3 \end{array}$$

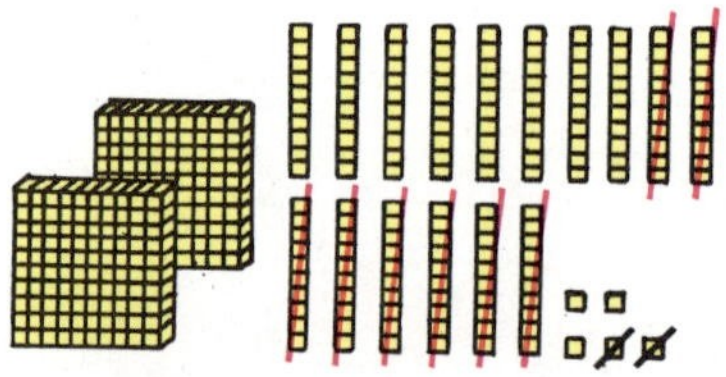

**Step 4.** Subtract hundreds.

$$\begin{array}{r} \overset{2}{\not 3}\,{}^{1}6\,5 \\ -1\,8\,2 \\ \hline 1\,8\,3 \end{array}$$

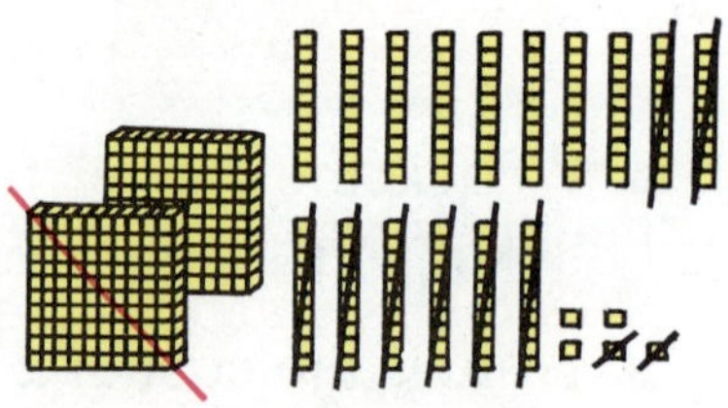

In this example we have to regroup twice.

**Step 1.** Regroup. Subtract ones.

$$\begin{array}{r} 4\,\overset{0}{\not 1}\,{}^{1}3 \\ -2\,5\,8 \\ \hline 5 \end{array}$$

**Step 2.** Regroup. Subtract tens.

$$\begin{array}{r} \overset{3}{\not 4}\,{}^{1}\overset{0}{\not 1}\,{}^{1}3 \\ -2\,5\,8 \\ \hline 5\,5 \end{array}$$

**Step 3.** Subtract hundreds.

$$\begin{array}{r} \overset{3}{\not 4}\,{}^{1}\overset{0}{\not 1}\,{}^{1}3 \\ -2\,5\,8 \\ \hline 1\,5\,5 \end{array}$$

## EXERCISES

**First estimate the difference. Then subtract.**

| | | | |
|---|---|---|---|
| 1. $\begin{array}{r} 537 \\ -248 \\ \hline \end{array}$ | 2. $\begin{array}{r} 714 \\ -80 \\ \hline \end{array}$ | 3. $\begin{array}{r} 558 \\ -362 \\ \hline \end{array}$ | 4. $\begin{array}{r} 693 \\ -157 \\ \hline \end{array}$ |
| 5. $\begin{array}{r} 946 \\ -58 \\ \hline \end{array}$ | 6. $\begin{array}{r} 725 \\ -258 \\ \hline \end{array}$ | 7. $\begin{array}{r} 423 \\ -55 \\ \hline \end{array}$ | 8. $\begin{array}{r} 628 \\ -346 \\ \hline \end{array}$ |
| 9. $\begin{array}{r} 532 \\ -274 \\ \hline \end{array}$ | 10. $\begin{array}{r} 641 \\ -185 \\ \hline \end{array}$ | 11. $\begin{array}{r} 946 \\ -275 \\ \hline \end{array}$ | 12. $\begin{array}{r} 835 \\ -59 \\ \hline \end{array}$ |
| 13. $\begin{array}{r} 624 \\ -495 \\ \hline \end{array}$ | 14. $\begin{array}{r} 753 \\ -299 \\ \hline \end{array}$ | 15. $\begin{array}{r} 542 \\ -76 \\ \hline \end{array}$ | 16. $\begin{array}{r} 512 \\ -348 \\ \hline \end{array}$ |
| 17. $\begin{array}{r} 731 \\ -256 \\ \hline \end{array}$ | 18. $\begin{array}{r} 623 \\ -48 \\ \hline \end{array}$ | 19. $\begin{array}{r} 514 \\ -175 \\ \hline \end{array}$ | 20. $\begin{array}{r} 862 \\ -496 \\ \hline \end{array}$ |
| 21. $\begin{array}{r} \$5.48 \\ -2.56 \\ \hline \end{array}$ | 22. $\begin{array}{r} \$6.25 \\ -3.58 \\ \hline \end{array}$ | 23. $\begin{array}{r} \$5.34 \\ -.68 \\ \hline \end{array}$ | 24. $\begin{array}{r} \$3.56 \\ -2.99 \\ \hline \end{array}$ |

**Solve.**

25. Anita had $4.35. She spent $2.75 for a Fun Park ticket. How much money did she have left?

26. The Silver Streak roller coaster could hold 56 people. Anita counted 84 people waiting in line. How many would not be able to go on the next ride?

### *Challenge!*

**Complete the table.**

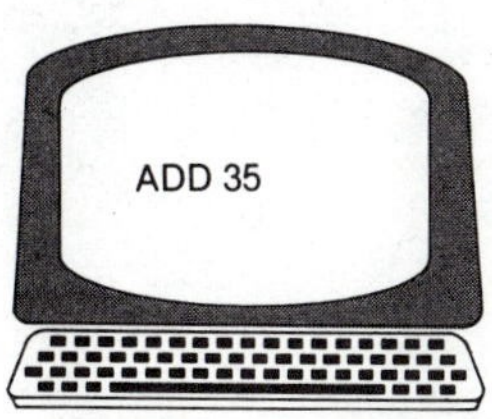


| | Input | Output |
|---|---|---|
| | 50 | 85 |
| 27. | 64 | ? |
| 28. | ? | 125 |
| 29. | 42 | ? |
| 30. | ? | 114 |

## KEEPING SKILLS SHARP

1. $\begin{array}{r} 38 \\ 24 \\ +15 \\ \hline \end{array}$

2. $\begin{array}{r} 821 \\ 465 \\ +378 \\ \hline \end{array}$

3. $\begin{array}{r} 4625 \\ 718 \\ +2469 \\ \hline \end{array}$

4. $\begin{array}{r} 36 \\ 78 \\ 95 \\ +43 \\ \hline \end{array}$

5. $\begin{array}{r} 159 \\ 146 \\ 380 \\ +56 \\ \hline \end{array}$

6. $\begin{array}{r} 3521 \\ 78 \\ 365 \\ +4175 \\ \hline \end{array}$

# Zeros in subtraction

Not enough ones.

$$\begin{array}{r} 802 \\ -149 \\ \hline \end{array}$$

There are no tens.
So, regroup 1 hundred
for 10 tens.

$$\begin{array}{r} \overset{7}{\cancel{8}}\,\overset{10}{0}\,2 \\ -1\,4\,9 \\ \hline \end{array}$$

Regroup 1 ten
for 10 ones.

$$\begin{array}{r} \overset{7}{\cancel{8}}\,\overset{9}{\cancel{10}}\,\overset{12}{2} \\ -1\,4\,9 \\ \hline \end{array}$$

Subtract.

$$\begin{array}{r} \overset{7}{\cancel{8}}\,\overset{9}{\cancel{10}}\,\overset{12}{2} \\ -1\,4\,9 \\ \hline 6\,5\,3 \end{array}$$

Estimate the difference.
Does the answer seem reasonable?

**EXERCISES**
**Subtract.**

1. 701 − 425
2. 503 − 367
3. 602 − 564
4. 405 − 177
5. 806 − 288
6. 604 − 156
7. 700 − 385
8. 505 − 347
9. 301 − 216
10. 400 − 287

| | | | | |
|---|---|---|---|---|
| 11. $\begin{array}{r} 713 \\ -516 \\ \hline \end{array}$ | 12. $\begin{array}{r} 621 \\ -148 \\ \hline \end{array}$ | 13. $\begin{array}{r} 502 \\ -209 \\ \hline \end{array}$ | 14. $\begin{array}{r} 600 \\ -207 \\ \hline \end{array}$ | 15. $\begin{array}{r} 306 \\ -199 \\ \hline \end{array}$ |
| 16. $\begin{array}{r} 602 \\ -348 \\ \hline \end{array}$ | 17. $\begin{array}{r} 537 \\ -229 \\ \hline \end{array}$ | 18. $\begin{array}{r} 807 \\ -499 \\ \hline \end{array}$ | 19. $\begin{array}{r} 694 \\ -396 \\ \hline \end{array}$ | 20. $\begin{array}{r} 847 \\ -548 \\ \hline \end{array}$ |
| 21. $\begin{array}{r} 674 \\ -497 \\ \hline \end{array}$ | 22. $\begin{array}{r} 800 \\ -537 \\ \hline \end{array}$ | 23. $\begin{array}{r} 708 \\ -186 \\ \hline \end{array}$ | 24. $\begin{array}{r} 507 \\ -309 \\ \hline \end{array}$ | 25. $\begin{array}{r} 635 \\ -207 \\ \hline \end{array}$ |
| 26. $\begin{array}{r} \$7.85 \\ -4.81 \\ \hline \end{array}$ | 27. $\begin{array}{r} \$6.03 \\ -3.49 \\ \hline \end{array}$ | 28. $\begin{array}{r} \$7.01 \\ -1.85 \\ \hline \end{array}$ | 29. $\begin{array}{r} \$8.05 \\ -7.39 \\ \hline \end{array}$ | 30. $\begin{array}{r} \$5.00 \\ -1.73 \\ \hline \end{array}$ |

**Solve.**

**31.** Shirley paid $1.75 to get admitted to the rink and $1.35 to rent ice skates. How much did she spend?

**32.** Bart had his ice skates sharpened for $1.65. How much change did he get from $5?

**33.** Dorothy practiced a Lutz jump for 45 minutes and some school figures for 55 minutes. How many minutes did she practice?

**34.** On Saturday 638 people skated at the rink. On Sunday 901 people skated at the rink. How many more people skated on Sunday?

**35.** Five hundred six students skated on a Youth Night. Two hundred eighty-eight were girls. How many were boys?

★ **36.** Shirley skated for 2 hours. Bart skated for 1 hour 40 minutes. How many more minutes did Shirley skate than Bart?

★ **37.** Andrew had $5. He spent $1.75 to get admitted to the rink and $.65 for hot chocolate. How much money did he have left?

★ **38.** On a Youth Night there were 271 boys and 27 fewer girls than boys. What was the total number of boys and girls?

## *Challenge!*

**39.** Here are the first and last tickets sold for a Saturday matinee. How many tickets were sold?

# Subtracting larger numbers

You use the same method to subtract larger numbers that you use to subtract smaller numbers.

**Step 1.** Regroup and subtract ones.

$$\begin{array}{r} {\scriptstyle 5\ \ } \\ 83\not{6}\,{}^{1}4 \\ -2519 \\ \hline 5 \end{array}$$

**Step 2.** Subtract tens.

$$\begin{array}{r} {\scriptstyle 5\ \ } \\ 83\not{6}\,{}^{1}4 \\ -2519 \\ \hline 45 \end{array}$$

**Step 3.** Regroup and subtract hundreds.

$$\begin{array}{r} {\scriptstyle 7\ \ \ \ 5\ \ } \\ \not{8}\,{}^{1}3\not{6}\,{}^{1}4 \\ -2519 \\ \hline 845 \end{array}$$

**Step 4.** Subtract thousands.

$$\begin{array}{r} {\scriptstyle 7\ \ \ \ 5\ \ } \\ \not{8}\,{}^{1}3\not{6}\,{}^{1}4 \\ -2519 \\ \hline 5845 \end{array}$$

**EXERCISES**

**Give each difference.**

1. 7659 − 2384
2. 8134 − 529
3. 6052 − 1080
4. 3904 − 2465
5. 3829 − 1469
6. 5302 − 1469
7. 8204 − 3691
8. 7002 − 816
9. 42,148 − 8,029
10. 71,253 − 14,065
11. 61,974 − 3,895
12. 82,345 − 17,586
13. 823,517 − 219,486
14. 593,475 − 16,829
15. 436,219 − 85,774
16. 369,003 − 151,274
17. \$532.46 − 83.19
18. \$414.83 − 79.61
19. \$3618.05 − 221.68
20. \$5231.64 − 856.75

# Computers in banking

The numbers printed at the bottom of checks can be read by computers. Banks use the computer to sort the checks and compute each customer's bank balance.

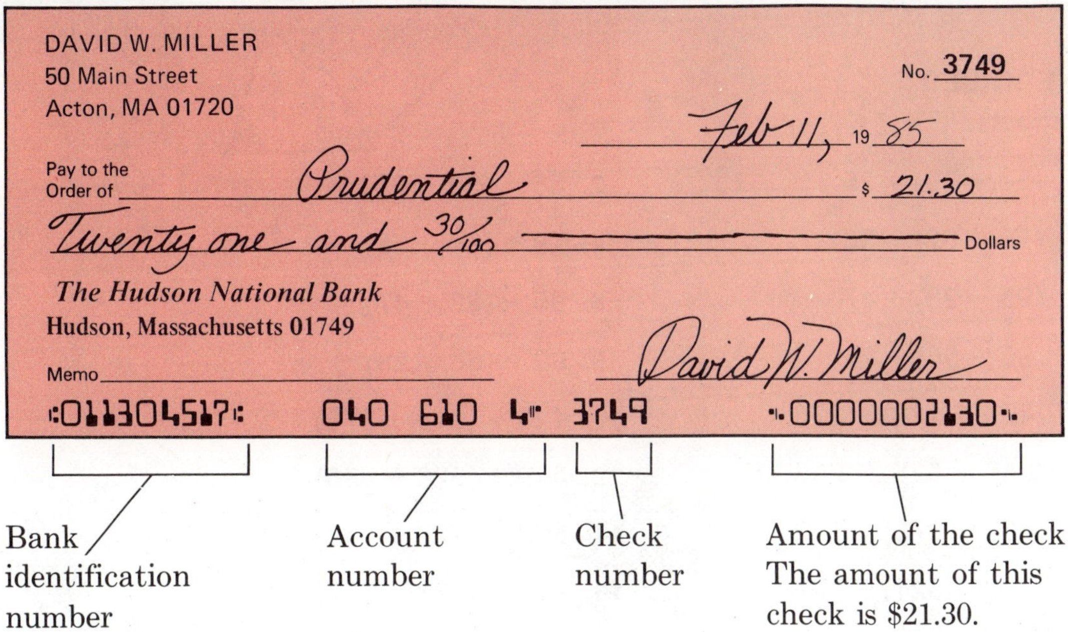


**Solve.**

1. What is the amount of the check?

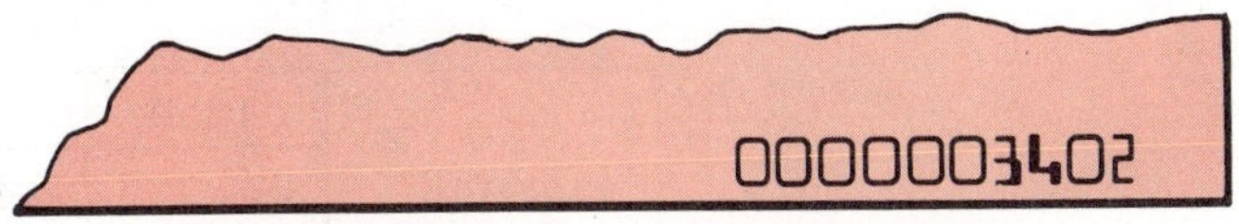


2. Joanne Kilpatrick's beginning balance was $519.36. She wrote these checks. The computer subtracted the amount of each check from her beginning balance. What was her ending balance?

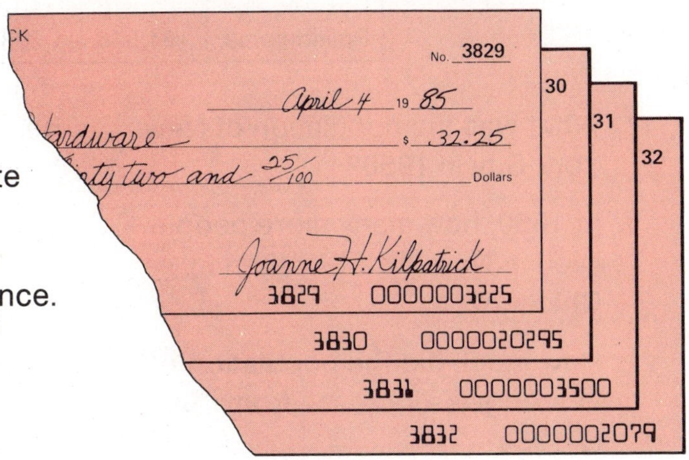

# Addition and subtraction

**(1603 − 452) + 897** = 2048

Remember to work inside the grouping symbols first.

**EXERCISES**

**Compute.**

1. (58 + 39) + 26
2. 58 + (39 + 26)
3. (79 + 82) − 56
4. 79 + (82 − 56)
5. (95 − 23) + 47
6. 95 − (23 + 47)
7. (90 − 56) − 28
8. 90 − (56 − 28)
9. (420 + 118) + 276
10. 420 + (118 + 276)
11. (349 + 526) − 235
12. 349 + (526 − 235)
13. (802 − 278) + 149
14. 802 − (278 + 149)
15. (712 − 321) − 257
16. 712 − (321 − 257)

**Solve.**

| *City* | *Population* | | |
|---|---|---|---|
| | 1850 | 1900 | 1950 |
| New York | 696,115 | 3,437,202 | 7,891,957 |
| Chicago | 29,963 | 1,698,575 | 3,620,962 |
| Los Angeles | 1,610 | 102,479 | 1,970,358 |
| Philadelphia | 121,376 | 1,293,697 | 2,071,605 |

17. What was the population of New York City in 1950?
18. What was the total population of the four cities in 1900?
19. In 1850, how many more people lived in New York City than in Chicago?
20. In 1900, how many more people lived in Philadelphia than in Los Angeles?
21. How much did the population of Los Angeles increase from 1900 to 1950?
22. What was the population of Chicago in 1900, rounded to the nearest thousand?

SKILL GAME

**How close did these players come to the target number?**

1. Beth
   (753 + 258) − 495 → 500

2. Alan
   (826 + 435) − 527 → 500

3. Juan
   (366 + 402) − 265 → 500

4. Dave
   (829 + 648) − 897 → 500

**Play the game.**

1. Make a digit card for each of the digits 0 through 9.
2. Choose a leader.
3. Draw a table like this:

   (☐☐☐ + ☐☐☐) − ☐☐☐ → 500

4. Without looking, the leader picks a card. Each player writes the digit in any square of the table.
5. The card is put back, and step 4 is repeated until all squares have been filled in.
6. The player who gets closest to the target number, 500, wins the game.

# Problem solving

| | |
|---|---|
| 1. Study the problem until you understand it. | Brian had $20. He rented skis for $5.75. How much money did he have left? |
| Find the question. | How much money did he have left? |
| Find the information you need to answer the question. | had $20<br>rented skis for $5.75 |
| 2. Plan what to do and do it. | Take $5.75 away from $20.<br>Subtract. $\begin{array}{r} \$20.00 \\ -5.75 \\ \hline \$14.25 \end{array}$ |
| 3. Answer the question and check your answer. | He had $14.25 left.<br>The answer seems right because $5.75 is about $6 and $20 − $6 = $14. |

## EXERCISES

**Solve.**

1. 43 students went on a ski trip to Wildcat Mountain. 29 students took their own skis. How many did not take their own skis?

2. Lauren went on a 40-kilometer cross-country ski trip. She skied 23 kilometers before noon. How many kilometers did she have to ski in the afternoon?

3. Ms. Cooper bought new skis that cost $149 and new ski boots that cost $119. How much did she spend?

4. Mateo skied down the hill in 101 seconds. George skied down the hill in 88 seconds. How many seconds faster was George?

5. Barry bought a ski mask that cost $3.79 plus $.19 sales tax. What was the total cost?

6. An all-day lift ticket cost $11.50. An afternoon lift ticket cost $8.75. How much more was the all-day ticket?

7. Jennifer wanted to buy skis that cost $125. She had $89. How much more did she need?

8. The ski bus traveled 65 kilometers from the mountain to a restaurant. Then it traveled 89 kilometers from the restaurant to the school. How many kilometers did the bus travel?

# Problem solving—too much information

1. Study and understand.
2. Plan and do.
3. Answer and check.

Carol's school had a bike-a-thon to raise money for the school library. She earned 28¢ for each kilometer that she rode. Before she rested, she rode 17 kilometers, stopped to put air in a tire, and then rode 12 kilometers. How far did she ride before she rested?

*Question:* How far did she ride before she rested?

*Facts needed:* rode 17 kilometers, rode 12 kilometers

$$\begin{array}{r} 17 \\ +12 \\ \hline 29 \end{array}$$

She rode 29 kilometers before she rested.

**EXERCISES**

**Solve.**

**1.** There were 52 students in the bike-a-thon. 19 of them were in fourth grade, 20 were in fifth grade, and 13 were in sixth grade.

- **a.** How many fifth-grade and sixth-grade students were in the bike-a-thon?
- **b.** Which grade had the most students in the bike-a-thon?

**2.** The bike route covered 127 kilometers. There were 4 rest stops for the bikers. Janet rode 48 kilometers and stopped at the second rest stop.

- **a.** How much farther did Janet have to ride to complete the entire route?
- **b.** Had she gone at least half the distance?

3. Carol's class wanted to collect a total of $300. They collected $206.75 from the bike-a-thon and $78.40 from a car wash.
   a. How much money did they collect?
   b. Would they collect more money from 3 car washes than from 1 bike-a-thon?

4. 20 students had lunch together at the third rest stop. They had traveled 70 kilometers. 13 students drank milk with lunch and the rest drank grape juice.
   a. How many students drank grape juice?
   b. How many students did not drink grape juice?

5. Bill, Kiku, and Sarah collected contributions from their neighbors. Bill collected $13.78, Kiku collected $16.85, and Sarah collected $12.34.
   a. How much more did Kiku collect than Bill?
   b. How much did Kiku and Sarah collect together?

6. To get her bike ready, Carol bought a new tire for $5.29 and 2 reflectors for $1.35. Her mother gave her $10 to buy them.
   a. How much did she spend for the tire and reflectors?
   b. How much more was the tire than the reflectors?

# CHAPTER CHECKUP

**Add.** [pages 24–36, 48–49]

1. 58 + 21
2. 69 + 18
3. 78 + 35
4. 593 + 298
5. 658 + 597
6. 3824 + 1536
7. 85,674 + 2,981
8. 483,673 + 297,575
9. 28, 37, + 23
10. 596, 784, + 356
11. 2814, 326, 7821, + 599
12. 537, 261, 83, 145, + 21

**Subtract.** [pages 38–46, 48–49]

13. 98 − 46
14. 82 − 36
15. 473 − 47
16. 603 − 459
17. 800 − 258
18. 4314 − 1682
19. 54,175 − 17,986
20. 836,052 − 38,791

**Solve.** [pages 37, 50–53]

21. 

What is the total cost?

22. 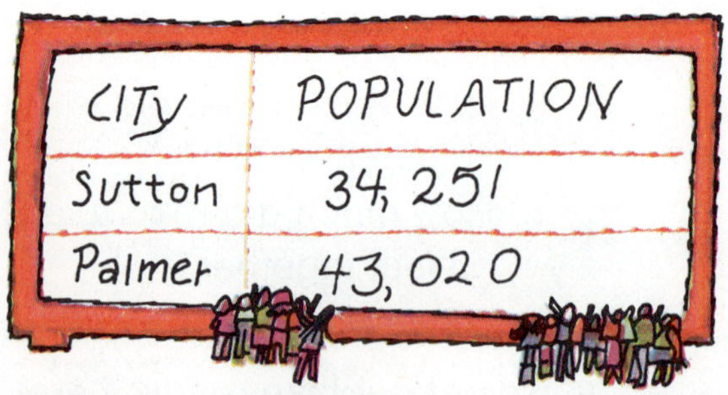


How many more people live in Palmer than in Sutton?

23. Sandy had 219 football cards. She bought 57 more at a garage sale. How many did she have then?

24. David had 206 baseball cards. He bought 67 more on Monday and 95 more on Tuesday. How many did he buy?

# CHAPTER PROJECT

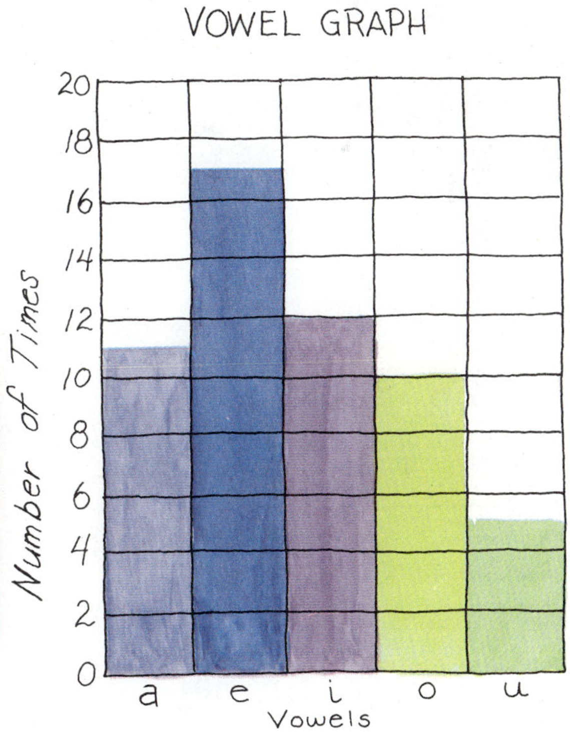


Think of something you would like to make a bar graph of. Use one of the ideas above or make up your own. Collect the information and make a graph.

# CHAPTER REVIEW

Add.
Regroup once.

$$\begin{array}{r} \overset{1}{7}461 \\ +1294 \\ \hline 8755 \end{array}$$

1. $\begin{array}{r} 542 \\ +63 \\ \hline \end{array}$

2. $\begin{array}{r} 823 \\ +58 \\ \hline \end{array}$

3. $\begin{array}{r} 3765 \\ +2712 \\ \hline \end{array}$

Add.
Regroup more than once.

$$\begin{array}{r} 1\ 1\ 1\quad 1\ \ \\ 283{,}714 \\ +59{,}648 \\ \hline 343{,}362 \end{array}$$

4. $\begin{array}{r} 8432 \\ +2591 \\ \hline \end{array}$

5. $\begin{array}{r} 76{,}345 \\ +4{,}738 \\ \hline \end{array}$

6. $\begin{array}{r} 370{,}956 \\ +839{,}614 \\ \hline \end{array}$

Add more than two numbers.

$$\begin{array}{r} 1\ 2\ 2\ \ \\ 8465 \\ 374 \\ 158 \\ +9535 \\ \hline 18{,}532 \end{array}$$

7. $\begin{array}{r} 35 \\ 26 \\ 38 \\ +15 \\ \hline \end{array}$

8. $\begin{array}{r} 263 \\ 75 \\ 184 \\ +942 \\ \hline \end{array}$

9. $\begin{array}{r} 3521 \\ 759 \\ 2163 \\ +97 \\ \hline \end{array}$

Subtract.
Regroup once.

$$\begin{array}{r} 4\quad\ \ \ \\ \not{5}\,{}^{1}286 \\ -1943 \\ \hline 3343 \end{array}$$

10. $\begin{array}{r} 596 \\ -48 \\ \hline \end{array}$

11. $\begin{array}{r} 742 \\ -117 \\ \hline \end{array}$

12. $\begin{array}{r} 826 \\ -353 \\ \hline \end{array}$

Subtract.
Regroup more than once.

$$\begin{array}{r} 2\ 9\ 4\ \ \\ \not{3}\,{}^{1}\not{0}\,\not{5}\,{}^{1}4 \\ -1387 \\ \hline 1667 \end{array}$$

13. $\begin{array}{r} 361 \\ -194 \\ \hline \end{array}$

14. $\begin{array}{r} 9835 \\ -2379 \\ \hline \end{array}$

15. $\begin{array}{r} 6308 \\ -149 \\ \hline \end{array}$

16. $\begin{array}{r} 36{,}482 \\ -19{,}536 \\ \hline \end{array}$

17. $\begin{array}{r} 530{,}491 \\ -206{,}838 \\ \hline \end{array}$

18. $\begin{array}{r} 730{,}046 \\ -415{,}348 \\ \hline \end{array}$

# CHAPTER CHALLENGE

Complete each number pattern.

1. 2, 4, 8, 16, _?_
2. 14, 12, 10, 8, _?_
3. 1, 2, 1, 3, 1, _?_
4. 41, 38, 35, 32, 29, _?_
5. 1, 3, 6, _?_, 15
6. 29, 25, 22, _?_, 19
7. 5, 25, 6, 24, 7, 23, 8, _?_
8. 64, 32, _?_, 8, 4
9. 15, 17, 16, _?_, 17, 19, 18
10. 40, 33, 26, 19, _?_
11. 5, 4, 10, 9, 15, _?_, 20, 19
12. _?_, 35, 46, 57, 68
13. 28, 20, 14, 10, _?_
14. 23, 28, 25, _?_, 27, 32
15. 9, 10, 18, 19, 27, _?_, 36, 37
16. _?_, 6, 5, 9, 8, 12, 11, 15

Copy this square and fill in your answers. If the sum of your answers along each row, column, and diagonal is the same, your answers are correct!

You may want to use a calculator to find the sums.

| 1. | 2. | 3. | 4. |
|---|---|---|---|
| 5. | 6. | 7. | 8. |
| 9. | 10. | 11. | 12. |
| 13. | 14. | 15. | 16. |

a b c d [ ] [ ] [ ] [ ] 2. a b c d [ ] [ ] [ ] [ ] a b c d a b c d 5. a b c d [ ] [ ] [ ] [ ] 6. a b c [ ] [ ] [ ]

a b c d [ ] [ ] [ ] [ ] 8. a b c d [ ] [ ] [ ] [ ] 9. [ ] [ ] [ ] [ ] 10. [ ] [ ] [ ] [ ] 11. a b c d [ ] [ ] [ ] [ ] 12. a b c [ ] [ ] [ ]

# MAJOR CHECKUP

## STANDARDIZED FORMAT

**Choose the correct letter.**

**1.** In 5834, the 8 stands for

- **a.** 80
- **b.** 8000
- **c.** 800
- **d.** none of these

**2.** In 267,810, the digit in the ten thousands place is

- **a.** 6
- **b.** 7
- **c.** 8
- **d.** none of these

**3.** The standard numeral for four hundred thousand, forty-four is

- **a.** 444,000
- **b.** 400,044
- **c.** 400,440
- **d.** none of these

**4.** Which number is greatest?

- **a.** 416,382
- **b.** 418,015
- **c.** 389,996
- **d.** 58,361

**5.** 547,376 rounded to the nearest thousand is

- **a.** 547,000
- **b.** 548,000
- **c.** 547,400
- **d.** none of these

**6.** In 356,749,182 the digit in the millions place is

- **a.** 6
- **b.** 3
- **c.** 7
- **d.** none of these

**7.** Add.

$$\begin{array}{r} 47{,}839 \\ +32{,}816 \\ \hline \end{array}$$

- **a.** 79,645
- **b.** 79,655
- **c.** 80,645
- **d.** none of these

**8.** Add.

$$\begin{array}{r} 382 \\ 167 \\ 593 \\ +142 \\ \hline \end{array}$$

- **a.** 1284
- **b.** 1274
- **c.** 1084
- **d.** none of these

**9.** Subtract.

$$\begin{array}{r} 59{,}281 \\ -36{,}948 \\ \hline \end{array}$$

- **a.** 23,747
- **b.** 23,333
- **c.** 22,333
- **d.** none of these

**10.** Subtract.

$$\begin{array}{r} 650{,}378 \\ -149{,}739 \\ \hline \end{array}$$

- **a.** 519,441
- **b.** 500,639
- **c.** 501,639
- **d.** none of these

**11.** Helen spent $30 for a clock and a sweater. The clock cost $18. What did the sweater cost?

- **a.** $12
- **b.** $14
- **c.** $48
- **d.** none of these

**12.** Mike had $12. He bought a book for $4 and a record for $5. How much did he spend?

- **a.** $7
- **b.** $8
- **c.** $9
- **d.** $21

# Multiplication

# 3

# The basic multiplication facts

The numbers that you multiply are called factors. The answer is called the product. It is important that you memorize all the basic multiplication facts.

$2 \times 6 = 12$ (2 and 6 are the factors; 12 is the product)

**The Order Property**
Changing the order of the factors does not change the product.

$$3 \times 2 = 2 \times 3$$

$$4 \times 5 = 5 \times 4$$

**The Multiplying by 1 Property**
The product of any number and 1 is that same number.

$$4 \times 1 = 4$$

$$7 \times 1 = 7$$

**The Multiplying by 0 Property**
The product of any number and 0 is 0.

$$5 \times 0 = 0$$

$$8 \times 0 = 0$$

**The Grouping Property**
Changing the grouping of the factors does not change the product.

$$(4 \times 2) \times 3 = 4 \times (2 \times 3)$$

$$(3 \times 1) \times 5 = 3 \times (1 \times 5)$$

## EXERCISES

**Give each product.**

1. If $4 \times 2 = 8$, then $2 \times 4 = \underline{?}$.
2. If $8 \times 7 = 56$, then $7 \times 8 = \underline{?}$.
3. If $5 \times 4 = 20$, then $4 \times 5 = \underline{?}$.
4. If $9 \times 8 = 72$, then $8 \times 9 = \underline{?}$.
5. If $3 \times 7 = 21$, then $7 \times 3 = \underline{?}$.
6. If $7 \times 9 = \underline{?}$, then $9 \times 7 = \underline{?}$.
7. If $7 \times 6 = 42$, then $6 \times 7 = \underline{?}$.
8. If $6 \times 8 = \underline{?}$, then $8 \times 6 = \underline{?}$.

**9.** $6 \times 1 = \underline{?}$ **10.** $9 \times 0 = \underline{?}$ **11.** $1 \times 8 = \underline{?}$ **12.** $0 \times 7 = \underline{?}$

**13.** $5 \times 0 = \underline{?}$ **14.** $1 \times 9 = \underline{?}$ **15.** $0 \times 0 = \underline{?}$ **16.** $1 \times 1 = \underline{?}$

**Complete.**

**17.** a. $(3 \times 2) \times 4 = \underline{?}$
b. $3 \times (2 \times 4) = \underline{?}$

**18.** a. $(3 \times 3) \times 2 = \underline{?}$
b. $3 \times (3 \times 2) = \underline{?}$

**19.** a. $(4 \times 2) \times 1 = \underline{?}$
b. $4 \times (2 \times 1) = \underline{?}$

**20.** a. $(2 \times 2) \times 4 = \underline{?}$
b. $2 \times (2 \times 4) = \underline{?}$

**Study the shortcut. Then use the shortcut to find each product.**

$7 \times 6 =$ (6 less)
$8 \times 6 = 48$
$9 \times 6 =$ (6 greater)

21. If $7 \times 7 = 49$, then $8 \times 7 = \underline{?}$.
22. If $6 \times 9 = 54$, then $7 \times 9 = \underline{?}$.
23. If $8 \times 4 = 32$, then $9 \times 4 = \underline{?}$.
24. If $8 \times 8 = 64$, then $9 \times 8 = \underline{?}$.
25. If $6 \times 5 = 30$, then $5 \times 5 = \underline{?}$.
26. If $7 \times 9 = 63$, then $6 \times 9 = \underline{?}$.

# Practicing the basic multiplication facts

**WHAT'S YOUR SCORE?**

**All the basic multiplication facts are listed below. See how quickly you can list the products.**

1. $9 \times 2$ 2. $2 \times 1$ 3. $7 \times 5$ 4. $0 \times 2$ 5. $2 \times 8$ 6. $1 \times 5$ 7. $0 \times 8$ 8. $2 \times 5$

9. $1 \times 0$ 10. $0 \times 3$ 11. $1 \times 4$ 12. $8 \times 2$ 13. $5 \times 4$ 14. $4 \times 8$ 15. $0 \times 0$ 16. $0 \times 1$

17. $3 \times 8$ 18. $7 \times 7$ 19. $5 \times 9$ 20. $8 \times 5$ 21. $5 \times 0$ 22. $6 \times 1$ 23. $2 \times 0$ 24. $9 \times 8$

25. $0 \times 6$ 26. $9 \times 3$ 27. $1 \times 3$ 28. $9 \times 1$ 29. $2 \times 9$ 30. $6 \times 4$ 31. $4 \times 6$ 32. $8 \times 4$

33. $8 \times 0$ 34. $2 \times 6$ 35. $4 \times 7$ 36. $3 \times 4$ 37. $5 \times 8$ 38. $9 \times 5$ 39. $7 \times 9$ 40. $7 \times 0$

41. $8 \times 8$ 42. $8 \times 6$ 43. $3 \times 6$ 44. $9 \times 0$ 45. $1 \times 6$ 46. $4 \times 1$ 47. $2 \times 2$ 48. $0 \times 9$

49. $3 \times 3$ 50. $5 \times 3$ 51. $2 \times 3$ 52. $5 \times 2$ 53. $9 \times 7$ 54. $2 \times 4$ 55. $1 \times 1$ 56. $8 \times 7$

57. $1 \times 9$ 58. $6 \times 3$ 59. $8 \times 3$ 60. $4 \times 9$ 61. $0 \times 7$ 62. $1 \times 8$ 63. $3 \times 9$ 64. $4 \times 0$

65. $9 \times 6$ 66. $1 \times 2$ 67. $5 \times 5$ 68. $7 \times 8$ 69. $2 \times 7$ 70. $6 \times 0$ 71. $3 \times 1$ 72. $9 \times 9$

73. $1 \times 7$ 74. $4 \times 4$ 75. $3 \times 7$ 76. $7 \times 1$ 77. $5 \times 6$ 78. $4 \times 5$ 79. $7 \times 6$ 80. $6 \times 7$

81. $7 \times 2$ 82. $3 \times 2$ 83. $4 \times 2$ 84. $6 \times 2$ 85. $0 \times 4$ 86. $3 \times 5$ 87. $5 \times 1$ 88. $6 \times 9$

89. $6 \times 6$ 90. $8 \times 1$ 91. $6 \times 5$ 92. $0 \times 5$ 93. $4 \times 3$ 94. $5 \times 7$ 95. $6 \times 8$ 96. $8 \times 9$

97. $7 \times 4$ 98. $7 \times 3$ 99. $3 \times 0$ 100. $9 \times 4$

**Check your answers.**

| | | | | | | | |
|---|---|---|---|---|---|---|---|
| **1.** 18 | **2.** 2 | **3.** 35 | **4.** 0 | **5.** 16 | **6.** 5 | **7.** 0 | **8.** 10 |
| **9.** 0 | **10.** 0 | **11.** 4 | **12.** 16 | **13.** 20 | **14.** 32 | **15.** 0 | **16.** 0 |
| **17.** 24 | **18.** 49 | **19.** 45 | **20.** 40 | **21.** 0 | **22.** 6 | **23.** 0 | **24.** 72 |
| **25.** 0 | **26.** 27 | **27.** 3 | **28.** 9 | **29.** 18 | **30.** 24 | **31.** 24 | **32.** 32 |
| **33.** 0 | **34.** 12 | **35.** 28 | **36.** 12 | **37.** 40 | **38.** 45 | **39.** 63 | **40.** 0 |
| **41.** 64 | **42.** 48 | **43.** 18 | **44.** 0 | **45.** 6 | **46.** 4 | **47.** 4 | **48.** 0 |
| **49.** 9 | **50.** 15 | **51.** 6 | **52.** 10 | **53.** 63 | **54.** 8 | **55.** 1 | **56.** 56 |
| **57.** 9 | **58.** 18 | **59.** 24 | **60.** 36 | **61.** 0 | **62.** 8 | **63.** 27 | **64.** 0 |
| **65.** 54 | **66.** 2 | **67.** 25 | **68.** 56 | **69.** 14 | **70.** 0 | **71.** 3 | **72.** 81 |
| **73.** 7 | **74.** 16 | **75.** 21 | **76.** 7 | **77.** 30 | **78.** 20 | **79.** 42 | **80.** 42 |
| **81.** 14 | **82.** 6 | **83.** 8 | **84.** 12 | **85.** 0 | **86.** 15 | **87.** 5 | **88.** 54 |
| **89.** 36 | **90.** 8 | **91.** 30 | **92.** 0 | **93.** 12 | **94.** 35 | **95.** 48 | **96.** 72 |
| | | **97.** 28 | **98.** 21 | **99.** 0 | **100.** 36 | | |

**101.** Practice the facts that you missed.

**102.** Work with a classmate. See how quickly you can give the products orally. Have your partner keep a tally of your mistakes.

**103.** To find your score, add the number of seconds and the number of mistakes. Try for a better (lower) score.

**104.** Practice your facts for the next several days and keep a line graph of your scores as shown.

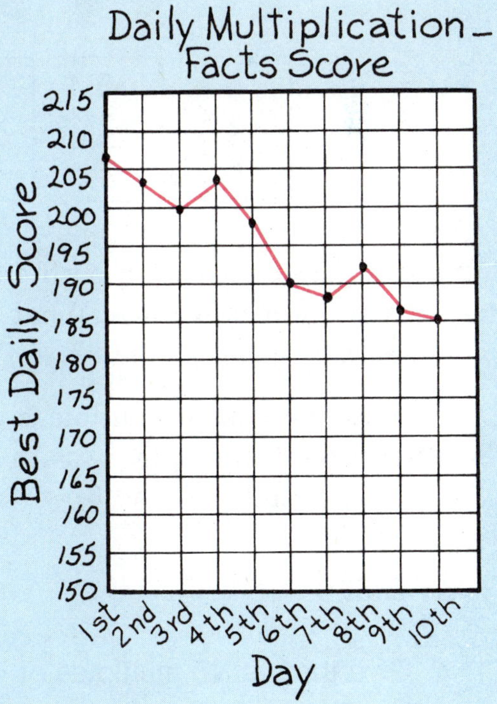

# Multiples and common multiples

4, 8, 12, 16, 20, and 24 are the first few **multiples** of 4.

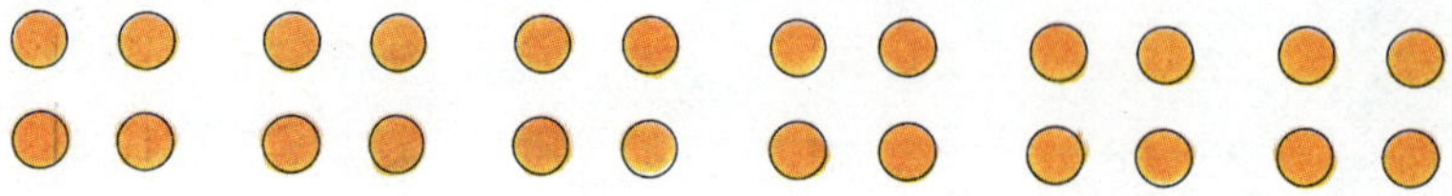

| $4 \times 1 = 4$ | $4 \times 2 = 8$ | $4 \times 3 = 12$ | $4 \times 4 = 16$ | $4 \times 5 = 20$ | $4 \times 6 = 24$ |
|---|---|---|---|---|---|

6, 12, 18, 24, and 30 are the first few multiples of 6.

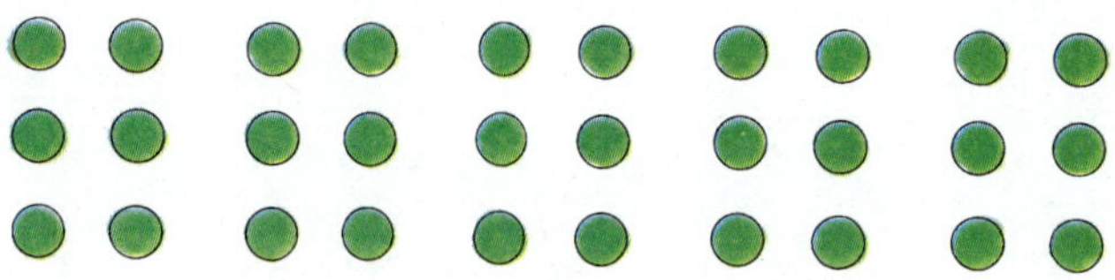

| $6 \times 1 = 6$ | $6 \times 2 = 12$ | $6 \times 3 = 18$ | $6 \times 4 = 24$ | $6 \times 5 = 30$ |
|---|---|---|---|---|

12 and 24 are multiples of both 4 and 6. They are called **common multiples.** What is the next common multiple of 4 and 6?

The smallest common multiple is called the **least common multiple.** 12 is the least common multiple of 4 and 6.

## EXERCISES

1. a. Give the first 15 multiples of 4.
   b. Give the first 15 multiples of 5.
   c. What are the first 3 common multiples of 4 and 5?
   d. What is the least common multiple of 4 and 5?

2. a. Give the first 10 multiples of 9.
   b. Give the first 10 multiples of 6.
   c. What are the first 3 common multiples of 9 and 6?
   d. What is the least common multiple of 9 and 6?

**Give the least common multiple of the two numbers.**

| | | | |
|---|---|---|---|
| 3. 2, 4 | 4. 3, 4 | 5. 2, 9 | 6. 4, 5 |
| 7. 4, 6 | 8. 10, 6 | 9. 5, 3 | 10. 9, 3 |
| 11. 5, 6 | 12. 3, 12 | 13. 4, 10 | 14. 7, 8 |

0 and multiples of 2 are called even numbers. The other whole numbers are called odd numbers.

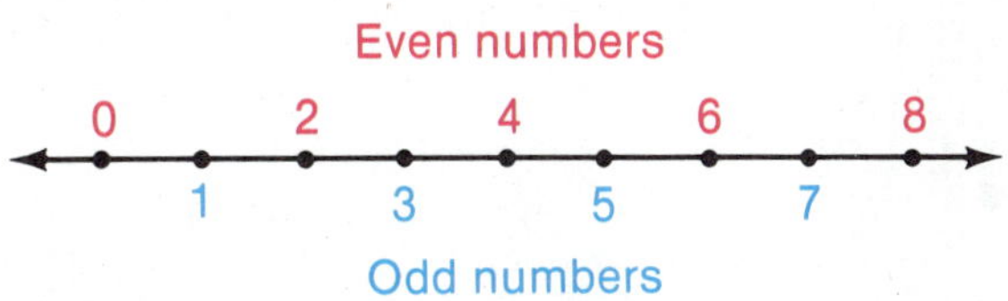


**Even or odd?**

15. 10 16. 17 17. 28 18. 23 19. 39

20. If you add two even numbers, the sum is _?_.
21. The sum of two odd numbers is _?_.
22. The difference of two even numbers is _?_.
23. The product of two odd numbers is _?_.
24. If you multiply two even numbers, the product is _?_.

## *Challenge!*

**The ones digit has been painted over.**
**Tell what the number could be.**

25. 26. 27. 28.

# Multiplying a 2-digit number

There are 2 sets of blocks with 34 blocks in each set. To find the total number of blocks, multiply.

$$\begin{array}{r} 34 \\ \times 2 \\ \hline \end{array}$$

**Step 1.** Multiply to find how many ones.

$$\begin{array}{r} 34 \\ \times 2 \\ \hline 8 \end{array}$$

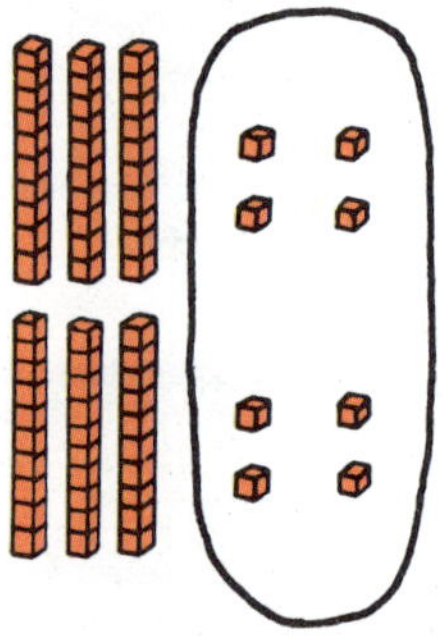

**Step 2.** Multiply to find how many tens.

$$\begin{array}{r} 34 \\ \times 2 \\ \hline 68 \end{array}$$

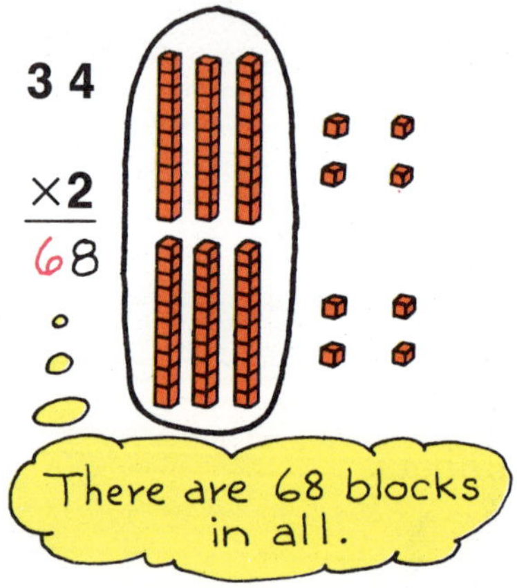


In this example, 10 ones are regrouped for 1 ten.

**Step 1.** Multiply to find how many ones.

$$\begin{array}{r} 26 \\ \times 3 \\ \hline \end{array}$$

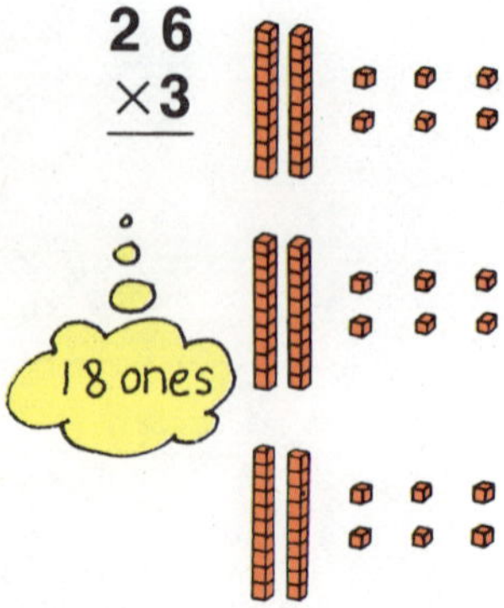


**Step 2.** Regroup.

$$\begin{array}{r} {}^{1}\phantom{6} \\ 26 \\ \times 3 \\ \hline 8 \end{array}$$

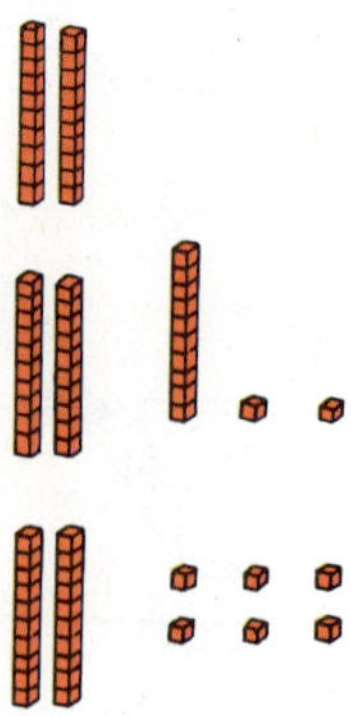

**Step 3.** Multiply and add to find how many tens.

$$\begin{array}{r} {}^{1}\phantom{6} \\ 26 \\ \times 3 \\ \hline 78 \end{array}$$

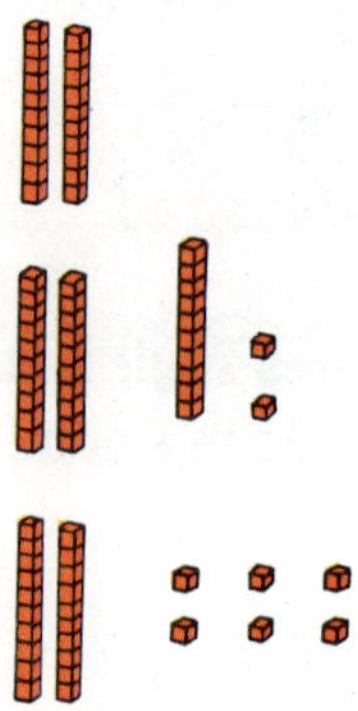

## EXERCISES

**Multiply.**

| | | | | | |
|---|---|---|---|---|---|
| 1. 23 ×3 | 2. 22 ×4 | 3. 10 ×5 | 4. 21 ×4 | 5. 32 ×3 | 6. 10 ×6 |
| 7. 20 ×4 | 8. 32 ×3 | 9. 12 ×4 | 10. 23 ×2 | 11. 10 ×7 | 12. 11 ×6 |
| 13. 26 ×3 | 14. 35 ×2 | 15. 45 ×2 | 16. 28 ×3 | 17. 16 ×4 | 18. 18 ×5 |
| 19. 68 ×2 | 20. 46 ×5 | 21. 38 ×6 | 22. 49 ×4 | 23. 76 ×8 | 24. 53 ×9 |
| 25. 74 ×7 | 26. 58 ×4 | 27. 66 ×9 | 28. 93 ×6 | 29. 87 ×8 | 30. 75 ×5 |
| 31. 53 ×5 | 32. 74 ×8 | 33. 85 ×9 | 34. 67 ×5 | 35. 39 ×7 | 36. 58 ×6 |

| | | | | |
|---|---|---|---|---|
| 37. \$.28 ×2 | 38. \$.38 ×6 | 39. \$.41 ×7 | 40. \$.68 ×9 | 41. \$.57 ×3 |
| 42. \$.85 ×6 | 43. \$.92 ×8 | 44. \$.49 ×5 | 45. \$.79 ×8 | 46. \$.57 ×9 |

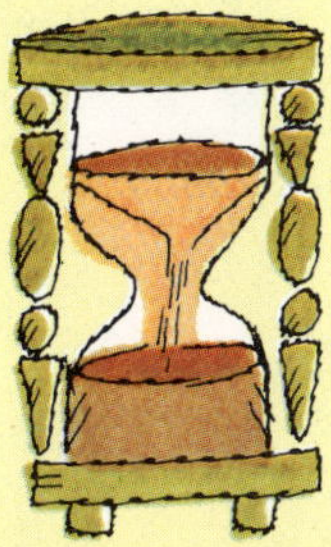

47. How many seconds are in 8 minutes?

48. How many minutes are in 7 hours?

49. How many hours are in 6 days?

# Multiplying a 3-digit number

**EXAMPLE.**
$$\begin{array}{r} 264 \\ \times 3 \\ \hline \end{array} \quad \text{rounds to} \quad \begin{array}{r} 300 \\ \times 3 \\ \hline 900 \end{array}$$

Is the product less than or greater than 900?

**Step 1.** Multiply to find how many ones.

$$\begin{array}{r} 264 \\ \times 3 \\ \hline \end{array}$$
12 ones

**Step 2.** Regroup.

$$\begin{array}{r} \scriptsize{1}\phantom{4} \\ 264 \\ \times 3 \\ \hline 2 \end{array}$$
12 ones = 1 ten + 2 ones

**Step 3.** Multiply and add to find how many tens.

$$\begin{array}{r} \scriptsize{1}\phantom{4} \\ 264 \\ \times 3 \\ \hline 2 \end{array}$$
19 tens

**Step 4.** Regroup.

$$\begin{array}{r} \scriptsize{1\,1}\phantom{4} \\ 264 \\ \times 3 \\ \hline 92 \end{array}$$
19 tens = 1 hundred + 9 tens

**Step 5.** Multiply and add to find how many hundreds.

$$\begin{array}{r} \scriptsize{1\,1}\phantom{4} \\ 264 \\ \times 3 \\ \hline 792 \end{array}$$

## EXERCISES

**First estimate the product.**
**Then multiply.**

1. 124 × 3
2. 238 × 4
3. 316 × 3
4. 278 × 2
5. 156 × 4
6. 387 × 5
7. 425 × 6
8. 800 × 8
9. 742 × 5
10. 502 × 9
11. 726 × 4
12. 656 × 3
13. 529 × 5
14. 599 × 7
15. 680 × 6
16. 538 × 5
17. 742 × 8
18. 613 × 7
19. 467 × 9
20. 962 × 5
21. \$2.58 × 7
22. \$4.93 × 3
23. \$9.02 × 4
24. \$7.83 × 6
25. \$6.75 × 8

**Solve.**

26. 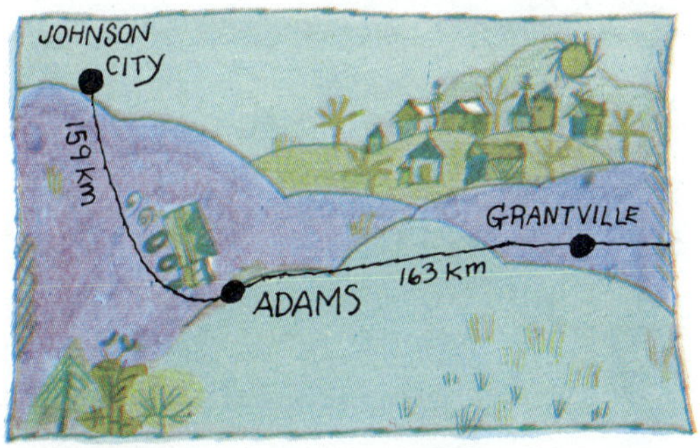


Each week, a truck driver drives 5 round trips from Johnson City to Grantville. How far is that?

27. There are 144 screws in 1 box. How many screws are in 6 boxes?

28. A certain kind of bus seats 46 people. How many people will 9 such buses seat?

29. Each member of a youth club wants to collect 375 kilograms of paper. There are 8 members. How much should they collect in all?

### *Challenge!*

**Find the missing digits.**

30. □8□ × 4 = 3140

31. □94 × □ = 4164

32. □□□ × 8 = 2864

# Multiplying larger numbers

You multiply a larger number the same way you multiply a smaller number.

**Step 1.** Multiply and regroup.

$$\begin{array}{r} {}^{3}\;\; \\ 9364 \\ \times 8 \\ \hline 2 \end{array}$$

**Step 2.** Multiply, add, and regroup.

$$\begin{array}{r} {}^{5\,3}\;\; \\ 9364 \\ \times 8 \\ \hline 12 \end{array}$$

**Step 3.** Multiply, add, and regroup.

$$\begin{array}{r} {}^{2\,5\,3}\;\; \\ 9364 \\ \times 8 \\ \hline 912 \end{array}$$

**Step 4.** Multiply and add.

$$\begin{array}{r} {}^{2\,5\,3}\;\; \\ 9364 \\ \times 8 \\ \hline 74{,}912 \end{array}$$

*Other examples:*

$$\begin{array}{r} {}^{4\,4\,1\,2}\;\; \\ 37{,}824 \\ \times 6 \\ \hline 226{,}944 \end{array} \qquad \begin{array}{r} {}^{2\,1\,1\,3\,2}\;\; \\ 753{,}296 \\ \times 4 \\ \hline 3{,}013{,}184 \end{array}$$

## EXERCISES

**Multiply.**

1. 3274 ×5
2. 9106 ×4
3. 3582 ×6
4. 7136 ×3
5. 52,816 ×4
6. 39,428 ×2
7. 74,385 ×3
8. 60,342 ×5
9. 216,342 ×3
10. 516,394 ×8
11. 491,763 ×6
12. 378,219 ×5
13. 258 × 6
14. 423 × 9
15. 8 × 7421
16. 5 × 8416
17. 3506 × 7
18. 5106 × 5
19. 4 × 39,174
20. 37,468 × 3

**Solve.**

21. How many pictures can you take with 6 rolls of this film?
22. What is the total price of 4 rolls of film?

### *Challenge!*

**Multiply.**

23. 12,345,679 ×9
24. 12,345,679 ×18
25. 12,345,679 ×27

### KEEPING SKILLS SHARP

1. 925 −258
2. 845 −674
3. 8210 −374
4. 6107 −539
5. 2000 −1374
6. 7123 −5160
7. 1111 −66
8. 2525 −1517
9. 4903 −4895
10. 6157 −359

# Multiplication practice

**Multiply.**

| | | | | |
|---|---|---|---|---|
| 1. 38 ×2 | 2. 42 ×4 | 3. 57 ×3 | 4. 93 ×8 | 5. 67 ×2 |
| 6. 49 ×5 | 7. 72 ×9 | 8. 80 ×3 | 9. 19 ×9 | 10. 53 ×8 |
| 11. 427 ×9 | 12. 674 ×2 | 13. 174 ×9 | 14. 459 ×4 | 15. 768 ×8 |
| 16. 201 ×9 | 17. 562 ×6 | 18. 750 ×3 | 19. 832 ×4 | 20. 593 ×4 |
| 21. 1038 ×7 | 22. 4061 ×4 | 23. 6783 ×5 | 24. 7354 ×6 | 25. 5216 ×7 |
| 26. 7159 ×4 | 27. 5938 ×6 | 28. 2654 ×7 | 29. 2614 ×3 | 30. 8092 ×5 |
| 31. 16,054 ×6 | 32. 23,091 ×4 | 33. 18,319 ×3 | 34. 40,038 ×5 | 35. 36,174 ×7 |
| 36. 59,308 ×3 | 37. 61,704 ×6 | 38. 23,852 ×7 | 39. 43,250 ×2 | 40. 63,172 ×4 |

**WATCH THE SIGNS**

**Copy and complete.**

41. (16 − 8) + 4 = ?
42. 16 − (8 + 4) = ?
43. (42 − 16) − 5 = ?
44. 42 − (16 − 5) = ?
45. (18 × 9) − 3 = ?
46. 18 × (9 − 3) = ?
47. (132 × 4) + 3 = ?
48. 132 × (4 + 3) = ?

# Problem solving—using computers

Jack's Bowladrome uses a computer to find out how much people must pay for bowling. The cashier enters into the computer the time of day and the age group of each player. Then the computer assigns an alley to the bowlers. When the bowlers finish playing, the computer tells how many games were played. Then it uses the time of day and the age group of each player to compute the total cost.

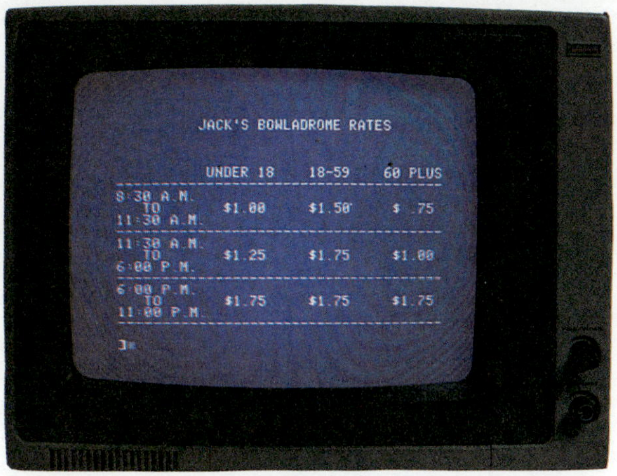


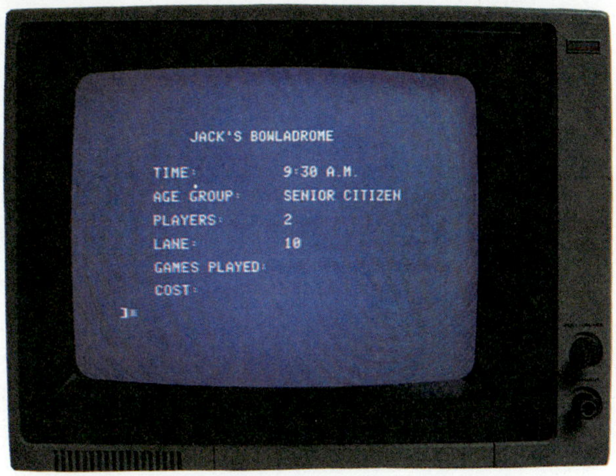


**Use the screens.**

1. When Mr. and Mrs. Netto went bowling, the computer printed the information shown on the second screen. They each bowled 1 game. How will the computer complete the last two lines when they finish bowling?

2. Irene is in fifth grade. One day she went bowling at 2:00 P.M. and bowled 3 games. How much will the computer tell her to pay?

3. Vincent went bowling with three classmates. They went at 7:30 P.M. and each bowled 2 games. How much will the computer tell them to pay?

★ 4. Mr. James and two other people went bowling at 10:30 A.M. They paid a total of $13.50. How many games did they bowl altogether?

★ 5. Melissa and four friends went bowling at 4:30 P.M. They each bowled one game. The total cost was $8.75. In what age group were Melissa and her friends?

# Problem solving

Tickets Sold for School Fun Night

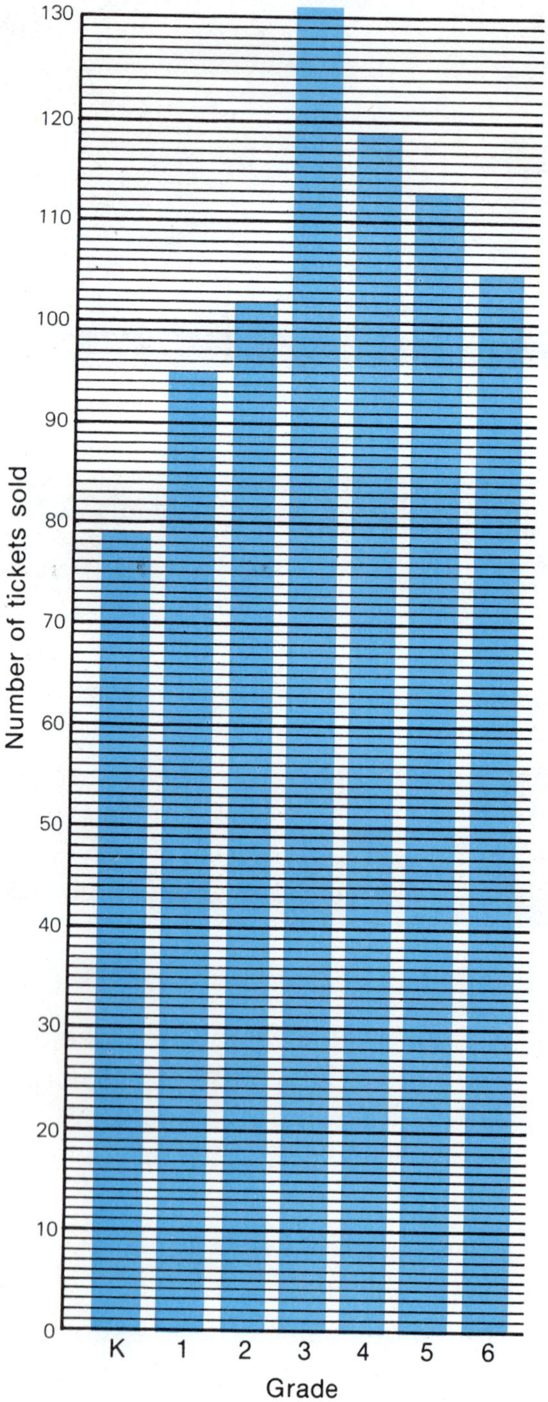

The bar graph shows the number of tickets sold by each grade for the school Fun Night.

**EXERCISES**

**Solve.**

1. How many tickets were sold by the kindergarten?
2. Which grade sold the most tickets?
3. How many grades sold more than 110 tickets?
4. How many grades sold less than 100 tickets?
5. Which grade sold exactly 102 tickets?
6. How many grades sold more than the sixth grade?

7. The sixth-grade goal was to sell 120 tickets. By how many tickets were they short of their goal?

8. The third-grade goal was to sell 125 tickets. By how many tickets were they over their goal?

9. Find the total number of tickets sold by grades K, 1, and 2.

10. What was the total number of tickets sold?

11. The fourth grade sold how many more tickets than the fifth?

12. Tickets cost $2.75 each. Joan's father bought 4 tickets. How much did they cost?

13. A book of tickets contained 16 tickets. Alan sold 9 of them but had to return the rest. How many tickets did he return?

14. A book of tickets contained 16 tickets. Althea and Paul together sold 3 books of tickets. How many tickets did they sell?

15. Of the 744 people who bought tickets, 658 went to Fun Night. How many did not go to Fun Night?

16. The school made $372 on tickets and $243.20 on refreshments, and spent $78.75 for door prizes. How much profit did they make?

# Multiplying by 10 and 100

There are 10 bottle caps in each stack. To find the total number of bottle caps, you can multiply.

$$\begin{array}{r} 10 \\ \times 7 \\ \hline 70 \end{array} \qquad \begin{array}{r} 7 \\ \times 10 \\ \hline 70 \end{array}$$

Study these products.

$$\begin{array}{r} 9 \\ \times 10 \\ \hline 90 \end{array} \qquad \begin{array}{r} 25 \\ \times 10 \\ \hline 250 \end{array} \qquad \begin{array}{r} 186 \\ \times 10 \\ \hline 1860 \end{array} \qquad \begin{array}{r} 2436 \\ \times 10 \\ \hline 24{,}360 \end{array}$$

There are 100 bottle caps in each pile.

$$\begin{array}{r} 100 \\ \times 3 \\ \hline 300 \end{array} \qquad \begin{array}{r} 3 \\ \times 100 \\ \hline 300 \end{array}$$

Study these products.

$$\begin{array}{r} 6 \\ \times 100 \\ \hline 600 \end{array} \qquad \begin{array}{r} 12 \\ \times 100 \\ \hline 1200 \end{array} \qquad \begin{array}{r} 152 \\ \times 100 \\ \hline 15{,}200 \end{array} \qquad \begin{array}{r} 283 \\ \times 100 \\ \hline 28{,}300 \end{array} \qquad \begin{array}{r} 3426 \\ \times 100 \\ \hline 342{,}600 \end{array}$$

What is an easy way to multiply a number by 100?

## EXERCISES

**Multiply.**

1. $16 \times 10$
2. $19 \times 10$
3. $25 \times 10$
4. $37 \times 10$
5. $68 \times 10$
6. $125 \times 10$
7. $150 \times 10$
8. $246 \times 10$
9. $373 \times 10$
10. $658 \times 10$
11. $5214 \times 10$
12. $3782 \times 10$
13. $5963 \times 10$
14. $4000 \times 10$
15. $7800 \times 10$
16. $18 \times 100$
17. $35 \times 100$
18. $64 \times 100$
19. $80 \times 100$
20. $92 \times 100$
21. $130 \times 100$
22. $152 \times 100$
23. $268 \times 100$
24. $642 \times 100$
25. $867 \times 100$
26. $3214 \times 100$
27. $2178 \times 100$
28. $4320 \times 100$
29. $5000 \times 100$
30. $3871 \times 100$

**Complete.**

31.

| Meters | 24 | 68 | 142 | 236 | 458 | 698 |
|---|---|---|---|---|---|---|
| Centimeters | ? | ? | ? | ? | ? | ? |

*Remember:*
1 meter = 100 centimeters

**Solve.**

32. There were 15 players on a Little League team. Each player sold 10 tickets. How many tickets did the team sell?

33. To raise money, 825 students at Lowry Elementary School each ran 100 meters in a school relay. How many meters was the relay?

### *Challenge!*

34. Guess the product of these ten numbers.
    1 2 3 4 5 6 7 8 9 10

35. Find the product. How close was your guess?

# Multiplying by multiples of 10 and 100

Notice that there are 24 stamps on each sheet, 3 sheets in each stack, and 10 stacks.

Here is how to find the total number of stamps.

| | |
|---:|---|
| 24 | stamps on each sheet |
| ×3 | sheets in each stack |
| 72 | stamps in each stack |
| ×10 | stacks |
| 720 | stamps |

$$\begin{array}{r} 24 \\ \times 30 \\ \hline 720 \end{array}$$

$$\begin{array}{r} 154 \\ \times 80 \\ \hline 12{,}320 \end{array} \qquad \begin{array}{r} 256 \\ \times 200 \\ \hline 51{,}200 \end{array} \qquad \begin{array}{r} 593 \\ \times 400 \\ \hline 237{,}200 \end{array} \qquad \begin{array}{r} 742 \\ \times 600 \\ \hline 445{,}200 \end{array}$$

## EXERCISES

**Multiply.**

| | | | | |
|---|---|---|---|---|
| **1.** $24 \times 20$ | **2.** $42 \times 50$ | **3.** $63 \times 70$ | **4.** $82 \times 40$ | **5.** $56 \times 60$ |
| **6.** $125 \times 50$ | **7.** $156 \times 30$ | **8.** $240 \times 70$ | **9.** $356 \times 80$ | **10.** $456 \times 60$ |
| **11.** $571 \times 50$ | **12.** $602 \times 60$ | **13.** $800 \times 90$ | **14.** $456 \times 70$ | **15.** $783 \times 80$ |
| **16.** $295 \times 200$ | **17.** $746 \times 400$ | **18.** $519 \times 300$ | **19.** $920 \times 600$ | **20.** $693 \times 500$ |
| **21.** $527 \times 800$ | **22.** $640 \times 700$ | **23.** $806 \times 600$ | **24.** $758 \times 900$ | **25.** $846 \times 500$ |

**Solve.**

**26.**

How many crayons in 30 boxes?

**27.**

How many sheets of paper in 200 packages?

**28.** A truck farmer planted 260 tomato plants in each of 60 rows. How many tomato plants did he plant?

**29.** To build a section of fence for her dog, Miss Bailey needs 512 nails. She plans to build 20 sections. How many nails does she need?

***Challenge!***

**Multiply across. Multiply down.**

**30.**

| | | |
|---|---|---|
| 24 | 4 | ? |
| 20 | 5 | ? |
| ? | ? | ? |

**31.**

| | | |
|---|---|---|
| 52 | 6 | ? |
| 8 | 5 | ? |
| ? | ? | ? |

**32.**

| | | |
|---|---|---|
| 68 | 5 | ? |
| 20 | 4 | ? |
| ? | ? | ? |

# Multiplying by a 2-digit number

Twenty-one cases of soup have been delivered to the supermarket. How many cans of soup is that?

There are 24 cans in each case. We can find the total number of cans by using the **distributive property:**

$$21 \times 24 = (20 \times 24) + (1 \times 24)$$

**Step 1.** Multiply by 1.

$$\begin{array}{r} 24 \\ \times 21 \\ \hline 24 \end{array} \text{ cans in 1 case}$$

**Step 2.** Multiply by 20.

$$\begin{array}{r} 24 \\ \times 21 \\ \hline 24 \\ 480 \\ \hline \end{array} \text{ cans in 20 cases}$$

**Step 3.** Add.

$$\begin{array}{r} 24 \\ \times 21 \\ \hline 24 \\ 480 \\ \hline 504 \end{array} \text{ cans in all 21 cases}$$

Another example:

$$\begin{array}{r} 3\,2\phantom{0} \\ 5\,3\phantom{0} \\ 574 \\ \times 58 \\ \hline 4592 \\ 28700 \\ \hline 33{,}292 \end{array} \quad \begin{array}{l} \\ \\ \\ \\ \\ (8 \times 574) \\ (50 \times 574) \\ \\ \end{array}$$

## EXERCISES

**Give each product.**

| | | | | |
|---|---|---|---|---|
| 1. 53 ×24 | 2. 37 ×25 | 3. 73 ×32 | 4. 95 ×43 | 5. 60 ×38 |
| 6. 49 ×49 | 7. 83 ×45 | 8. 54 ×53 | 9. 78 ×64 | 10. 95 ×58 |
| 11. \$.96 ×25 | 12. \$.57 ×36 | 13. \$.74 ×62 | 14. \$.62 ×40 | 15. \$.81 ×74 |
| 16. 158 ×31 | 17. 296 ×45 | 18. 342 ×62 | 19. 506 ×51 | 20. 821 ×78 |
| 21. \$1.42 ×46 | 22. \$6.95 ×70 | 23. \$9.24 ×43 | 24. \$5.38 ×97 | 25. \$8.26 ×64 |

**Solve.**

26. A can of soup costs $.57. How much will 24 cans cost?

27. A 20-ounce loaf of bread costs $1.29 and a 16-ounce loaf of bread costs $1.06. How much do two 16-ounce loaves cost?

28. A ham weighs 12 pounds 9 ounces. How many ounces does it weigh? (*Hint:* 1 pound = 16 ounces.)

29. Ground beef is on sale for $1.65 a pound. You have $20. Can you buy 12 pounds of ground beef?

### *Challenge!*

**Guess my number.**

30. My number is 68 greater than the product of 42 and 16.

31. My number is 54 less than the product of 232 and 34.

# Multiplication practice

## EXERCISES

Give each product.

| | | | | |
|---|---|---|---|---|
| 1. 34 ×12 | 2. 85 ×14 | 3. 91 ×11 | 4. 62 ×23 | 5. 52 ×31 |
| 6. 47 ×45 | 7. 79 ×25 | 8. 44 ×33 | 9. 87 ×46 | 10. 43 ×18 |
| 11. 46 ×21 | 12. 55 ×48 | 13. 94 ×24 | 14. 65 ×52 | 15. 74 ×26 |
| 16. 68 ×32 | 17. 77 ×42 | 18. 35 ×18 | 19. 60 ×44 | 20. 63 ×35 |
| 21. 42 ×38 | 22. 54 ×53 | 23. 90 ×72 | 24. 86 ×65 | 25. 36 ×17 |
| 26. 69 ×43 | 27. 76 ×19 | 28. 45 ×35 | 29. 75 ×34 | 30. 53 ×46 |
| 31. 136 ×23 | 32. 320 ×41 | 33. 411 ×35 | 34. 256 ×60 | 35. 521 ×33 |
| 36. 753 ×45 | 37. 222 ×82 | 38. 158 ×29 | 39. 468 ×53 | 40. 243 ×64 |
| 41. 190 ×42 | 42. 824 ×43 | 43. 578 ×73 | 44. 172 ×26 | 45. 303 ×42 |

## SKILL GAME

**This game is for two people. Play the game.**

1. Copy the table of products shown below.
2. Player A picks **2** factors from the factor board and marks the product with an X.
3. Player B picks **2** factors from the factor board and marks the product with an O.
4. The first player to get 4 marks in a row, column, or diagonal wins.

## PRODUCTS

| | | | | | |
|---|---|---|---|---|---|
| 3010 | 160 | 240 | 3150 | 504 | 6364 |
| 688 | 3526 | 2590 | 1000 | 2050 | 492 |
| 1750 | 400 | 1720 | 328 | 1032 | 5418 |
| 1260 | 4300 | 600 | 3034 | 96 | 820 |
| 700 | 280 | 2205 | 592 | 1435 | 4662 |
| 2583 | 420 | 888 | 1480 | 756 | 3700 |

# Multiplying by a 3-digit number

This example shows how to multiply by a 3-digit number. If you estimate first, you can tell whether your answer is reasonable.

$$\begin{array}{r} 364 \\ \times 123 \\ \hline \end{array} \quad \begin{array}{l} \text{rounds to} \\ \text{rounds to} \end{array} \quad \begin{array}{r} 400 \\ \times 100 \\ \hline 40{,}000 \end{array}$$

**Step 1.**
Multiply by 3.

$$\begin{array}{r} 364 \\ \times 123 \\ \hline 1092 \end{array}$$

**Step 2.**
Multiply by 20.

$$\begin{array}{r} 364 \\ \times 123 \\ \hline 1092 \\ 7280 \end{array}$$

**Step 3.**
Multiply by 100.

$$\begin{array}{r} 364 \\ \times 123 \\ \hline 1092 \\ 7280 \\ 36400 \end{array}$$

**Step 4.**
Add.

$$\begin{array}{r} 364 \\ \times 123 \\ \hline 1092 \\ 7280 \\ 36400 \\ \hline 44{,}772 \end{array}$$

Here is another example.

$$\begin{array}{r} 2564 \\ \times 397 \\ \hline 17948 \\ 230760 \\ 769200 \\ \hline 1{,}017{,}908 \end{array}$$

## EXERCISES

**First estimate the product. Then multiply.**

1. $\begin{array}{r} 52 \\ \times 38 \\ \hline \end{array}$ $\begin{array}{r} 50 \\ \times 40 \\ \hline \end{array}$
2. $\begin{array}{r} 61 \\ \times 52 \\ \hline \end{array}$
3. $\begin{array}{r} 78 \\ \times 39 \\ \hline \end{array}$
4. $\begin{array}{r} 185 \\ \times 64 \\ \hline \end{array}$
5. $\begin{array}{r} 298 \\ \times 42 \\ \hline \end{array}$
6. $\begin{array}{r} 321 \\ \times 39 \\ \hline \end{array}$
7. $\begin{array}{r} 468 \\ \times 126 \\ \hline \end{array}$
8. $\begin{array}{r} 705 \\ \times 283 \\ \hline \end{array}$
9. $\begin{array}{r} 883 \\ \times 327 \\ \hline \end{array}$

**Multiply.**

10. $\begin{array}{r} 374 \\ \times 152 \\ \hline \end{array}$
11. $\begin{array}{r} 495 \\ \times 344 \\ \hline \end{array}$
12. $\begin{array}{r} 784 \\ \times 285 \\ \hline \end{array}$
13. $\begin{array}{r} 714 \\ \times 329 \\ \hline \end{array}$
14. $\begin{array}{r} 866 \\ \times 354 \\ \hline \end{array}$
15. $\begin{array}{r} 589 \\ \times 296 \\ \hline \end{array}$

★ 16. $\begin{array}{r} 7419 \\ \times 340 \\ \hline \end{array}$

★ 17. $\begin{array}{r} 3659 \\ \times 428 \\ \hline \end{array}$

★ 18. $\begin{array}{r} 8953 \\ \times 765 \\ \hline \end{array}$

**Solve.**

19. A computer graphics program prints out 168 rows of characters with 92 characters in each row. How many characters are in the display?
20. Ms. Allison's computer is on 7 hours a day, 5 days a week. How many hours is the computer on in 48 weeks?

## KEEPING SKILLS SHARP

1. $\begin{array}{r} 93 \\ -21 \\ \hline \end{array}$
2. $\begin{array}{r} 80 \\ -43 \\ \hline \end{array}$
3. $\begin{array}{r} 453 \\ -271 \\ \hline \end{array}$
4. $\begin{array}{r} 834 \\ -456 \\ \hline \end{array}$
5. $\begin{array}{r} 903 \\ -275 \\ \hline \end{array}$
6. $\begin{array}{r} 600 \\ -458 \\ \hline \end{array}$
7. $\begin{array}{r} 2319 \\ -528 \\ \hline \end{array}$
8. $\begin{array}{r} 5678 \\ -1759 \\ \hline \end{array}$
9. $\begin{array}{r} 32{,}005 \\ -6{,}384 \\ \hline \end{array}$
10. $\begin{array}{r} 91{,}063 \\ -30{,}847 \\ \hline \end{array}$
11. $\begin{array}{r} 827{,}609 \\ -59{,}318 \\ \hline \end{array}$
12. $\begin{array}{r} 508{,}394 \\ -186{,}503 \\ \hline \end{array}$

# More about multiplying

Richard Dorr delivers cases of milk to grocery stores. Each day he drives 103 kilometers on his route. He works 235 days a year. How many kilometers does he drive?

Estimate.

$$\begin{array}{r} 235 \\ \times 103 \\ \hline \end{array} \qquad \begin{array}{r} 200 \\ \times 100 \\ \hline 20{,}000 \end{array}$$

Is the product less than or greater than 20,000?

**Step 1.**
Multiply by 3.

$$\begin{array}{r} 235 \\ \times 103 \\ \hline 705 \end{array}$$

**Step 2.**
Multiply by 100.

$$\begin{array}{r} 235 \\ \times 103 \\ \hline 705 \\ 23500 \end{array}$$

**Step 3.**
Add.

$$\begin{array}{r} 235 \\ \times 103 \\ \hline 705 \\ 23500 \\ \hline 24{,}205 \end{array}$$

The product seems right.

He drives 24,205 kilometers.

**EXAMPLES.**

$$\begin{array}{r} 937 \\ \times 325 \\ \hline 4685 \\ 1874\phantom{0} \\ 2811\phantom{00} \\ \hline 304{,}525 \end{array} \qquad \begin{array}{r} 624 \\ \times 507 \\ \hline 4368 \\ 3120\phantom{00} \\ \hline 316{,}368 \end{array} \qquad \begin{array}{r} 827 \\ \times 150 \\ \hline 41350 \\ 827\phantom{00} \\ \hline 124{,}050 \end{array}$$

## EXERCISES

**First estimate the product. Then multiply.**

1. $261 \times 304$
2. $384 \times 607$
3. $703 \times 205$
4. $812 \times 600$
5. $432 \times 130$
6. $756 \times 240$
7. $860 \times 400$
8. $571 \times 310$
9. $629 \times 420$
10. $826 \times 573$
11. $946 \times 704$
12. $860 \times 306$

★ 13. $6734 \times 507$

★ 14. $8341 \times 608$

★ 15. $5943 \times 290$

**Solve.**

16. Each morning the Dairy gives 800 liters of milk to each of 13 drivers. How many liters of milk is that?

17. The Dairy sold 24 cartons of milk and 16 cartons of butter to a grocery store. The store paid $1.79 for each carton of milk and $1.89 for each carton of butter. What was the total cost?

## KEEPING SKILLS SHARP

**Give the least common multiple of the two numbers.**

1. 3 and 5
2. 3 and 6
3. 3 and 7
4. 4 and 6
5. 4 and 7
6. 4 and 8
7. 6 and 7
8. 6 and 8
9. 6 and 9
10. 7 and 1
11. 7 and 2
12. 7 and 3
13. 10 and 3
14. 10 and 4
15. 10 and 5
16. 10 and 6
17. 10 and 7
18. 10 and 8

# Practice exercises

Give each product.

| | | | | |
|---|---|---|---|---|
| 1. 77 ×86 | 2. 49 ×32 | 3. 57 ×40 | 4. 80 ×67 | 5. 50 ×40 |
| 6. 372 ×26 | 7. 406 ×37 | 8. 293 ×83 | 9. 756 ×48 | 10. 872 ×70 |
| 11. 935 ×45 | 12. 225 ×88 | 13. 248 ×25 | 14. 394 ×27 | 15. 630 ×19 |
| 16. 803 ×59 | 17. 159 ×73 | 18. 562 ×143 | 19. 781 ×222 | 20. 457 ×300 |
| 21. 283 ×500 | 22. 743 ×302 | 23. 643 ×507 | 24. 231 ×421 | 25. 775 ×806 |
| 26. 504 ×276 | 27. 703 ×543 | 28. 805 ×607 | 29. 906 ×505 | 30. 619 ×342 |
| 31. 460 ×527 | 32. 908 ×709 | 33. 638 ×419 | 34. 456 ×911 | 35. 920 ×105 |

# Computers store information

An important part of a computer is its **memory.** The memory is made of electronic circuits that can store information. The basic unit for storage of information is the **byte** (pronounced *bite*). You can think of a byte as the amount of memory required to store a single letter, digit, space, or punctuation mark. It takes 26 bytes to store

ACTON, MASSACHUSETTS 01720

because there are 18 letters, 5 numerals, 1 comma, and 2 spaces.

| Information | Bytes |
|---|---|
| 212-555-9381 | 12 |
| 917 S. Randolph | 15 |
| Nov. 20, 1983 | 13 |
| J. P. Jones | 11 |

**EXERCISES**

1. How many bytes of memory are needed to store your full name?

2. How much memory is needed to store the full names of all students in your class?

3. Get a telephone directory and open it to one of the white pages. How many listings are recorded on the page? (Can you find out without actually counting them all?)

4. Suppose each telephone listing uses 45 spaces. How many bytes of memory are needed to store the page of the telephone directory?

★ 5. Estimate the amount of memory needed to store all of the white-page listings.

# Problem solving—two or more steps

One year Jerry West scored a total of 2309 points. He scored 831 field goals worth 2 points each. The rest of the points were for free throws, worth 1 point each. How many free throws did he score?

**Step 1.**

Multiply to find the number of points he scored for field goals.

| | |
|---:|---|
| **831** | field goals |
| **×2** | points each |
| 1662 | points for field goals |

**Step 2.**

Subtract to find the number of points he scored for free throws.

| | |
|---:|---|
| **2309** | total points |
| **−1662** | points for field goals |
| 647 | points for free throws |

The number of points he scored for free throws is the same as the number of free throws.

He scored 647 free throws.

**Solve.**

1. One year Rick Barry scored 1011 field goals and 753 free throws. Each field goal was worth 2 points and each free throw 1 point. How many points did he score?

2. In one basketball stadium a section contains 32 rows with 26 seats in each row. At a game there were only 5 empty seats in the section. How many people sat in the section?

3. In one basketball stadium a section contains 32 rows with 26 seats in each row. There are 3 of these sections. How many seats are in the stadium?

4. A basketball stadium has 3 sections: A, B, and C. Section A has 3842 seats. Section B has 468 more seats than section A. There are 13,284 seats in all. How many seats are in section C?

5. Season tickets for 41 home games cost $270. Single tickets cost $6.50 each. How much more expensive are season tickets?

6. One year Larry Bird scored 635 field goals worth 2 points each and 58 field goals worth 3 points each. How many points did he score on field goals?

7. One year Julius "Doctor J" Erving scored a total of 1680 points. He scored 834 field goals worth 2 points each. The rest of the points were for 3-point field goals. How many 3-point field goals did he score?

8. One program seller sold 748 programs. The programs cost her $.40 each and she sold them for $.75 each. What was her total profit?

# CHAPTER CHECKUP

**Give the least common multiple.** [pages 64–65]

1. 4, 6
2. 3, 7
3. 6, 8
4. 9, 6
5. 8, 4

**Find each product.** [pages 60–63, 66–73, 76–88]

6. $42 \times 2$
7. $28 \times 4$
8. $279 \times 6$
9. $1496 \times 7$
10. $84{,}252 \times 8$
11. $342 \times 10$
12. $3482 \times 100$
13. $3938 \times 60$
14. $253 \times 52$
15. $792 \times 68$
16. $384 \times 200$
17. $391 \times 125$

**Solve.** [pages 74–75, 90–91]

18.

How many paper clips in 9 boxes?

19.

How many tacks in 36 boxes?

20. Ms. Davis owns a school supply store. She has 78 ball-point pens in one box and 120 ball-point pens in another box. How many pens does she have?

21. Ms. Davis has 124 boxes of pencils with 144 pencils in each box. How many pencils does she have?

# CHAPTER PROJECT

1. What was the date of your birth?
2. How many years old are you?
3. How many days old are you?
4. Determine about how many hours you have lived.
5. About how many minutes have you lived?
6. About how many seconds have you lived?

*Remember:*

60 seconds = 1 minute
60 minutes = 1 hour
24 hours = 1 day
365 days = 1 year
366 days = 1 leap year

# CHAPTER REVIEW

**Multiply.**

| | | | | |
|---|---|---|---|---|
| 36<br>×3<br>108 | **1.** 45<br>×7 | **2.** 254<br>×9 | **3.** 6.92<br>×5 | **4.** 1326<br>×7 |
| 42<br>×10<br>420 | **5.** 68<br>×10 | **6.** 256<br>×10 | **7.** 384<br>×100 | **8.** 4623<br>×100 |
| 1<br>83<br>×60<br>4980 | **9.** 78<br>×40 | **10.** 584<br>×50 | **11.** 453<br>×200 | **12.** 732<br>×600 |
| 85<br>×23<br>255 (3 X 85)<br>1700 (20 X 85)<br>1955 | **13.** 79<br>×46 | **14.** 58<br>×85 | **15.** 164<br>×51 | **16.** 602<br>×37 |
| 523<br>×146<br>3138 (6 X 523)<br>20920 (40 X 523)<br>52300 (100 X 523)<br>76,358 | **17.** 642<br>×234 | **18.** 859<br>×357 | **19.** 468<br>×209 | **20.** 208<br>×706 |

# CHAPTER CHALLENGE

Sometimes we have to use the same factor several times.

exponent

$$6 \times 6 \times 6 \times 6 = 6^4$$

The exponent 4 tells us how many times the number 6 is used as a factor.

*Other examples:*

$5 \times 5 \times 5 = 5^3$ $3 \times 3 \times 3 \times 3 \times 3 = 3^5$

**Give the exponent.**

1. $2 \times 2 \times 2 = 2^?$
2. $3 \times 3 = 3^?$
3. $5 \times 5 \times 5 \times 5 = 5^?$
4. $8 \times 8 \times 8 \times 8 \times 8 = 8^?$
5. $10 \times 10 \times 10 = 10^?$
6. $4 \times 4 \times 4 \times 4 \times 4 \times 4 = 4^?$

**Write, using an exponent.**

7. $3 \times 3 \times 3 \times 3$
8. $5 \times 5$
9. $4 \times 4 \times 4 \times 4 \times 4$
10. $8 \times 8 \times 8$
11. $2 \times 2 \times 2 \times 2 \times 2 \times 2$
12. $6 \times 6$

**Write the standard numeral.**

$2 \times 2 \times 2 = 8$

13. $2^3$
14. $2^2$
15. $2^4$
16. $3^2$
17. $1^5$
18. $5^2$
19. $4^3$
20. $5^3$
21. $3^3$
22. $4^2$
23. $2^6$
24. $6^2$
25. $10^1$
26. $10^2$
27. $10^3$
28. $10^4$
29. $10^5$
30. $10^6$

1. a b c d 2. a b c d 3. a b c d 4. a b c d 5. a b c d 6. a b c d
7. a b c d 8. a b c d 9. a b c d 10. a b c d 11. a b c d 12. a b c d

# MAJOR CHECKUP

## STANDARDIZED FORMAT

**Choose the correct letter.**

1. In 94,167, the digit in the ten thousands place is
   - a. 1
   - b. 9
   - c. 4
   - d. none of these

2. 743,500 rounded to the nearest thousand is
   - a. 743,500
   - b. 743,000
   - c. 744,000
   - d. none of these

3. In 438,715,602, the 3 stands for
   - a. 3 ten thousands
   - b. 3 million
   - c. 3 ten millions
   - d. none of these

4. Add.
   $$\begin{array}{r} 35{,}281 \\ +7{,}896 \\ \hline \end{array}$$
   - a. 32,077
   - b. 42,177
   - c. 33,177
   - d. none of these

5. Add.
   $$\begin{array}{r} 521 \\ 78 \\ 256 \\ +642 \\ \hline \end{array}$$
   - a. 1497
   - b. 1487
   - c. 1397
   - d. none of these

6. Subtract.
   $$\begin{array}{r} 705 \\ -156 \\ \hline \end{array}$$
   - a. 651
   - b. 559
   - c. 549
   - d. none of these

7. Subtract.
   $$\begin{array}{r} 753{,}214 \\ -85{,}392 \\ \hline \end{array}$$
   - a. 732,182
   - b. 667,822
   - c. 668,822
   - d. none of these

8. The least common multiple of 8 and 6 is
   - a. 2
   - b. 48
   - c. 24
   - d. none of these

9. Multiply.
   $$\begin{array}{r} 3821 \\ \times 7 \\ \hline \end{array}$$
   - a. 26,747
   - b. 26,757
   - c. 21,647
   - d. none of these

10. Multiply.
    $$\begin{array}{r} 164 \\ \times 258 \\ \hline \end{array}$$
    - a. 33,312
    - b. 42,312
    - c. 42,302
    - d. none of these

11. Each of 7 pads has 63 paper sheets. What is the total number of paper sheets?
    - a. 56
    - b. 70
    - c. 441
    - d. none of these

12. Greg had 75¢. He bought 4 pencils for 12¢ each. How much money did he have then?
    - a. 63¢
    - b. 59¢
    - c. 48¢
    - d. 27¢

# Division

# 4

# The basic division facts

Division is finding a missing factor. If you remember the multiplication facts, you will know the division facts.

If you remember $6 \times 3 = 18$

then you know $18 \div 6 = 3$

and

$18 \div 3 = 6$

(3: divisor; 6: quotient)

Multiplication and division are related.

To solve a division problem, think about a multiplication fact.

(9 × ? = 72)

$72 \div 9 =$

Here is another way to write a division problem.

$$9\overline{)36}\ \text{with quotient } 4$$

divisor → 9; 4 ← quotient

(36 ÷ 9 = 4)

In some division problems there is a remainder.

$$\begin{array}{r} 4\ R2 \\ 9\overline{)38} \\ -36 \\ \hline 2 \end{array}$$

remainder → R2, 2

## EXERCISES

**Give each missing factor.**

| | | |
|---|---|---|
| **1.** $6 \times \underline{?} = 30$ | **2.** $3 \times \underline{?} = 27$ | **3.** $4 \times \underline{?} = 24$ |
| **4.** $8 \times \underline{?} = 56$ | **5.** $4 \times \underline{?} = 32$ | **6.** $7 \times \underline{?} = 63$ |
| **7.** $6 \times \underline{?} = 36$ | **8.** $5 \times \underline{?} = 40$ | **9.** $6 \times \underline{?} = 42$ |
| **10.** $8 \times \underline{?} = 72$ | **11.** $6 \times \underline{?} = 48$ | **12.** $9 \times \underline{?} = 81$ |

**Give each quotient.**

| | | |
|---|---|---|
| $6 \times 7 = 42$ | $8 \times 6 = 48$ | $4 \times 9 = 36$ |
| **13.** $42 \div 6 = \underline{?}$ | **14.** $48 \div 8 = \underline{?}$ | **15.** $36 \div 4 = \underline{?}$ |
| $5 \times ? = 45$ | $9 \times ? = 54$ | $9 \times ? = 63$ |
| **16.** $45 \div 5 = \underline{?}$ | **17.** $54 \div 9 = \underline{?}$ | **18.** $63 \div 9 = \underline{?}$ |
| **19.** $45 \div 9 = \underline{?}$ | **20.** $63 \div 7 = \underline{?}$ | **21.** $72 \div 8 = \underline{?}$ |
| **22.** $64 \div 8 = \underline{?}$ | **23.** $81 \div 9 = \underline{?}$ | **24.** $49 \div 7 = \underline{?}$ |

| | | | | |
|---|---|---|---|---|
| **25.** $4\overline{)24}$ | **26.** $3\overline{)27}$ | **27.** $7\overline{)49}$ | **28.** $2\overline{)18}$ | **29.** $5\overline{)40}$ |
| **30.** $4\overline{)32}$ | **31.** $8\overline{)32}$ | **32.** $7\overline{)21}$ | **33.** $8\overline{)24}$ | **34.** $9\overline{)27}$ |
| **35.** $6\overline{)48}$ | **36.** $9\overline{)63}$ | **37.** $8\overline{)72}$ | **38.** $5\overline{)35}$ | **39.** $6\overline{)42}$ |

**Give each quotient and remainder.**

| | | | | |
|---|---|---|---|---|
| **40.** $4\overline{)17}$ | **41.** $3\overline{)16}$ | **42.** $3\overline{)17}$ | **43.** $5\overline{)24}$ | **44.** $6\overline{)23}$ |
| **45.** $8\overline{)53}$ | **46.** $9\overline{)48}$ | **47.** $7\overline{)51}$ | **48.** $6\overline{)50}$ | **49.** $8\overline{)43}$ |
| **50.** $7\overline{)41}$ | **51.** $8\overline{)62}$ | **52.** $9\overline{)71}$ | **53.** $9\overline{)65}$ | **54.** $9\overline{)75}$ |
| **55.** $8\overline{)46}$ | **56.** $4\overline{)33}$ | **57.** $5\overline{)39}$ | **58.** $4\overline{)29}$ | **59.** $6\overline{)38}$ |

# Basic division facts

WHAT'S YOUR SCORE?

**All the basic division facts are listed below.**
**See how many seconds it takes you to write the quotients.**

**1.** $4\overline{)4}$ **2.** $9\overline{)0}$ **3.** $2\overline{)4}$ **4.** $7\overline{)49}$ **5.** $2\overline{)2}$

**6.** $7\overline{)21}$ **7.** $1\overline{)8}$ **8.** $9\overline{)18}$ **9.** $5\overline{)5}$ **10.** $6\overline{)6}$

**11.** $1\overline{)7}$ **12.** $8\overline{)16}$ **13.** $5\overline{)20}$ **14.** $3\overline{)24}$ **15.** $3\overline{)15}$

**16.** $7\overline{)0}$ **17.** $9\overline{)9}$ **18.** $7\overline{)56}$ **19.** $8\overline{)40}$ **20.** $7\overline{)42}$

**21.** $4\overline{)0}$ **22.** $6\overline{)12}$ **23.** $1\overline{)6}$ **24.** $6\overline{)48}$ **25.** $3\overline{)3}$

**26.** $5\overline{)35}$ **27.** $4\overline{)36}$ **28.** $9\overline{)27}$ **29.** $3\overline{)12}$ **30.** $2\overline{)6}$

**31.** $3\overline{)0}$ **32.** $4\overline{)24}$ **33.** $1\overline{)0}$ **34.** $6\overline{)36}$ **35.** $4\overline{)8}$

**36.** $8\overline{)24}$ **37.** $1\overline{)9}$ **38.** $1\overline{)3}$ **39.** $7\overline{)35}$ **40.** $4\overline{)16}$

**41.** $6\overline{)18}$ **42.** $7\overline{)7}$ **43.** $5\overline{)30}$ **44.** $2\overline{)0}$ **45.** $5\overline{)10}$

**46.** $8\overline{)8}$ **47.** $9\overline{)36}$ **48.** $4\overline{)32}$ **49.** $1\overline{)1}$ **50.** $5\overline{)15}$

**51.** $3\overline{)18}$ **52.** $8\overline{)72}$ **53.** $7\overline{)63}$ **54.** $9\overline{)45}$ **55.** $9\overline{)63}$

**56.** $5\overline{)45}$ **57.** $2\overline{)8}$ **58.** $8\overline{)32}$ **59.** $4\overline{)12}$ **60.** $9\overline{)81}$

**61.** $2\overline{)12}$ **62.** $8\overline{)64}$ **63.** $6\overline{)24}$ **64.** $8\overline{)0}$ **65.** $1\overline{)2}$

**66.** $3\overline{)6}$ **67.** $6\overline{)42}$ **68.** $3\overline{)27}$ **69.** $6\overline{)54}$ **70.** $7\overline{)14}$

**71.** $2\overline{)18}$ **72.** $5\overline{)0}$ **73.** $8\overline{)48}$ **74.** $2\overline{)14}$ **75.** $5\overline{)25}$

**76.** $7\overline{)28}$ **77.** $5\overline{)40}$ **78.** $2\overline{)10}$ **79.** $4\overline{)28}$ **80.** $1\overline{)4}$

**81.** $9\overline{)72}$ **82.** $9\overline{)54}$ **83.** $3\overline{)9}$ **84.** $4\overline{)20}$ **85.** $6\overline{)0}$

**86.** $1\overline{)5}$ **87.** $8\overline{)56}$ **88.** $2\overline{)16}$ **89.** $3\overline{)21}$ **90.** $6\overline{)30}$

**Check your answers.**

| | | | | |
|---|---|---|---|---|
| **1.** 1 | **2.** 0 | **3.** 2 | **4.** 7 | **5.** 1 |
| **6.** 3 | **7.** 8 | **8.** 2 | **9.** 1 | **10.** 1 |
| **11.** 7 | **12.** 2 | **13.** 4 | **14.** 8 | **15.** 5 |
| **16.** 0 | **17.** 1 | **18.** 8 | **19.** 5 | **20.** 6 |
| **21.** 0 | **22.** 2 | **23.** 6 | **24.** 8 | **25.** 1 |
| **26.** 7 | **27.** 9 | **28.** 3 | **29.** 4 | **30.** 3 |
| **31.** 0 | **32.** 6 | **33.** 0 | **34.** 6 | **35.** 2 |
| **36.** 3 | **37.** 9 | **38.** 3 | **39.** 5 | **40.** 4 |
| **41.** 3 | **42.** 1 | **43.** 6 | **44.** 0 | **45.** 2 |
| **46.** 1 | **47.** 4 | **48.** 8 | **49.** 1 | **50.** 3 |
| **51.** 6 | **52.** 9 | **53.** 9 | **54.** 5 | **55.** 7 |
| **56.** 9 | **57.** 4 | **58.** 4 | **59.** 3 | **60.** 9 |
| **61.** 6 | **62.** 8 | **63.** 4 | **64.** 0 | **65.** 2 |
| **66.** 2 | **67.** 7 | **68.** 9 | **69.** 9 | **70.** 2 |
| **71.** 9 | **72.** 0 | **73.** 6 | **74.** 7 | **75.** 5 |
| **76.** 4 | **77.** 8 | **78.** 5 | **79.** 7 | **80.** 4 |
| **81.** 8 | **82.** 6 | **83.** 3 | **84.** 5 | **85.** 0 |
| **86.** 5 | **87.** 7 | **88.** 8 | **89.** 7 | **90.** 5 |

**91.** To find your score, multiply the number of mistakes by 2 and add that number to the number of seconds.

**92.** Practice the facts that you missed. Try for a better (lower) score.

**93.** Work with a classmate. Give the facts orally. Keep a record of your best daily score.

**94.** Use graph paper. Make a graph of your scores like the one on the computer screen.

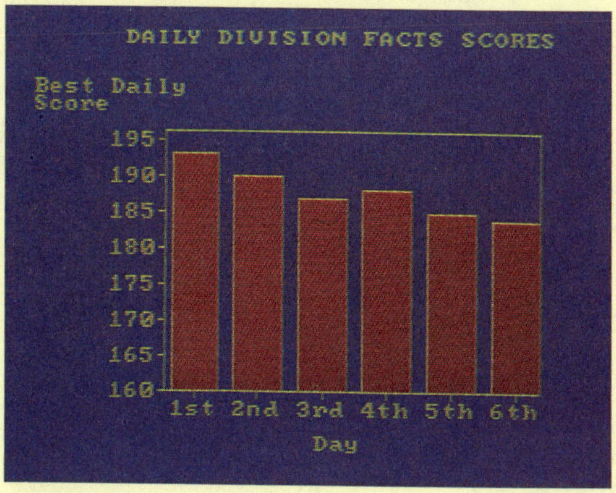

# Factors and common factors

Because $6 \times 3 = 18$, 6 and 3 are **factors,** or **divisors,** of 18.

5 is not a factor of 18.
Why not?

$\begin{array}{r} 6 \\ \times 3 \\ \hline 18 \end{array}$ $\begin{array}{r} 9 \\ \times 2 \\ \hline 18 \end{array}$ $\begin{array}{r} 18 \\ \times 1 \\ \hline 18 \end{array}$

$\begin{array}{r} 6 \\ \times 4 \\ \hline 24 \end{array}$ $\begin{array}{r} 8 \\ \times 3 \\ \hline 24 \end{array}$ $\begin{array}{r} 12 \\ \times 2 \\ \hline 24 \end{array}$ $\begin{array}{r} 24 \\ \times 1 \\ \hline 24 \end{array}$

1, 2, 3, and 6 are factors of both 18 and 24. They are called **common factors,** or **common divisors,** of 18 and 24. The **greatest common factor,** or **greatest common divisor,** of 18 and 24 is 6.

**EXERCISES**

**List all factors.**

**1.** 3 **2.** 8 **3.** 9 **4.** 12 **5.** 17 **6.** 16
**7.** 18 **8.** 21 **9.** 20 **10.** 24 **11.** 23 **12.** 25
**13.** 26 **14.** 27 **15.** 28 **16.** 30 **17.** 32 **18.** 36

**List all common factors.**

**19.** 8, 12 **20.** 6, 8 **21.** 7, 10 **22.** 8, 24 **23.** 15, 25
**24.** 10, 12 **25.** 16, 18 **26.** 9, 12 **27.** 15, 18 **28.** 10, 20
**29.** 16, 24 **30.** 12, 16 **31.** 18, 24 **32.** 24, 36 **33.** 12, 18

**True or false?**

**34.** 1 is a divisor of every number.

**35.** 1 is a common divisor of any two numbers.

**36.** 2 is a divisor of every even number.

**37.** 3 is a divisor of every odd number.

**38.** Every number has at least two divisors.

**Give the greatest common factor.**

| | | | | |
|---|---|---|---|---|
| **39.** 9, 12 | **40.** 10, 35 | **41.** 8, 12 | **42.** 9, 24 | **43.** 14, 35 |
| **44.** 18, 36 | **45.** 16, 20 | **46.** 16, 48 | **47.** 18, 45 | **48.** 18, 42 |

★ **Find the greatest common factor.**

**49.** 12, 18, 21    **50.** 20, 40, 35

**51.** 16, 28, 7    **52.** 28, 40, 32, 18

## *Challenge!*

**a or b?**

**53.** Complete: $\underline{?} \times 0 = 7$

**a.** no number    **b.** any number

**54.** Complete: $\underline{?} \times 0 = 0$

**a.** no number    **b.** any number

**55.** Complete: $6 \div 0 = \underline{?}$ (?×0=6)

**a.** no number    **b.** any number

**56.** Complete: $0 \div 0 = \underline{?}$ (?×0=0)

**a.** no number    **b.** any number

# Problem solving—choose the operation

1. Study and understand.
2. Plan and do.
3. Answer and check.

■ people had to be on the roller coaster before the ride could begin. ■ people were already on the roller coaster. How many more people were needed?

You need to compare two numbers. So, subtract.

■ people had to be in each of the ■ cars of the roller coaster before the ride could begin. ■ people were already on the roller coaster. How many more people were needed?

You need to compare two numbers. To find one of these numbers, you have to multiply two numbers. So, first multiply and then subtract.

## EXERCISES

**The numbers have been covered.**
**Add, subtract, multiply, or divide?**
**Decide what to do.**

1. A ride on the Big Wheel cost ■¢. Carlos bought ■ tickets. How much did he spend?

2. ■ people wanted to ride on the Bumper Car. ■ people had to be in each car. How many cars were needed?

3. Mio bought some cotton candy for ■¢ and some popcorn for ■¢. How much money did she spend?

4. Craig won ■ prizes. Barbara won ■ prizes. How many more prizes did Craig win?

5. Laura bought ■ tickets for each ride. There were ■ different rides. How many tickets did she buy?

6. Richard had ■¢. He wanted to buy a Jungle Cruise ticket for ■¢. How much more money did he need?

7. Charles had $■. He bought a Silver Streak ticket for $■. How much money did he have then?

8. Helen bought ■ Bumper Car tickets for ■¢. How much did each ticket cost?

**Here are some two-step problems.**
**Decide what to do. There may be**
**more than one way to do a problem.**

9. Steven had ■¢. He bought ■ Big Wheel tickets for ■¢ each. How much money did he have left?

10. Allison had ■ tickets. A friend gave her ■ more tickets. She used ■ of the tickets. How many tickets did she have then?

11. Kim bought ■ Silver Streak tickets for ■¢ each. She also bought a balloon for ■¢. How much money did she spend?

12. Sean spent $■. He bought a box of popcorn for ■¢ and ■ Bumper Car tickets. How much did each ticket cost?

13. Dolores had ■¢. She bought a Jungle Cruise ticket for ■¢ and a Big Wheel ticket for ■¢. How much money did she have left?

14. Kenneth bought ■ Creature Posters and ■ Rock Posters for ■¢ each. What was the total cost?

# Dividing a 2-digit number

Mark, Luis, and Sharon helped do the shopping.
The clerk gave them 63 Bonus stamps.
Example 1 shows how they divided the stamps.

**EXAMPLE 1.** Divide 63 stamps equally among 3 people.

**Step 1.** Divide tens. **Step 2.** Divide ones.

$3\overline{)63}$

$\begin{array}{r}2\phantom{3}\\3\overline{)63}\end{array}$

$\begin{array}{r}21\\3\overline{)63}\end{array}$

The quotient is 21.

**EXAMPLE 2.** Divide 80 stamps equally between 2 people.

**Step 1.** Divide tens. **Step 2.** Divide ones.

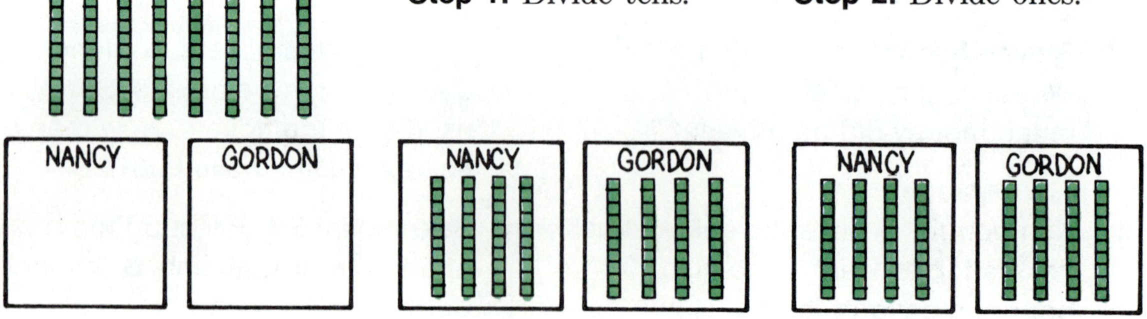


$2\overline{)80}$

$\begin{array}{r}4\phantom{0}\\2\overline{)80}\end{array}$

$\begin{array}{r}40\\2\overline{)80}\end{array}$

The quotient is 40.

## EXERCISES

**Divide.**

1. $2\overline{)68}$ 2. $3\overline{)36}$ 3. $4\overline{)40}$ 4. $4\overline{)44}$ 5. $3\overline{)39}$

6. $2\overline{)64}$ 7. $4\overline{)88}$ 8. $2\overline{)86}$ 9. $4\overline{)80}$ 10. $3\overline{)33}$

11. $2\overline{)42}$ 12. $2\overline{)84}$ 13. $4\overline{)84}$ 14. $2\overline{)44}$ 15. $2\overline{)62}$

16. $7\overline{)70}$ 17. $3\overline{)60}$ 18. $2\overline{)60}$ 19. $3\overline{)96}$ 20. $2\overline{)66}$

21. $3\overline{)69¢}$ 22. $4\overline{)80¢}$ 23. $3\overline{)93¢}$ 24. $5\overline{)50¢}$ 25. $9\overline{)99¢}$

26. $2\overline{)46¢}$ 27. $3\overline{)66¢}$ 28. $6\overline{)66¢}$ 29. $3\overline{)63¢}$ 30. $5\overline{)55¢}$

**Solve.**

31. Stella made some sandwiches for a hiking trip. She used 48 slices of bread. How many sandwiches did she make?

32. A club sandwich requires 3 slices of bread. If Stanley has 60 slices of bread, how many club sandwiches can he make?

33. Jerry baked 96 cookies. He wants to put the same number of cookies in each of 3 bags. How many should he put in each bag?

34. Jan collected 55 bottles. She was paid 5¢ each when she returned them to the store. How much did she earn?

### *Challenge!*

**Find the missing input or output.**

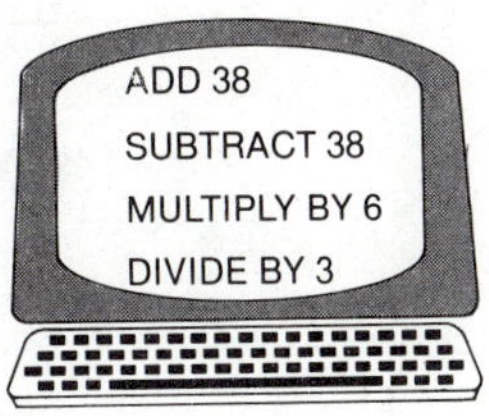


| | Input | Output |
|---|---|---|
| | 5 | 10 |
| 35. | 7 | ? |
| 36. | 25 | ? |
| 37. | ? | 104 |

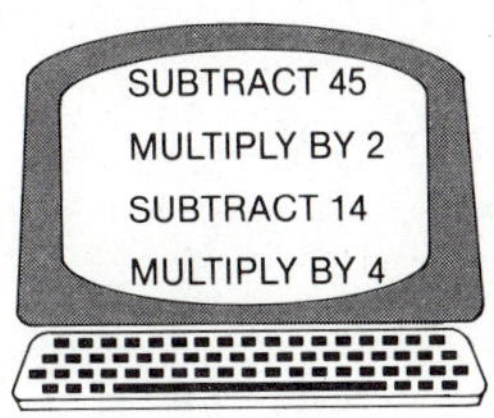


| | Input | Output |
|---|---|---|
| | 75 | 184 |
| 38. | 100 | ? |
| 39. | 80 | ? |
| 40. | ? | 64 |

# Division with regrouping

In some division problems you need to regroup.
Study the examples.

**EXAMPLE.** $4\overline{)98}$

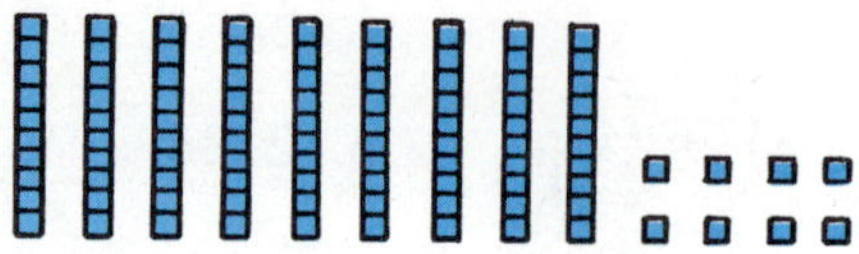

**Step 1.** Divide tens. Put 2 tens in each of the 4 sets. That leaves 1 ten.

Used 8 tens. There is 1 ten remaining.

$$\begin{array}{r} 2\phantom{8} \\ 4\overline{)98} \\ -8\phantom{8} \\ \hline 1\phantom{8} \end{array}$$

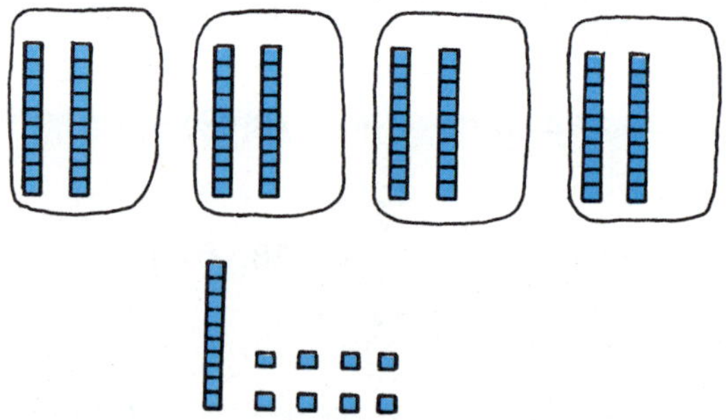

**Step 2.** Regroup 1 ten for 10 ones.

Write the 8 here and think of this as 18 ones.

$$\begin{array}{r} 2\phantom{8} \\ 4\overline{)98} \\ -8\phantom{8} \\ \hline 18 \end{array}$$

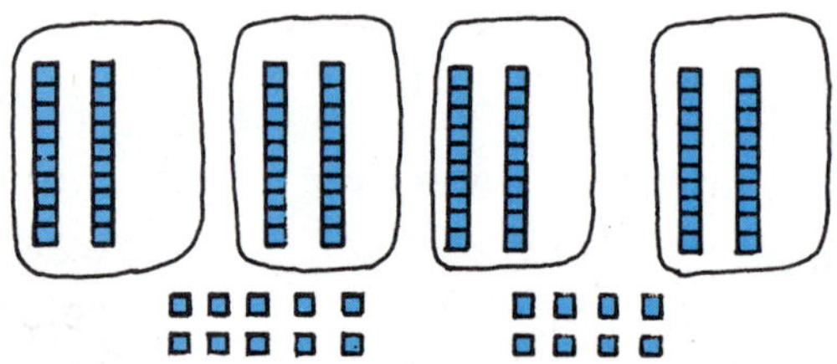

**Step 3.** Divide ones. Put 4 ones in each of the 4 sets.

Used 16 ones. There are 2 ones remaining.

$$\begin{array}{r} 24 \text{ R2} \\ 4\overline{)98}\phantom{\text{ R2}} \\ -8\phantom{8}\phantom{\text{ R2}} \\ \hline 18\phantom{\text{ R2}} \\ -16\phantom{\text{ R2}} \\ \hline 2\phantom{\text{ R2}} \end{array}$$

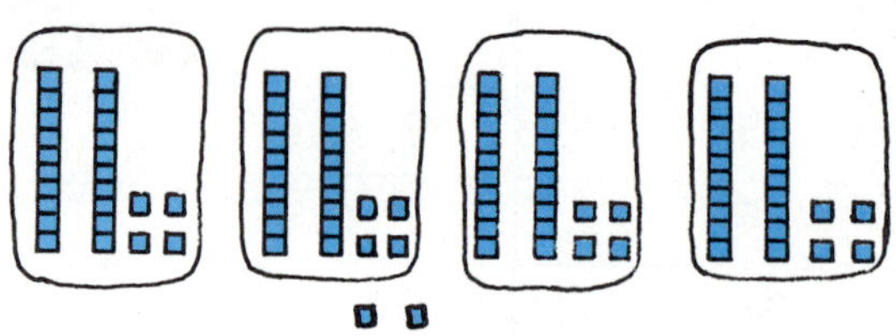

## EXERCISES

**Divide.**

| | | | | |
|---|---|---|---|---|
| 1. $4\overline{)88}$ | 2. $3\overline{)57}$ | 3. $2\overline{)75}$ | 4. $7\overline{)98}$ | 5. $7\overline{)86}$ |
| 6. $3\overline{)42}$ | 7. $6\overline{)88}$ | 8. $3\overline{)85}$ | 9. $6\overline{)78}$ | 10. $4\overline{)81}$ |
| 11. $2\overline{)52}$ | 12. $5\overline{)80}$ | 13. $7\overline{)84}$ | 14. $8\overline{)96}$ | 15. $5\overline{)90}$ |
| 16. $3\overline{)84}$ | 17. $3\overline{)54}$ | 18. $6\overline{)96}$ | 19. $7\overline{)91}$ | 20. $5\overline{)66}$ |
| 21. $2\overline{)64}$ | 22. $5\overline{)78}$ | 23. $5\overline{)60}$ | 24. $6\overline{)90}$ | 25. $2\overline{)96}$ |
| 26. $3\overline{)45¢}$ | 27. $5\overline{)95¢}$ | 28. $3\overline{)48¢}$ | 29. $4\overline{)76¢}$ | 30. $6\overline{)72¢}$ |

**Solve.**

31. There were 84 pepper plants in each row. There were 7 rows. How many pepper plants were there?

32. There were 84 pepper plants in a row. 7 of the plants were broken. How many plants were not broken?

33. There were 8 tomato plants in each box. There were 24 tomato plants. How many boxes were there?

34. There were 18 potato plants in each box. There were 9 boxes. How many potato plants were there?

35. 8 watermelon seeds were planted in each hill. There were 96 watermelon seeds. How many hills were there?

36. There were 8 cucumber seeds. There were 96 watermelon seeds. How many seeds were there?

### *Challenge!*

37. There are fewer than 45 tomato plants in a box. If you plant them 6 in a row, you will have 1 left over. If you plant them 5 in a row, you will have 4 left over. How many tomato plants are in the box?

38. You have 10 pepper plants. Draw a picture to show how you could plant them so that there are 4 plants in each of 5 rows. (*Hint:* Each plant is in two different rows.)

# Dividing larger numbers

You can divide larger numbers in the same way you divide 2-digit numbers.

**EXAMPLE 1.** $3\overline{)442}$

**Step 1.** Divide hundreds. Subtract.

```
   1
3)442
 -3
  1
```

**Step 2.** Regroup.

```
   1
3)442
 -3
  14
```

**Step 3.** Divide tens. Subtract.

```
   14
3)442
 -3
  14
 -12
   2
```

**Step 4.** Regroup.

```
   14
3)442
 -3
  14
 -12
   22
```

**Step 5.** Divide ones. Subtract.

```
   147 R1
3)442
 -3
  14
 -12
   22
  -21
    1
```

**EXAMPLE 2.** $4\overline{)3128}$

**Step 1.** Not enough thousands.

```
4)3128
```

**Step 2.** Regroup and divide hundreds. Subtract.

```
    7
4)3128
 -28
   3
```

**Step 3.** Regroup and divide tens. Subtract.

```
    78
4)3128
 -28
   32
  -32
```

**Step 4.** Regroup and divide ones. Subtract.

```
    782
4)3128
 -28
   32
  -32
     8
    -8
     0
```

## EXERCISES

**Divide.**

1. $3\overline{)521}$ 2. $6\overline{)719}$ 3. $7\overline{)926}$ 4. $3\overline{)742}$ 5. $2\overline{)725}$

6. $4\overline{)152}$ 7. $6\overline{)321}$ 8. $8\overline{)415}$ 9. $5\overline{)329}$ 10. $7\overline{)526}$

11. $4\overline{)664}$ 12. $3\overline{)500}$ 13. $8\overline{)976}$ 14. $6\overline{)394}$ 15. $9\overline{)498}$

16. $7\overline{)645}$ 17. $3\overline{)813}$ 18. $2\overline{)835}$ 19. $7\overline{)960}$ 20. $8\overline{)523}$

21. $6\overline{)\$8.34}$ 22. $4\overline{)\$6.44}$ 23. $8\overline{)\$9.04}$ 24. $6\overline{)\$5.04}$ 25. $3\overline{)\$2.52}$

26. $5\overline{)4173}$ 27. $6\overline{)3818}$ 28. $7\overline{)1675}$ 29. $6\overline{)5151}$ 30. $5\overline{)1269}$

31. $3\overline{)4276}$ 32. $8\overline{)1000}$ 33. $4\overline{)6195}$ 34. $5\overline{)1620}$ 35. $7\overline{)6000}$

**Solve.**

36. Pam bought 6 computer diskettes for $4.98 each. How much did she spend?

37. Arthur bought 3 copies of a book on programming in BASIC to give as gifts. He spent $15.78. How much did each book cost?

### KEEPING SKILLS SHARP

**Round to the nearest ten.**

1. 76 2. 54 3. 31 4. 65 5. 19

6. 153 7. 297 8. 812 9. 603 10. 408

**Round to the nearest hundred.**

11. 153 12. 297 13. 812 14. 603 15. 408

16. 6407 17. 1750 18. 8042 19. 5719 20. 3961

# Zeros in division

Here is an example that has a zero in the quotient.

5 ) 3 5 1 2

**Step 1.** Not enough thousands.

5 ) **3** 5 1 2

**Step 2.** Regroup and divide hundreds. Subtract.

```
    7
5 ) 3 5 1 2
  - 3 5
```

**Step 3.** Not enough tens.

```
    7 0
5 ) 3 5 1 2
  - 3 5
        1
```

**Step 4.** Regroup and divide ones. Subtract.

```
    7 0 2  R2
5 ) 3 5 1 2
  - 3 5
        1 2
      - 1 0
          2
```

Sometimes, people forget to write zeros in quotients. If you estimate first, you may not make that mistake. Here is a way to estimate a quotient.

**EXAMPLE 1.**

**Step 1.** Find the first digit of the quotient.

```
    9
3 ) 2 7 1 2
```

**Step 2.** Write 0's in *all* remaining places.

```
    9 0 0 ← estimate
3 ) 2 7 1 2
```

**EXAMPLE 2.**

**Step 1.** Find the first digit.

```
    6
4 ) 2 5 7
```

**Step 2.** Write 0's.

```
    6 0 ← estimate
4 ) 2 5 7
```

## EXERCISES

**Estimate. Then find each quotient.**

1. $5\overline{)512}$ 2. $4\overline{)831}$ 3. $6\overline{)362}$ 4. $2\overline{)157}$ 5. $3\overline{)919}$

6. $6\overline{)7153}$ 7. $2\overline{)9674}$ 8. $5\overline{)3517}$ 9. $8\overline{)2403}$ 10. $6\overline{)5418}$

11. $5\overline{)600}$ 12. $5\overline{)205}$ 13. $5\overline{)2535}$ 14. $8\overline{)6008}$ 15. $4\overline{)5000}$

**Divide.**

16. $5\overline{)653}$ 17. $8\overline{)726}$ 18. $6\overline{)948}$ 19. $6\overline{)100}$ 20. $4\overline{)784}$

21. $3\overline{)842}$ 22. $6\overline{)917}$ 23. $9\overline{)456}$ 24. $4\overline{)981}$ 25. $7\overline{)324}$

26. $2\overline{)\$8.50}$ 27. $4\overline{)\$8.32}$ 28. $6\overline{)\$8.52}$ 29. $3\overline{)\$6.21}$ 30. $5\overline{)\$7.00}$

31. $4\overline{)3042}$ 32. $6\overline{)3001}$ 33. $4\overline{)2482}$ 34. $2\overline{)3416}$ 35. $7\overline{)1000}$

36. $4\overline{)2815}$ 37. $8\overline{)4703}$ 38. $9\overline{)9314}$ 39. $4\overline{)7000}$ 40. $6\overline{)5436}$

**Solve.**

41. Pam has 8 pads of paper. Each pad has 120 sheets of paper. How many sheets of paper does she have?

42. There are 120 pictures in Pam's photo album. There are 8 pictures on each page. How many pages are there in the photo album?

43. Garrett had $5.04. He bought 4 ball-point pens for 29¢ each. How much money did he spend?

44. Garrett had $5.04. He bought 4 ball-point pens for 29¢ each. How much money did he have left?

# Finding an average, or mean

The computer table below shows the students who entered the rope-jumping contest.

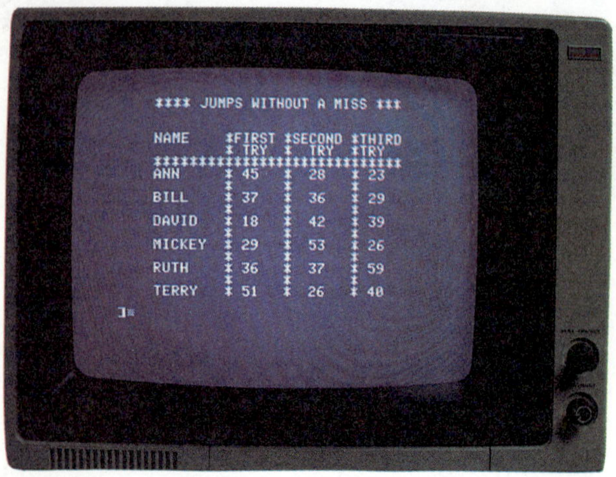

**** JUMPS WITHOUT A MISS ***

| NAME | FIRST TRY | SECOND TRY | THIRD TRY |
|---|---|---|---|
| ANN | 45 | 28 | 23 |
| BILL | 37 | 36 | 29 |
| DAVID | 18 | 42 | 39 |
| MICKEY | 29 | 53 | 26 |
| RUTH | 36 | 37 | 59 |
| TERRY | 51 | 26 | 40 |

Ann had a total of 96 jumps. If you divide this total by 3, you get the average number of jumps she made on each try.

Such an average is sometimes called the **mean.**

## EXERCISES

**Use the computer table on page 114 to solve these problems.**

1. What was Bill's average in the rope-jumping contest?
2. What was Ruth's average in the rope-jumping contest?
3. Find the average number of jumps for the first try.
4. Find the average number of jumps for the second try.
5. Who had the higher average, David or Terry? How much higher?
6. Who had the highest average? the lowest average?

**The computer table on the left shows the length of the long jumps in centimeters.**

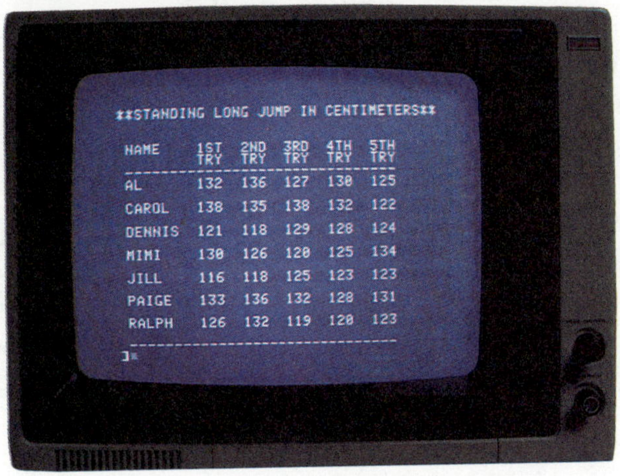

**STANDING LONG JUMP IN CENTIMETERS**

| NAME | 1ST TRY | 2ND TRY | 3RD TRY | 4TH TRY | 5TH TRY |
|---|---|---|---|---|---|
| AL | 132 | 136 | 127 | 130 | 125 |
| CAROL | 138 | 135 | 138 | 132 | 122 |
| DENNIS | 121 | 118 | 129 | 128 | 124 |
| MIMI | 130 | 126 | 120 | 125 | 134 |
| JILL | 116 | 118 | 125 | 123 | 123 |
| PAIGE | 133 | 136 | 132 | 128 | 131 |
| RALPH | 126 | 132 | 119 | 120 | 123 |

7. What was Carol's average long jump?
8. Who had the greater average, Al or Paige?
9. Find the average length of the jumps on the first try.
10. How does the first-jump average compare with the last-jump average?

**The bar graph shows the number of baskets made in the free-throw contest.**

11. What was the mean number of baskets?
12. Did anyone make the mean score? Who scored above the mean? Who scored below?

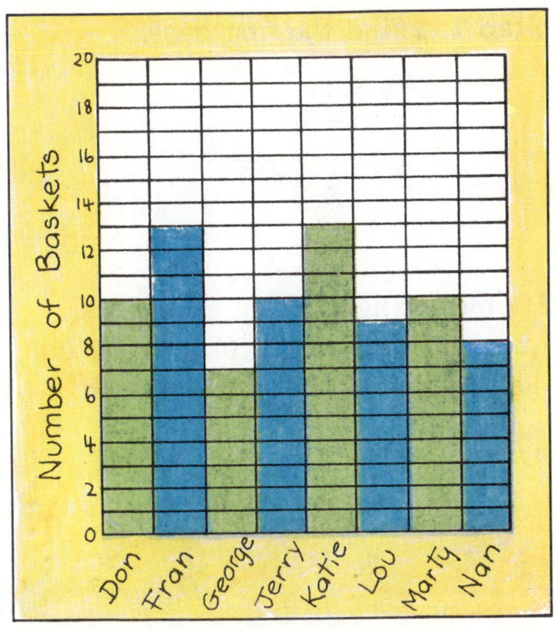

# Dividing by tens

We divide by a 2-digit number the same way that we divide by a 1-digit number. Here are the multiplication facts for 20.

| 20 | 20 | 20 | 20 | 20 |
|---|---|---|---|---|
| x0 | x1 | x2 | x3 | x4 |
| 0 | 20 | 40 | 60 | 80 |

| 20 | 20 | 20 | 20 | 20 |
|---|---|---|---|---|
| x5 | x6 | x7 | x8 | x9 |
| 100 | 120 | 140 | 160 | 180 |

Let's use these facts to do a division exercise.

**Step 1.** Not enough hundreds.

$20\overline{)956}$

**Step 2.** Regroup and divide. Think 95 tens.

$20 \times 4 = 80$

```
     4
20)956
  -80
   15
```

**Step 3.** Regroup and divide.

$20 \times 7 = 140$

```
     47 R16
20)956
  -80
   156
  -140
    16
```

Here is a way to estimate a quotient.

**Step 1.** Find the first digit.

```
    3
20)6341
```

**Step 2.** Write 0's in the remaining places.

```
    300 ← estimate
20)6341
```

If you do the division, should the quotient be less than or greater than 300? Should it be less than 400?

## EXERCISES

Use the given multiplication facts to help you divide.

| 40 | 40 | 40 | 40 | 40 |
|---|---|---|---|---|
| ×0 | ×1 | ×2 | ×3 | ×4 |
| 0 | 40 | 80 | 120 | 160 |

| 40 | 40 | 40 | 40 | 40 |
|---|---|---|---|---|
| ×5 | ×6 | ×7 | ×8 | ×9 |
| 200 | 240 | 280 | 320 | 360 |

1. $40\overline{)864}$
2. $40\overline{)395}$
3. $40\overline{)760}$
4. $40\overline{)942}$
5. $40\overline{)805}$
6. $40\overline{)677}$
7. $40\overline{)3248}$
8. $40\overline{)2974}$
9. $40\overline{)6359}$

First estimate the quotient. Then divide.

| 30 | 30 | 30 | 30 | 30 |
|---|---|---|---|---|
| ×0 | ×1 | ×2 | ×3 | ×4 |
| 0 | 30 | 60 | 90 | 120 |

| 30 | 30 | 30 | 30 | 30 |
|---|---|---|---|---|
| ×5 | ×6 | ×7 | ×8 | ×9 |
| 150 | 180 | 210 | 240 | 270 |

10. $30\overline{)958}$
11. $30\overline{)674}$
12. $30\overline{)840}$
13. $30\overline{)2356}$
14. $30\overline{)3048}$
15. $30\overline{)2965}$
16. $30\overline{)4712}$
17. $30\overline{)5836}$
18. $30\overline{)6542}$

### *Challenge!*

First list the multiplication facts for 57. Then divide.

| 57 | 57 | 57 | 57 | 57 |
|---|---|---|---|---|
| ×0 | ×1 | ×2 | ×3 | ×4 |
| | | | | |

| 57 | 57 | 57 | 57 | 57 |
|---|---|---|---|---|
| ×5 | ×6 | ×7 | ×8 | ×9 |
| | | | | |

19. $57\overline{)459}$
20. $57\overline{)564}$
21. $57\overline{)1077}$
22. $57\overline{)3675}$
23. $57\overline{)16,772}$
24. $57\overline{)39,564}$
25. $57\overline{)48,348}$
26. $57\overline{)84,165}$

# More about dividing by a 2-digit number

In the last lesson you were given all the multiplication facts. In this lesson you will have to work out your own multiplication facts as you divide. The examples show how to use rounding to decide which multiplication facts are needed.

**EXAMPLE 1.** **32 ) 8 7 0**

**Step 1.** Round the divisor to the nearest 10. Think about dividing by 30.

30

**32 ) 8 7 0**

| 32 |
|---|
| × 2 |
| 64 |

**Step 2.** Think about the multiplication facts for 30.

$30 \times 1 = 30$ $30 \times 2 = 60$ $30 \times 3 = 90$ ← Too big!

So, try 2.

$$
\begin{array}{r}
2\phantom{0} \\
32\overline{)870} \\
-64\phantom{0} \\
\hline
23\phantom{0}
\end{array}
$$

| 32 |
|---|
| × 7 |
| 224 |

**Step 3.** Think about the multiplication facts for 30.

$30 \times 6 = 180$ $30 \times 7 = 210$ $30 \times 8 = 240$

Try 7.

$$
\begin{array}{r}
27 \text{ R6} \\
32\overline{)870} \\
-64\phantom{0} \\
\hline
230 \\
-224 \\
\hline
6
\end{array}
$$

**EXAMPLE 2.** **36 ) 7 5 4**

**Step 1.** 40

$40 \times 1 = 40$ $40 \times 2 = 80$ ← Too big!

Try 1.

$$
\begin{array}{r}
1\phantom{0} \\
36\overline{)754} \\
-36\phantom{0} \\
\hline
39\phantom{0}
\end{array}
$$

| 36 |
|---|
| × 2 |
| 72 |

**Step 2.**

$$
\begin{array}{r}
2\phantom{0} \\
36\overline{)754} \\
-72\phantom{0} \\
\hline
3\phantom{0}
\end{array}
$$

**Step 3.**

$$
\begin{array}{r}
20 \text{ R34} \\
36\overline{)754} \\
-72\phantom{0} \\
\hline
34 \\
-\phantom{0}0 \\
\hline
34
\end{array}
$$

## EXERCISES

**To what number would you round the divisor?**

1. $23\overline{)783}$ 2. $49\overline{)956}$ 3. $71\overline{)803}$ 4. $36\overline{)916}$ 5. $54\overline{)642}$

**Tell which multiplication fact you would work out first.**

$\begin{array}{r}40\\ \times 1\\ \hline 40\end{array}$ $\begin{array}{r}40\\ \times 2\\ \hline 80\end{array}$ $\begin{array}{r}40\\ \times 3\\ \hline 120\end{array}$

6. $42\overline{)926}$

$42 \times 2$

$\begin{array}{r}20\\ \times 2\\ \hline 40\end{array}$ $\begin{array}{r}20\\ \times 3\\ \hline 60\end{array}$ $\begin{array}{r}20\\ \times 4\\ \hline 80\end{array}$

7. $19\overline{)683}$

$\begin{array}{r}30\\ \times 1\\ \hline 30\end{array}$ $\begin{array}{r}30\\ \times 2\\ \hline 60\end{array}$ $\begin{array}{r}30\\ \times 3\\ \hline 90\end{array}$

8. $33\overline{)697}$

**Divide.**

9. $18\overline{)927}$ 10. $40\overline{)900}$ 11. $61\overline{)765}$ 12. $16\overline{)699}$ 13. $34\overline{)742}$

14. $33\overline{)683}$ 15. $23\overline{)374}$ 16. $43\overline{)894}$ 17. $47\overline{)963}$ 18. $42\overline{)826}$

19. $27\overline{)958}$ 20. $22\overline{)695}$ 21. $31\overline{)9012}$ 22. $58\overline{)8376}$ 23. $24\overline{)7658}$

24. $43\overline{)8935}$ 25. $36\overline{)4916}$ 26. $50\overline{)7800}$ 27. $57\overline{)65{,}912}$ 28. $38\overline{)97{,}434}$

**Solve.**

29.

a. How many rolls of film would you have to buy to take 110 pictures?

b. What would be the total price?

30.

a. How many boxes of screws would you have to buy to get 600 screws?

b. What would be the total price?

### KEEPING SKILLS SHARP

1. $\begin{array}{r}39\\74\\+81\\ \hline\end{array}$ 2. $\begin{array}{r}253\\79\\+179\\ \hline\end{array}$ 3. $\begin{array}{r}688\\599\\+73\\ \hline\end{array}$ 4. $\begin{array}{r}652\\891\\+375\\ \hline\end{array}$ 5. $\begin{array}{r}3916\\803\\+2174\\ \hline\end{array}$ 6. $\begin{array}{r}5378\\929\\+1625\\ \hline\end{array}$

# Division practice

**EXAMPLE.** $72\overline{)59,863}$

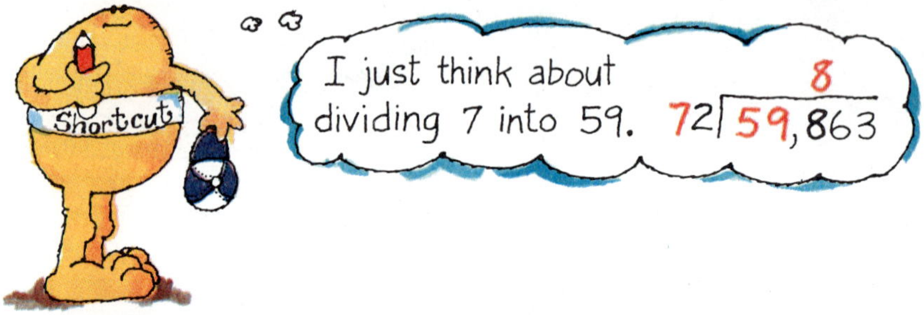


**Step 1.**

```
     8
72)59,863
  -576
    22
```

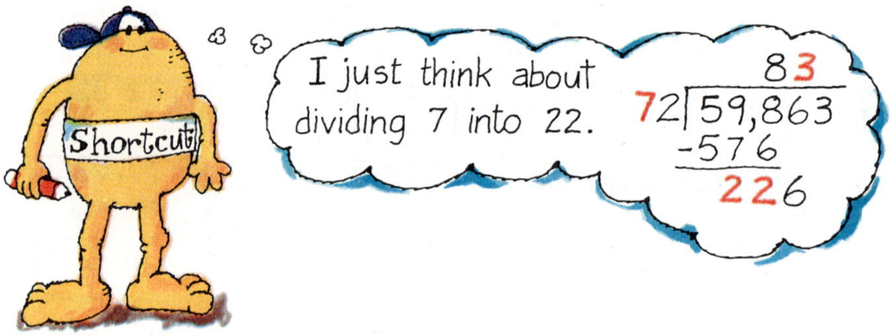


**Step 2.**

```
     83
72)59,863
  -576
    226
   -216
     10
```

**Step 3.**

```
     831 R31
72)59,863
  -576
    226
   -216
     103
    - 72
      31
```

We can check a division problem by multiplying the quotient by the divisor and then adding the remainder.

```
   831          59,832
  × 72        +     31 ← remainder
  1662          59,863  It checks!
 58170
 59,832
```

## EXERCISES

**Divide.**

1. $8\overline{)9367}$ 2. $5\overline{)8291}$ 3. $7\overline{)6403}$ 4. $9\overline{)2853}$

5. $4\overline{)28{,}345}$ 6. $3\overline{)70{,}652}$ 7. $8\overline{)916{,}304}$ 8. $6\overline{)523{,}068}$

9. $25\overline{)2834}$ 10. $32\overline{)7136}$ 11. $64\overline{)9284}$ 12. $42\overline{)9035}$

13. $37\overline{)38{,}216}$ 14. $59\overline{)74{,}555}$ 15. $48\overline{)92{,}307}$ 16. $27\overline{)56{,}283}$

**EXAMPLE.**

```
      $ 24.46
32 ) $ 7 8 2.7 2
     -6 4
       1 4 2
      -1 2 8
         1 4 7
        -1 2 8
           1 9 2
          -1 9 2
               0
```

17. $5\overline{)\$94.25}$ 18. $8\overline{)\$37.44}$

19. $7\overline{)\$666.33}$ 20. $9\overline{)\$597.87}$

21. $15\overline{)\$946.80}$ 22. $25\overline{)\$859.75}$

23. $47\overline{)\$496.79}$ 24. $39\overline{)\$654.42}$

**Solve.**

25. There are 180 school days in a year. How many school days are there in 12 years?

26. The 27 students in Bill's class gave \$8.64 to the United Fund. How much was the average gift?

## *Challenge!*

**Find the missing digits.**

27.
```
     6■7 R3
 6)3■2■
  -36
    22
   -18
    45
   -■■
     ■
```

28.
```
       6■■ R■
 14)91■5
   -■■
     ■7
    -70
     75
    -70
      ■
```

29.
```
        1 99■ R12
 ■■)■■,9■6
    -38
     37 ■
    -■■ ■
      3 73
     -3 42
        ■■■
       -304
         ■■
```

# Addition, subtraction, multiplication, and division

Remember to work inside the grouping symbols first.

$$(3905 - 193) \div 64 = 58$$

**EXERCISES**

**Complete.**

1. $(258 + 367) - 367 =$ ?
2. $(582 + 453) - 453 =$ ?
3. $(942 + 365) + 218 =$ ?
4. $942 + (365 + 218) =$ ?
5. $(782 - 526) - 142 =$ ?
6. $782 - (256 - 142) =$ ?
7. $(512 \div 16) \div 4 =$ ?
8. $512 \div (16 \div 4) =$ ?
9. $(256 \times 25) \div 25 =$ ?
10. $(874 \times 39) \div 39 =$ ?
11. $(4144 \div 74) + 125 =$ ?
12. $238 \times (1932 \div 28) =$ ?
13. $(1024 - 512) \div 32 =$ ?
14. $1024 - (512 \div 32) =$ ?
15. $(238 \times 5) + 17 =$ ?
16. $238 \times (5 + 17) =$ ?

**Solve.**

17. 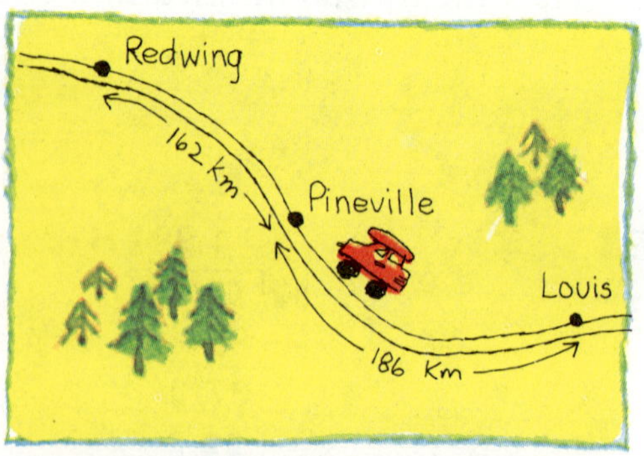


If you average 58 kilometers per hour, how long will it take you to drive from Redwing to Louis?

18. An apple grower ordered 1200 trees. If 42 trees are planted in each row, how many rows can be planted? How many trees will be left over?

★ 19. A farmer wants to build a straight fence that is 1026 meters long. If the fence posts are 3 meters apart, how many will be needed?

# SKILL GAME

**Tell which division has the greater remainder.**

1. 47)9135 14)5793
2. 60)1397 19)6307
3. 48)5271 47)2815
4. 36)2750 72)5306

**Play the game.**

1. Make a card for each of the digits.

   0 1 2 3 4 5 6 7 8 9

2. Choose a leader.
3. Each player draws a table.

   □□)□□□□

4. As the leader picks the cards, each player writes the digit in his or her table.
5. Repeat step 4 until your table is filled in.
6. The player who has the greatest remainder wins!

# Problem solving

The workers at Clark's Market grow and sell fresh fruit and vegetables. They sell gardening tools, seeds, and plants. They help people decide what food to give their plants and what sprays to use on them.

## EXERCISES

**Solve.**

1. The gardeners at Clark's Market planted 48 onions in each of 12 rows. How many onions were planted?

2. The gardeners planted 115 watermelon seeds in small hills. 5 seeds were planted in each hill. How many hills were used?

3. The gardeners planted 132 pepper plants. There were some rows of 8 plants each and one row with less than 8 plants. How many rows of 8 plants were there? How many plants were in the row that had less than 8 plants?

4. A clerk sold Michael a rake for $7.75 and a spade for $13.49. What was the total cost?

5. Judy spent $16.45 for seeds. She gave the clerk a $20 bill. How much change did she get?

Use the table below to compute each total price.

**CLARK'S FRESH VEGETABLES**

| | |
|---|---|
| BEANS | 69¢ lb |
| SPINACH | 79¢ lb |
| CABBAGE | 23¢ lb |
| CUCUMBERS | 17¢ each |
| PEAS | 57¢ lb |
| POTATOES | 25¢ lb |
| ONIONS | 48¢ lb |
| PEPPERS | 29¢ each |

6. CLARK'S MARKET

| | PRICE |
|---|---|
| 2 lb spinach | |
| 4 lb potatoes | |

7. CLARK'S MARKET

| | PRICE |
|---|---|
| 2 lb beans | |
| 5 peppers | |

8. CLARK'S MARKET

| | PRICE |
|---|---|
| 2 lb onions | |
| 3 lb peas | |
| 1 cucumber | |

9. CLARK'S MARKET

| | PRICE |
|---|---|
| 2 lb beans | |
| 2 lb cabbage | |
| 12 lb potatoes | |
| 6 peppers | |

# Problem solving—not enough information

1. Study and understand.
2. Plan and do.
3. Answer and check.

**EXAMPLE 1.**

A tour guide at Lee's Bike Company said that the workers expect to build 85,000 bicycles this year. How many wheels will be needed?

*Missing fact:* the number of wheels on a bicycle.

85,000 × (missing fact)

Everyone knows a bicycle has 2 wheels. So, you can solve the problem.

**EXAMPLE 2.**

The company makes 3000 type A wheels each day. How many spokes do they need each day to make the wheels?

*Missing fact:* the number of spokes on a wheel.

3000 × (missing fact)

Do you know the missing fact? If so, solve the problem.

**EXERCISES**

**What is the missing fact?**
**If you know the missing fact, solve the problem.**

1. Each bus could take 24 students to the bicycle factory. How many buses were needed?
2. The students had to pay $1.50 each for bus fare. Ellen had her fare in quarters. How many quarters did she have?

3. The buses left the school at 9:25. They arrived at the factory at 10:15. How long was the ride to the factory?

4. The tour guide said that the first bicycle was invented about 1790. How long ago was that?

5. One worker could build a frame in 50 minutes. At that rate, how many hours would it take her to build 18 frames?

6. The students saw about 20 yellow frames of the same size that were hanging from hooks. How much tubing was used for that many frames?

7. The guide said that a lightweight touring bicycle weighs about 14 kilograms. How much less does a racing bicycle weigh?

8. The average hourly wage at the factory was $8.32. At that rate, how much would a worker receive for a week's work?

9. The students left the factory at 12:15. How long were they at the factory?

10. After the tour, each student was given 2 decals. How many decals did the factory give away?

11. The driver of the last bus counted 3 empty seats when he closed the door. How many passengers were on the last bus?

12. After the tour, Edward decided he would use the $12 he had saved to buy a 10-speed bicycle. How much more money did he need?

# CHAPTER CHECKUP

**Give the greatest common factor (divisor).** [pages 102–103]

**1.** 6, 9　**2.** 8, 12　**3.** 12, 18　**4.** 15, 21　**5.** 16, 24

**Divide.** [pages 98–101, 106–113, 116–123]

**6.** $3\overline{)96}$　**7.** $2\overline{)84}$　**8.** $3\overline{)54}$　**9.** $5\overline{)75}$　**10.** $6\overline{)84}$

**11.** $8\overline{)93}$　**12.** $4\overline{)78}$　**13.** $5\overline{)653}$　**14.** $9\overline{)695}$　**15.** $7\overline{)853}$

**16.** $42\overline{)966}$　**17.** $35\overline{)980}$　**18.** $53\overline{)1961}$　**19.** $81\overline{)6000}$　**20.** $68\overline{)6835}$

**Solve.** [pages 104–105, 114–115, 124–127]

**21. a.** What is the average test score?

**b.** How many scored above the average? below the average?

**22.** 322 days
How many weeks?

**23.** 672 hours
How many days?

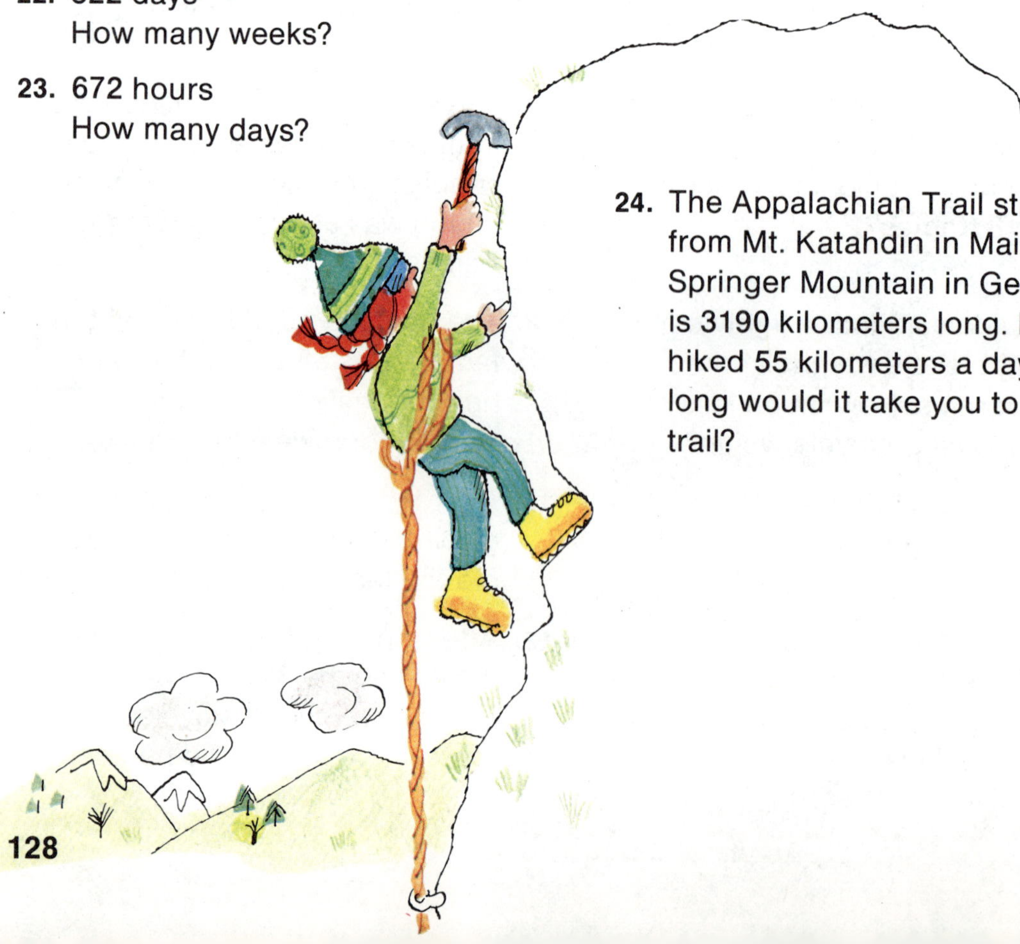

**24.** The Appalachian Trail stretches from Mt. Katahdin in Maine to Springer Mountain in Georgia. It is 3190 kilometers long. If you hiked 55 kilometers a day, how long would it take you to hike the trail?

# CHAPTER PROJECT

1. Copy and complete this dot graph.

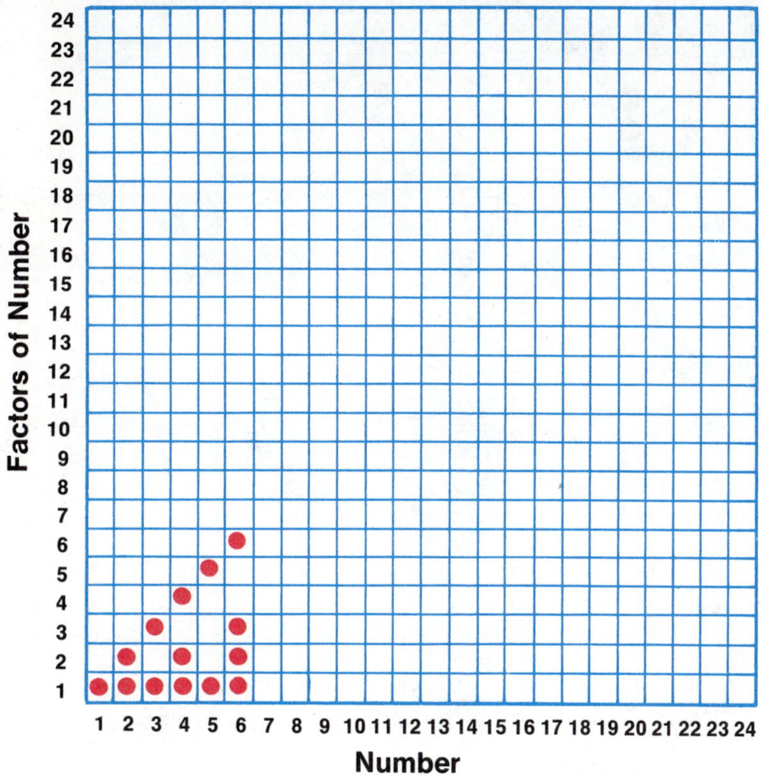


**Use your completed graph to answer the following questions.**

2. How many factors does 12 have? does 18 have?

3. Which number has the most factors?

4. Which numbers have exactly 3 factors?

5. A **prime number** has exactly two factors. Which numbers on your graph are prime numbers?

6. How can you use your graph to find the common factors of 12 and 18? the greatest common factor of 12 and 18?

# CHAPTER REVIEW

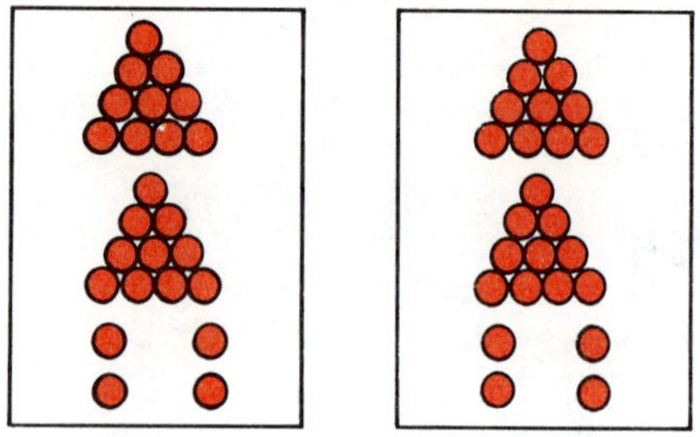

**Divide.**

$2\overline{)48}$ → 24

1. $3\overline{)93}$
2. $4\overline{)80}$

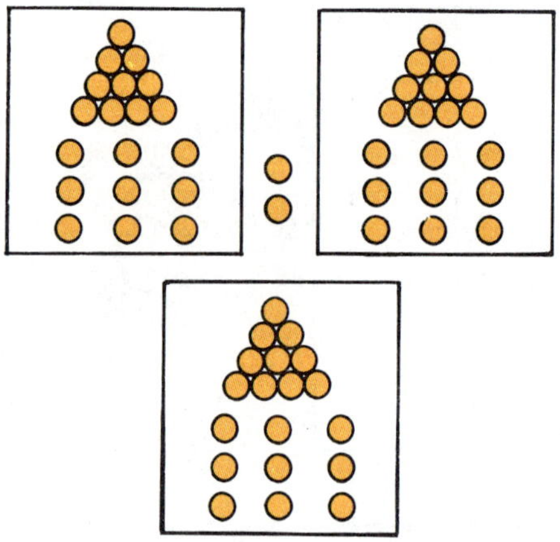

$6\overline{)78}$ → 13
$-6$, 18, $-18$, 0

3. $5\overline{)95}$
4. $7\overline{)91}$

$3\overline{)59}$ → 19 R2
$-3$, 29, $-27$, 2

5. $6\overline{)83}$
6. $8\overline{)99}$

$7\overline{)526}$ → 75 R1
$-49$, 36, $-35$, 1

7. $8\overline{)756}$
8. $9\overline{)700}$

$32\overline{)785}$ → 24 R17
$-64$, 145, $-128$, 17

9. $32\overline{)982}$
10. $32\overline{)896}$

$56\overline{)3974}$ → 70 R54
$-392$, 54, $-0$, 54

11. $61\overline{)8306}$
12. $78\overline{)9973}$

# CHAPTER CHALLENGE

When we divide the whole number 255 by 3, there is no remainder. We say that 255 is divisible by 3.

When we divide 255 by 4, there is a remainder. We say that 255 is not divisible by 4.

A whole number is divisible by 2 if its last digit is divisible by 2.

**Is the number divisible by 2?**

| | | |
|---|---|---|
| **1.** 74 | **2.** 63 | **3.** 80 |
| **4.** 258 | **5.** 327 | **6.** 405 |
| **7.** 1046 | **8.** 3509 | **9.** 2791 |

A whole number is divisible by 3 if the sum of its digits is divisible by 3.

**Is the number divisible by 3?**

| | | |
|---|---|---|
| **10.** 72 (7+2) | **11.** 84 (8+4) | **12.** 95 (9+5) |
| **13.** 153 | **14.** 263 | **15.** 570 |
| **16.** 3217 | **17.** 4614 | **18.** 5830 |

A whole number is divisible by 4 if its last two digits are divisible by 4.

**Is the number divisible by 4?**

| | | |
|---|---|---|
| **19.** 120 | **20.** 234 | **21.** 712 |
| **22.** 2316 | **23.** 5205 | **24.** 4236 |
| **25.** 3458 | **26.** 6039 | **27.** 5164 |

A whole number is divisible by 5 if its last digit is divisible by 5 (is either 5 or 0).

**Is the number divisible by 5?**

| | | |
|---|---|---|
| **28.** 80 | **29.** 74 | **30.** 65 |
| **31.** 328 | **32.** 260 | **33.** 435 |
| **34.** 2306 | **35.** 5120 | **36.** 4205 |

b c d 2. a b c d a b c d a b c d 5. a b c d 6. a b c

b c d 8. a b c d 9. a b c d 10. a b c d 11. a b c d 12. a b c

# MAJOR CHECKUP

## STANDARDIZED FORMAT

**Choose the correct letter.**

**1.** In 748,192, the digit in the hundred thousands place is

**a.** 8
**b.** 4
**c.** 7
**d.** none of these

**2.** 450,000 rounded to the nearest thousand is

**a.** 500,000
**b.** 450,000
**c.** 400,000
**d.** none of these

**3.** Which number is the least?

**a.** 5,963,849
**b.** 5,874,026
**c.** 43,281,752
**d.** 49,714,260

**4.** Add.

$$\begin{array}{r} 7813 \\ 942 \\ 1604 \\ +97 \\ \hline \end{array}$$

**a.** 10,456
**b.** 10,446
**c.** 8346
**d.** none of these

**5.** Subtract.

$$\begin{array}{r} 7421 \\ -3859 \\ \hline \end{array}$$

**a.** 4438
**b.** 4672
**c.** 3672
**d.** none of these

**6.** Which number is a common multiple of 6 and 9?

**a.** 3
**b.** 15
**c.** 36
**d.** none of these

**7.** Multiply.

$$\begin{array}{r} 613 \\ \times 8 \\ \hline \end{array}$$

**a.** 4884
**b.** 4804
**c.** 4904
**d.** none of these

**8.** Multiply.

$$\begin{array}{r} 7296 \\ \times 75 \\ \hline \end{array}$$

**a.** 88,552
**b.** 529,370
**c.** 547,200
**d.** none of these

**9.** Divide.

$4\overline{)2718}$

**a.** 879 R2
**b.** 679
**c.** 679 R2
**d.** none of these

**10.** Divide.

$38\overline{)3914}$

**a.** 103
**b.** 13
**c.** 1003
**d.** none of these

**11.** Apples cost 24¢ each. How much do 4 apples cost?

**a.** 96¢ **b.** 6¢
**c.** 8¢ **d.** none of these

**12.** 3 hot dogs cost $2.40. How much does 1 hot dog cost?

**a.** $7.20 **b.** $6.20
**c.** $.70 **d.** none of these

# Geometry

# 5

# READY OR NOT?

**Match each figure with its name.**

1. 

2. 

3. 

4. 

5. 

6. 

7. 

8. 

a. square
b. triangle
c. segment
d. line
e. circle
f. ray
g. point
h. rectangle

# Points and segments

Sylvester the Space Spider can travel any place in space. He marks his path with a thin spiderweb. Sometimes he takes the shortest path between two points, a **segment.**

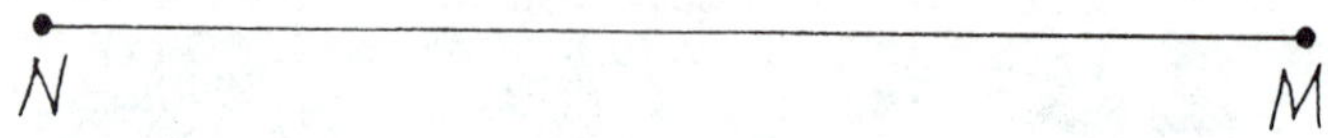


A segment is named by its **endpoints.** Here are two ways to name the segment:

$\overline{MN}$ (segment *MN*) $\overline{NM}$ (segment *NM*)

One day Sylvester was caught speeding. He had to spend the next week in a flat surface, a **plane.** He could go as far as he wanted in that plane, but he could not leave the plane. Here are some plane figures that Sylvester made:

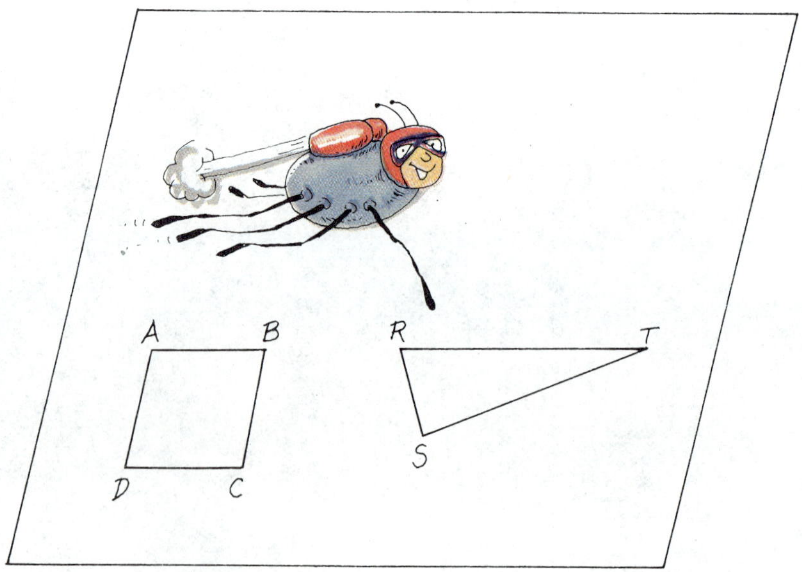


[Remember that the corner points are called **vertices** (plural of **vertex**) and the segments are called **sides.**]

## EXERCISES

**Which figures are plane figures? (*Hint:* Could Sylvester make the figure by staying on a flat surface?)**

**1.** 

sphere

**2.** 

cylinder

**3.** 

cone

**4.** 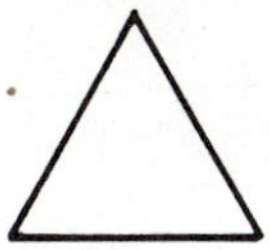

triangle

**5.** 

rectangular solid

**6.** 

rectangle

**7.** 

cube

**8.** 

segment

**9.** 

square

**10.** Give the endpoints of the segment.

**11.** Give two ways to name the segment.

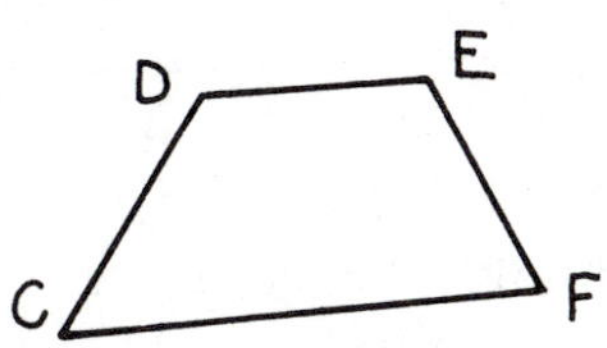


**12.** List the vertices of the plane figure.

**13.** How many sides does it have? List the sides.

**14.** Complete this table.

| Name | Figure | Number of vertices | Number of sides |
|---|---|---|---|
| triangle |  | | |
| square |  | | |
| pentagon |  | | |
| hexagon |  | | |

# Rays, angles, and lines

Hester, another spider, wanted to spend her whole life traveling. She started at point $A$, went straight through point $B$, and then continued along the straight path. Her path is a **ray.** It has an endpoint and goes on and on in one direction.

This is ray $AB$.

We write $\overrightarrow{AB}$.

Notice that we write the endpoint first.

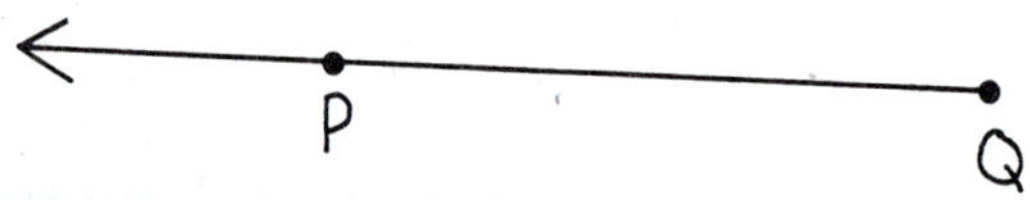


This is ray $QP$.

We write $\overrightarrow{QP}$.

$\overrightarrow{MX}$ and $\overrightarrow{MY}$ together form a **line.** Notice that a line has no endpoints. It goes on and on in both directions.

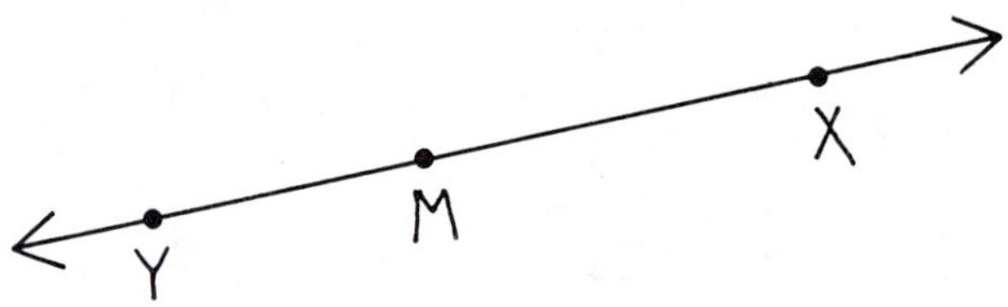


We name a line by using any two points on the line. Here are some names for the line above:

$\overleftrightarrow{XY}$ (line $XY$)   $\overleftrightarrow{YX}$   $\overleftrightarrow{XM}$

$\overrightarrow{SR}$ and $\overrightarrow{ST}$ together form an **angle.** Point $S$ is the endpoint of both rays. It is called the **vertex** of the angle. The two rays are the **sides** of the angle.

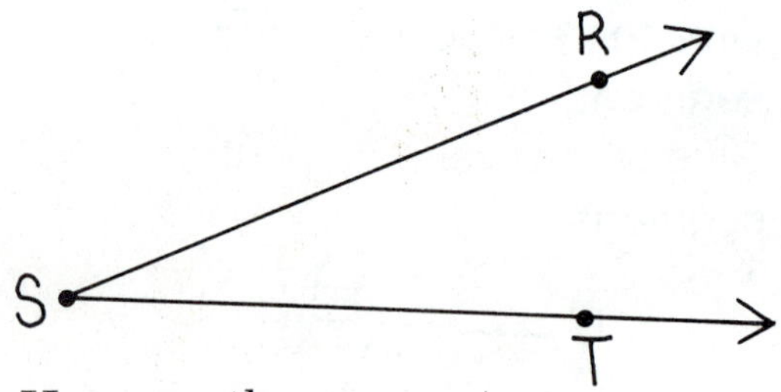


Here are three ways to name the angle:

$\angle S$ (angle $S$)   $\angle RST$   $\angle TSR$

The vertex is written in the middle.

## EXERCISES

**True or false?**

1. A segment has two endpoints.
2. A line has two endpoints.
3. A ray has one endpoint.
4. $E$ is the vertex of $\angle DEF$.
5. $\overrightarrow{FE}$ is a side of $\angle DEF$.
6. $\angle DEF$ can be called $\angle E$.
7. $\overleftrightarrow{ED}$ is a side of the angle.

**Use three letters to name each angle.**

8. 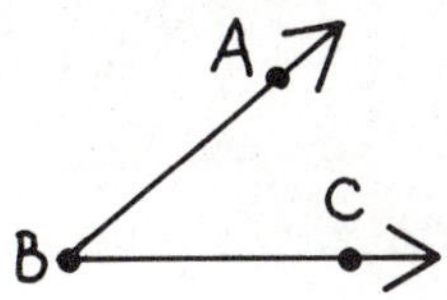


9. 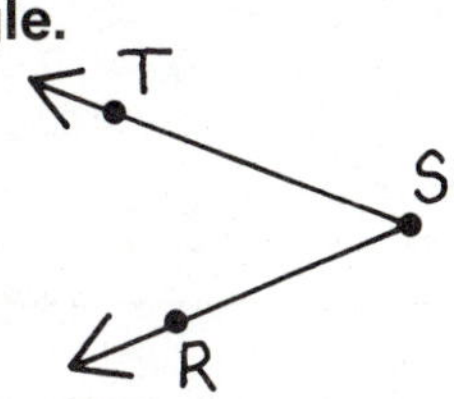


10. 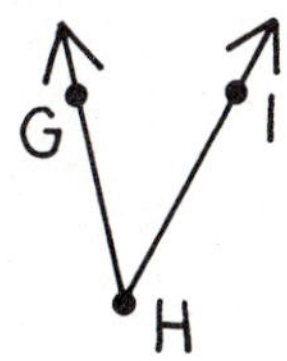


11. 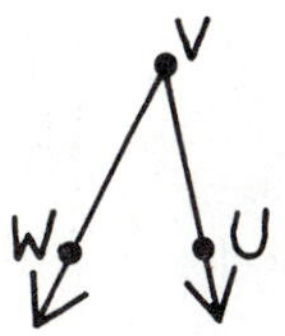


12. 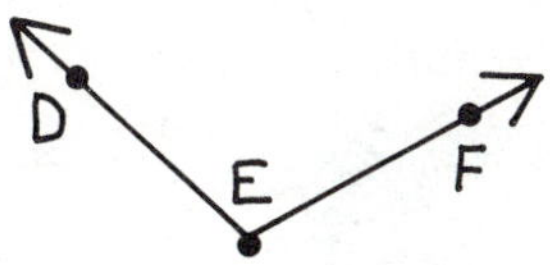


13. 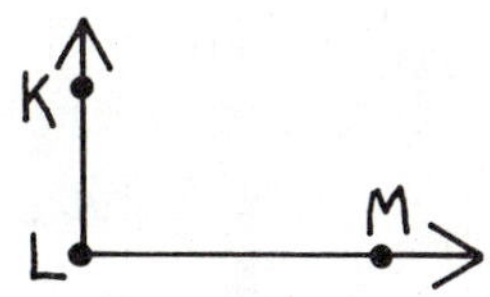


14. a. Use three letters to name each angle.

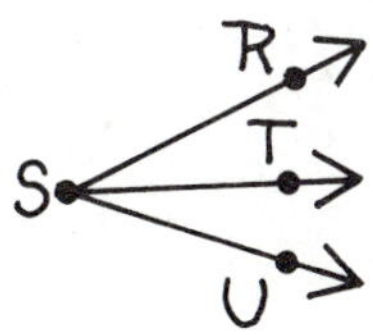


b. Why can't you use "$\angle S$" to name one of the angles?

15. Angle $R$ is called an angle of the triangle. Name the other angles of the triangle.

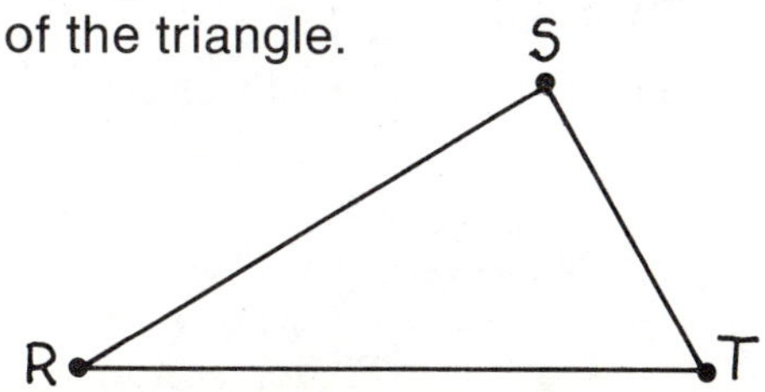


### KEEPING SKILLS SHARP

1. $56 \times 8$
2. $72 \times 5$
3. $193 \times 7$
4. $1523 \times 4$
5. $2580 \times 6$
6. $83 \times 15$
7. $74 \times 26$
8. $54 \times 91$
9. $158 \times 32$
10. $195 \times 146$
11. $783 \times 250$
12. $709 \times 623$

# Measuring angles

The **degree** is a unit used for measuring angles.

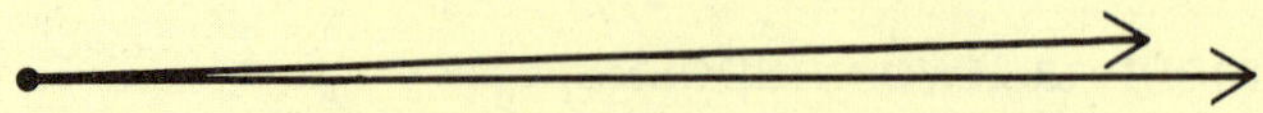

This is a 1° (one degree) angle.

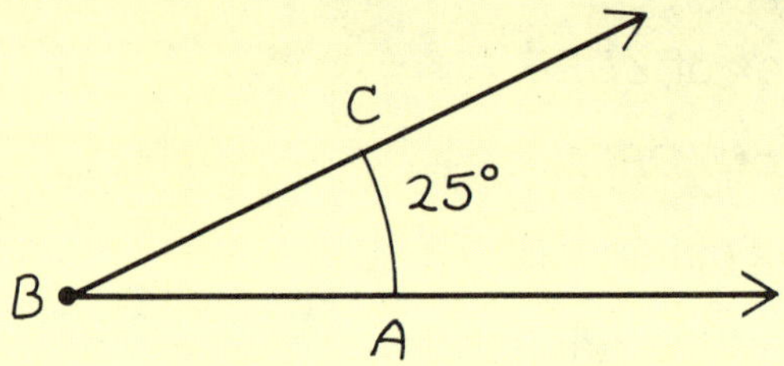


The measure of $\angle ABC$ is 25°.

A **protractor** is used to measure angles.

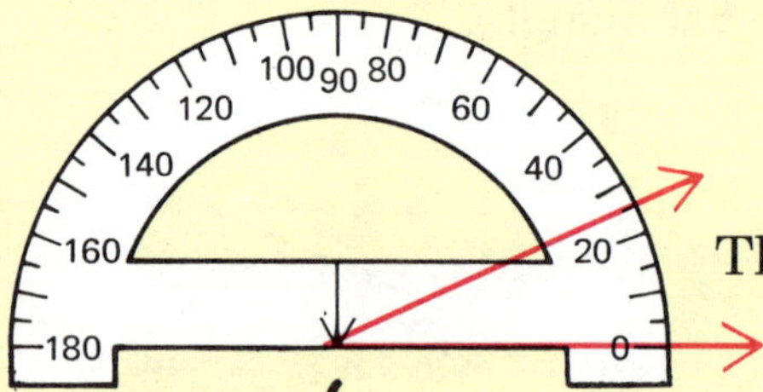


This is a 25° angle.

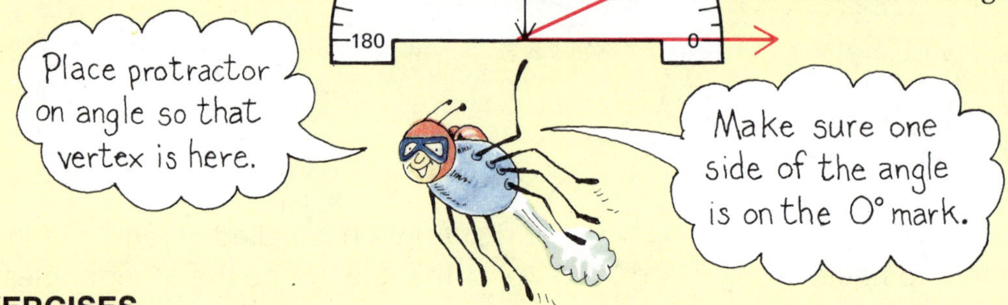


**EXERCISES**

**Give the measure of each angle.**

**1.**

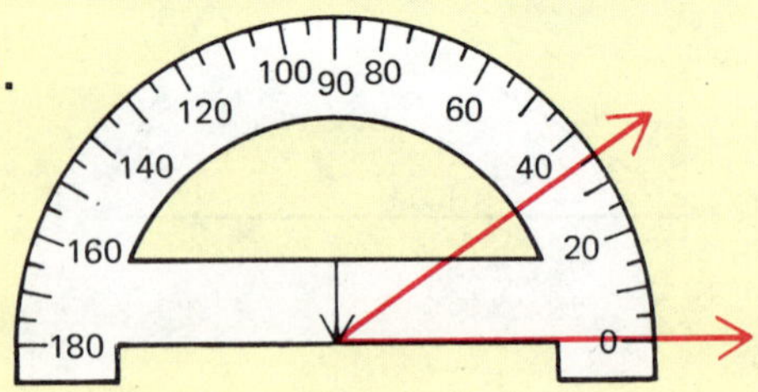

**2.**

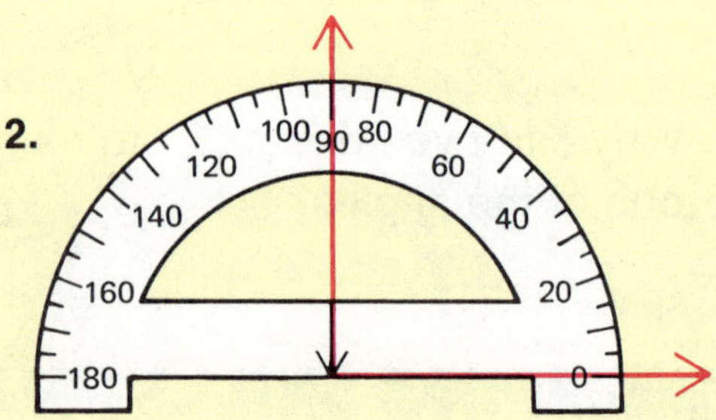

**3.**

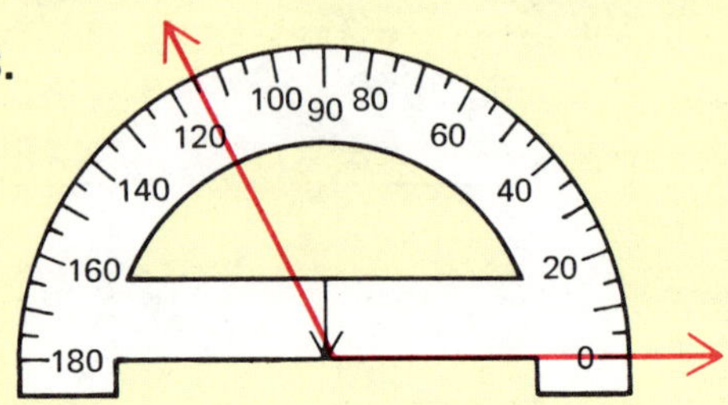

**4.**

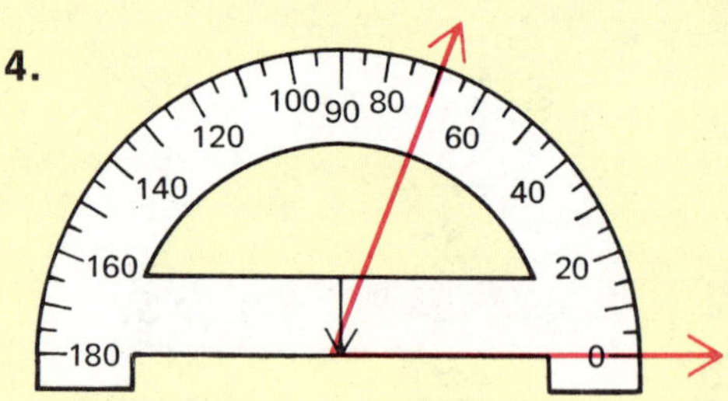

5.

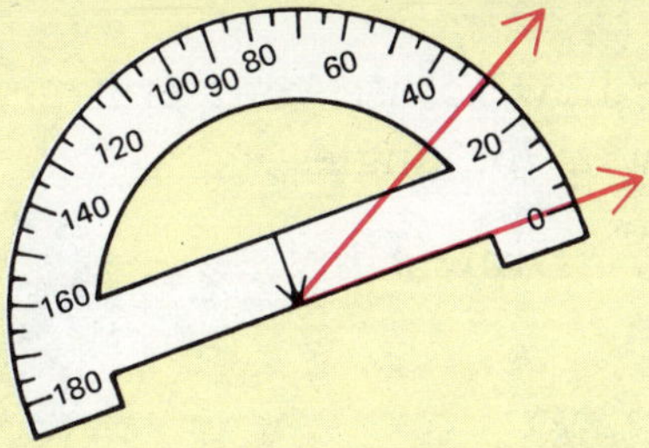

6.

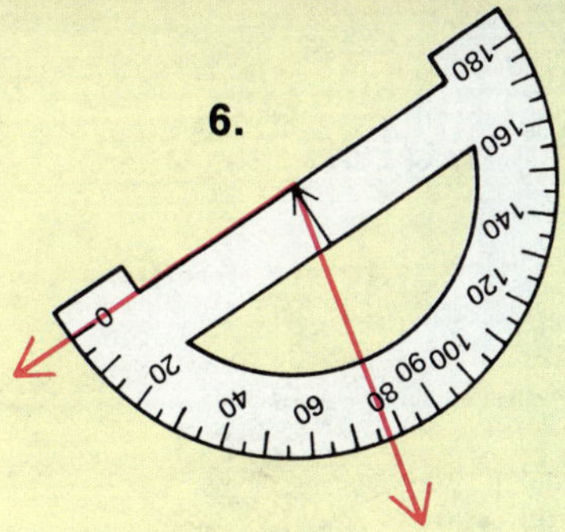

**Measure each angle with a protractor.**

7. 8. 9. 10.

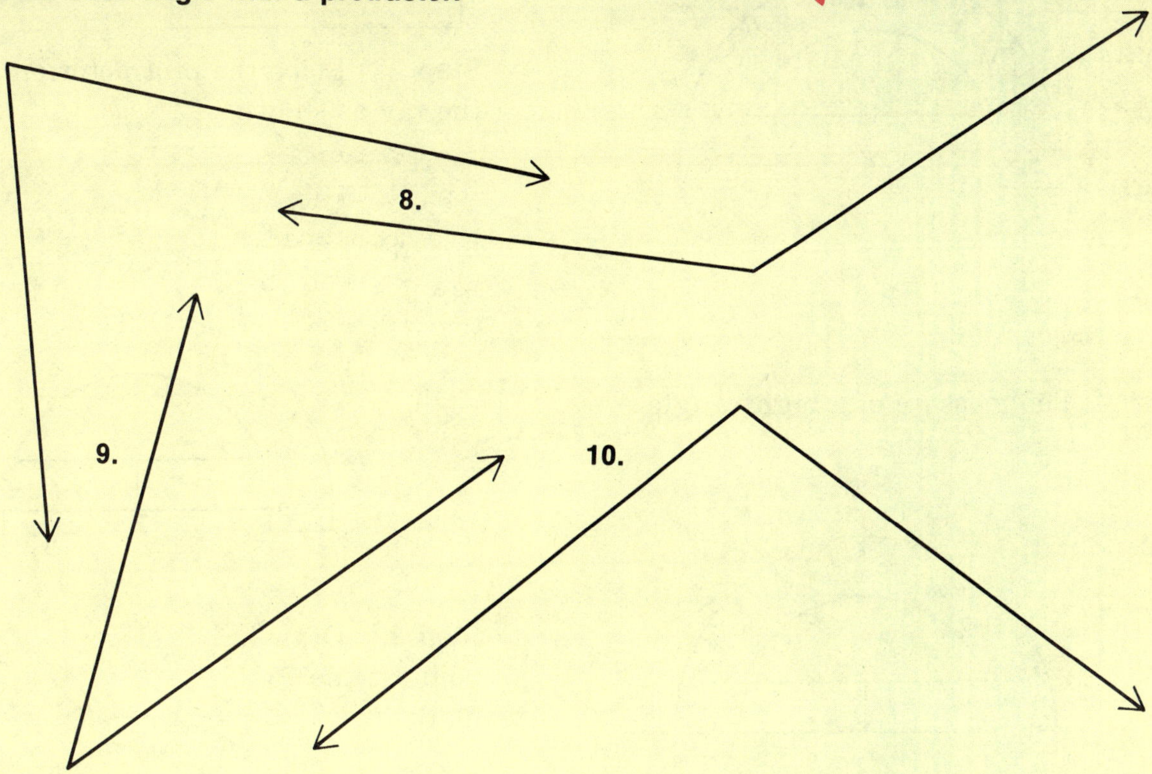

## *Challenge!*

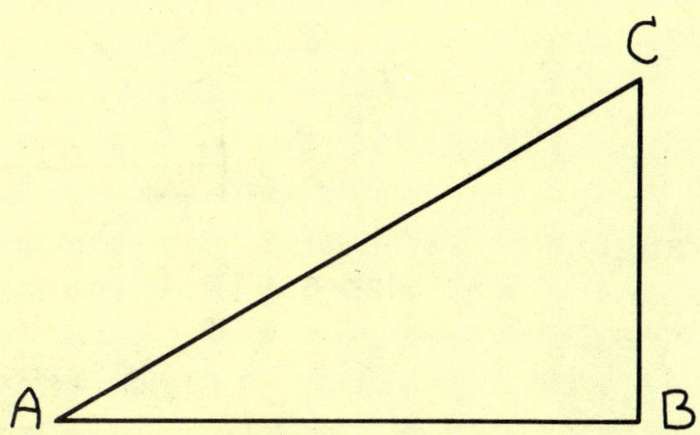


**11.** Measure

**a.** $\angle A$.

**b.** $\angle B$.

**c.** $\angle C$.

**12.** Add the measures of $\angle A$, $\angle B$, and $\angle C$.

**13.** Draw a triangle of your own. Find the sum of the angles. Did you get the same sum as you did in exercise 12?

# Kinds of angles

The measure of an **acute angle** is less than 90°.

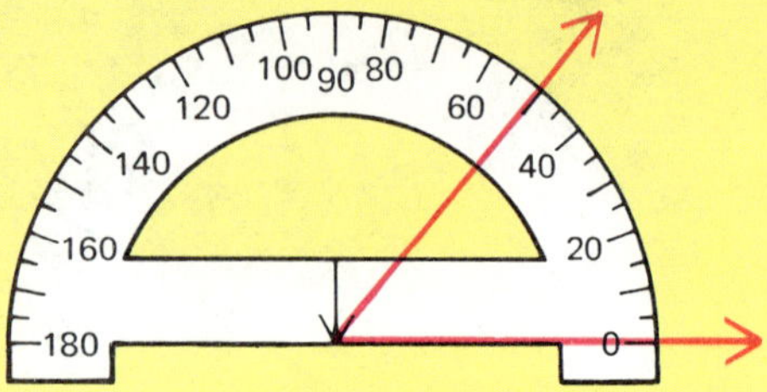


The measure of a **right angle** is 90°.

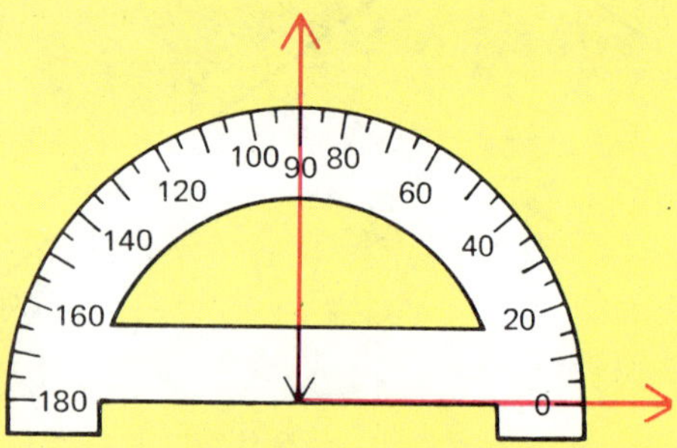


The measure of an **obtuse angle** is greater than 90°.

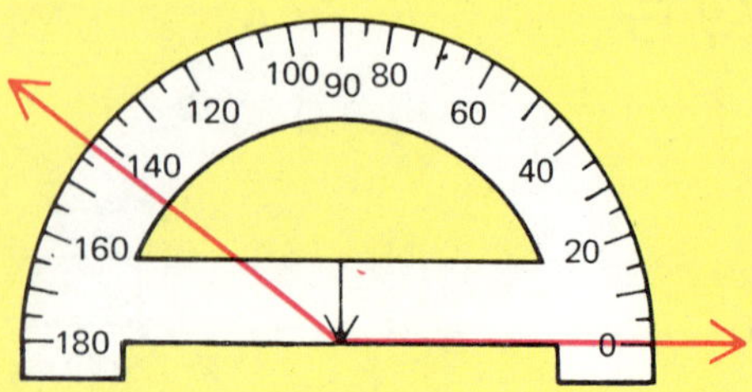


Here is how to use your protractor to draw a 75° angle.

**Step 1.** Draw a ray.

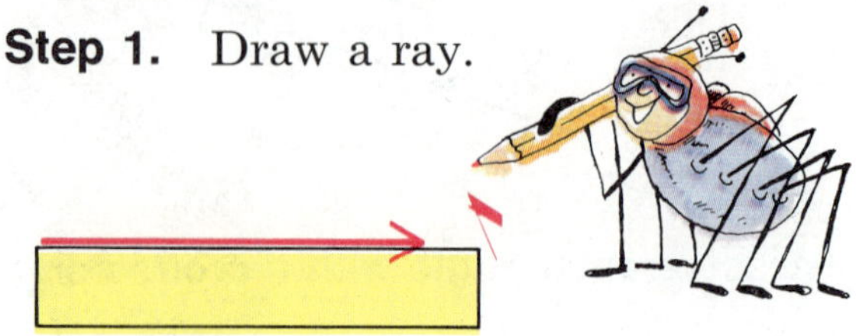

**Step 2.** Place the protractor on the ray as shown.

**Step 3.** Draw a point at the 75° mark.

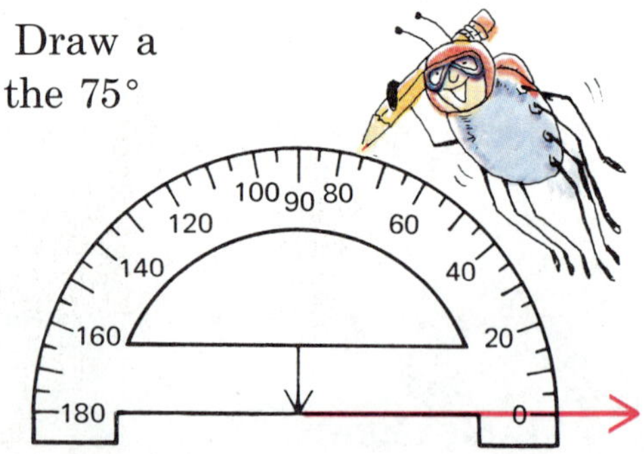


**Step 4.** Draw the second ray.

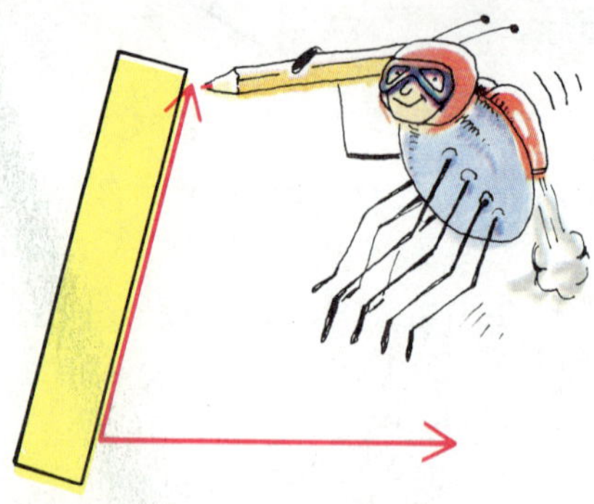

## EXERCISES

**Tell whether the angle is acute, right, or obtuse.**

1. 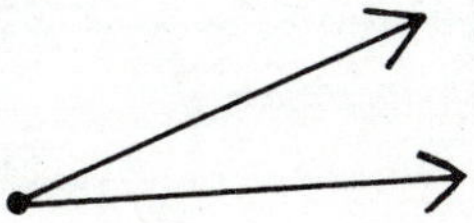

2. 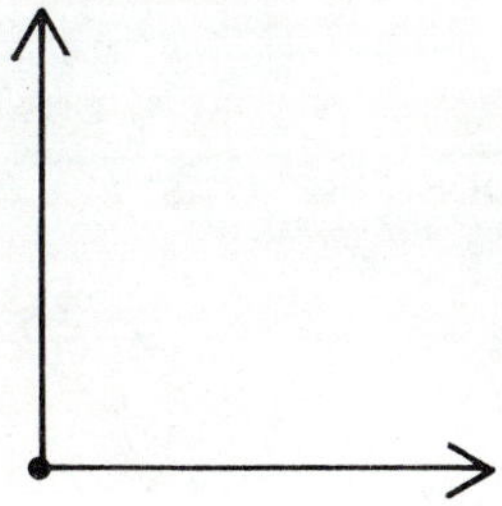

3. 

4. 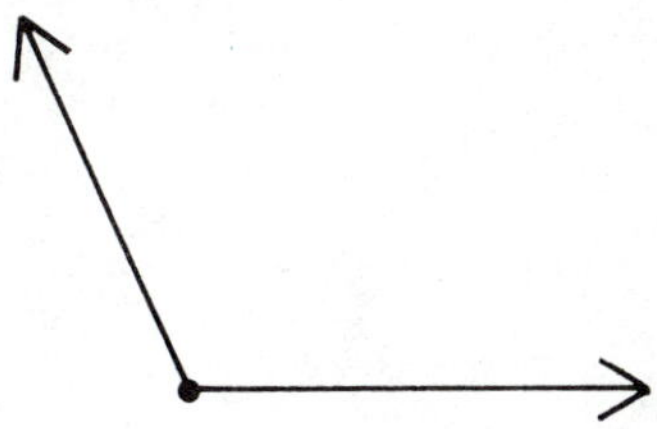

5. 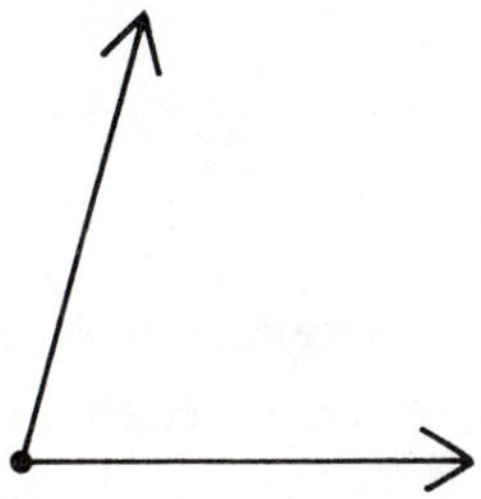

6. 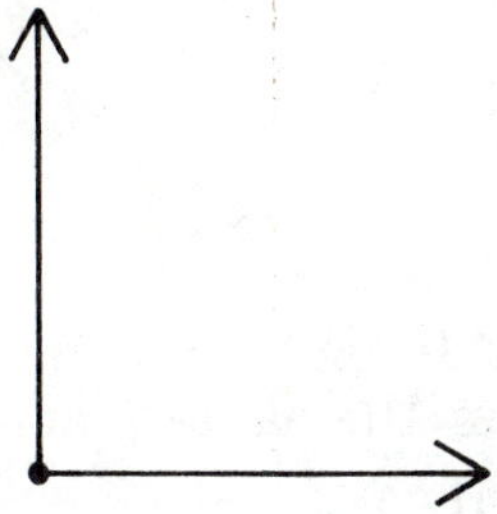

7. $\angle BAD$
8. $\angle BAC$
9. $\angle CAD$
10. $\angle CAE$
11. $\angle BAE$
12. $\angle CAF$

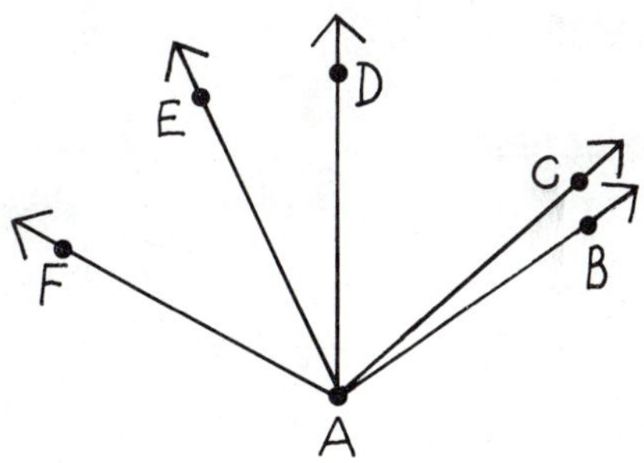


**Draw angles having these measures.**

13. 35°
14. 115°
15. 90°
16. 38°
17. 152°

**True or false? (If you need to, make a sketch.)**

18. A square has 4 right angles.
19. All rectangles have 4 right angles.
20. No triangle has 3 acute angles.
21. Some triangles have 1 obtuse angle.
22. No triangle has 2 obtuse angles.

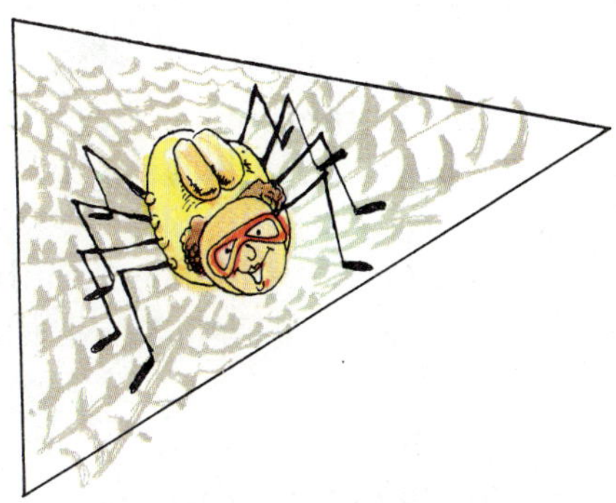

# Parallel and perpendicular lines

Sometimes Esther and Sylvester fly straight paths in formation. Their paths look like this:

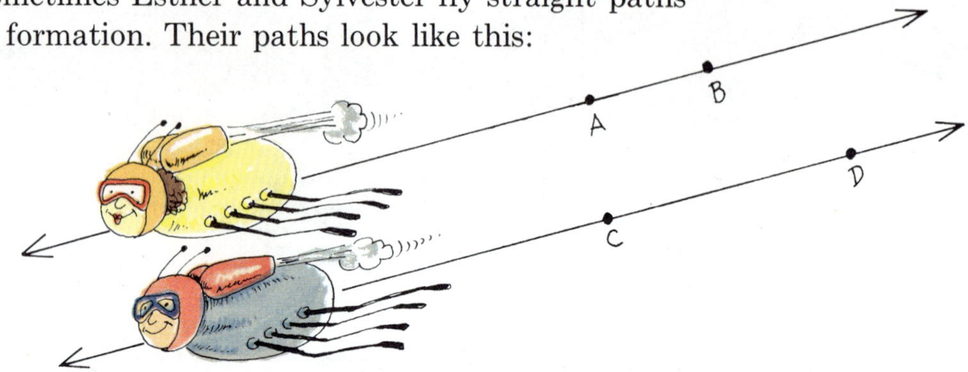


The two lines are called **parallel lines.** They are in the same plane, they never meet, and they are always the same distance apart.

$\overleftrightarrow{AB}$ is parallel to $\overleftrightarrow{CD}$.

We also say that two segments are parallel if they are parts of two parallel lines.

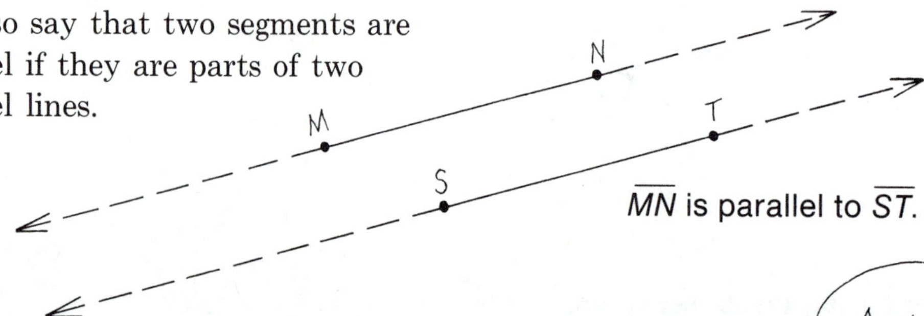


$\overline{MN}$ is parallel to $\overline{ST}$.

Two lines that intersect to form right angles are called **perpendicular** lines.

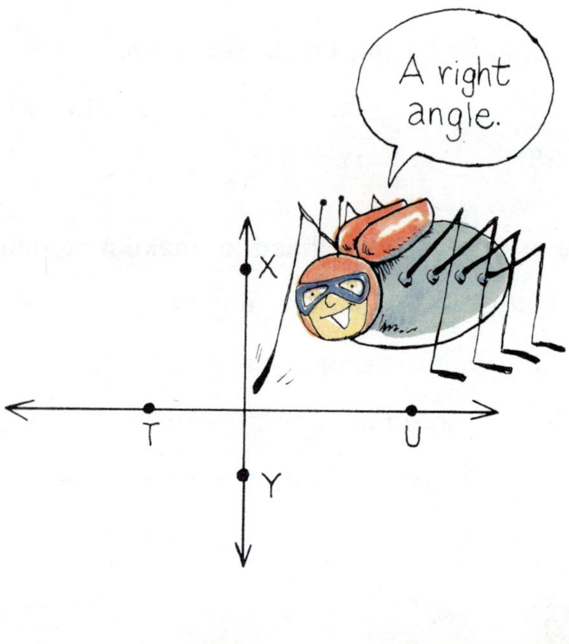


$\overleftrightarrow{XY}$ is perpendicular to $\overleftrightarrow{TU}$.

Two segments are perpendicular if they belong to perpendicular lines.

$\overline{XY}$ is perpendicular to $\overline{TU}$.

## EXERCISES

**The lines shown in each exercise were torn from a large piece of paper. Do you think that the lines are parallel, perpendicular, or neither?**

**1.** 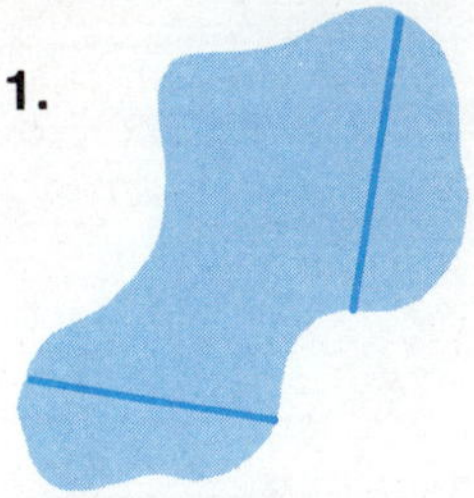

**2.** 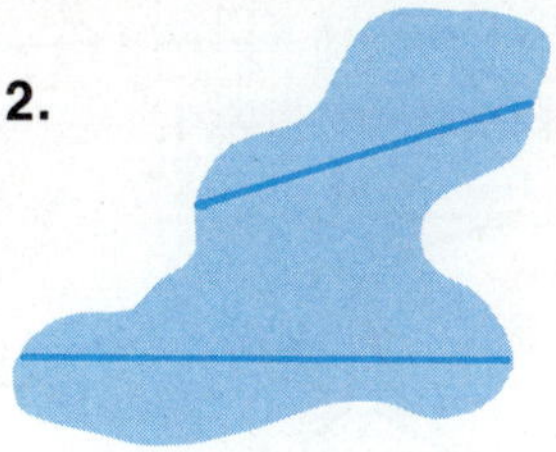

**3.** 

**Tell which segments are parallel and which are perpendicular.**

**4.** 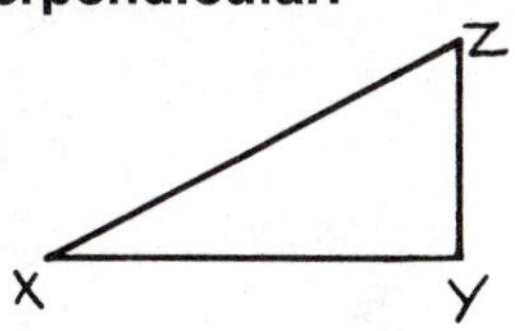


**5.** 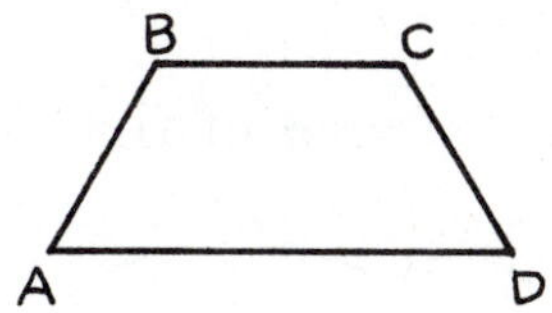


**6.** 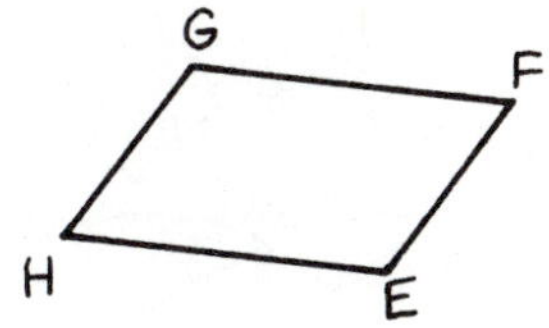


**7.** 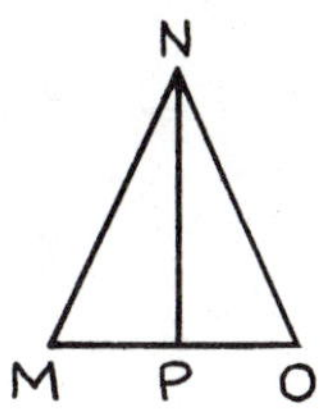


**8.** 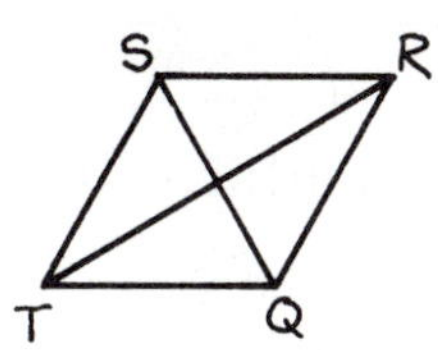


**9.** 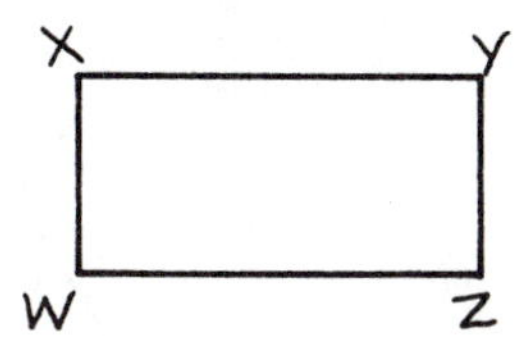

**Draw a figure to go with each sentence.**

**10.** $\overleftrightarrow{AB}$ is perpendicular to $\overleftrightarrow{CD}$.

**11.** $\overleftrightarrow{RS}$ is parallel to $\overleftrightarrow{XY}$.

**12.** $\overleftrightarrow{TU}$ intersects $\overleftrightarrow{WX}$ at point *P*.

**13.** $\overleftrightarrow{EF}$ is perpendicular to $\overleftrightarrow{EG}$.

★ **14.** $\overleftrightarrow{AB}$ is parallel to $\overleftrightarrow{CD}$, $\overleftrightarrow{RS}$ is perpendicular to $\overleftrightarrow{AB}$, $\overleftrightarrow{RS}$ intersects $\overleftrightarrow{AB}$ at point *M*, and $\overleftrightarrow{CD}$ intersects $\overleftrightarrow{RS}$ at point *P*.

# Circles

A compass is used to draw a circle. The point of the compass doesn't move, while the pencil in the compass makes a curve around the point. The center of the circle is where the point of the compass is.

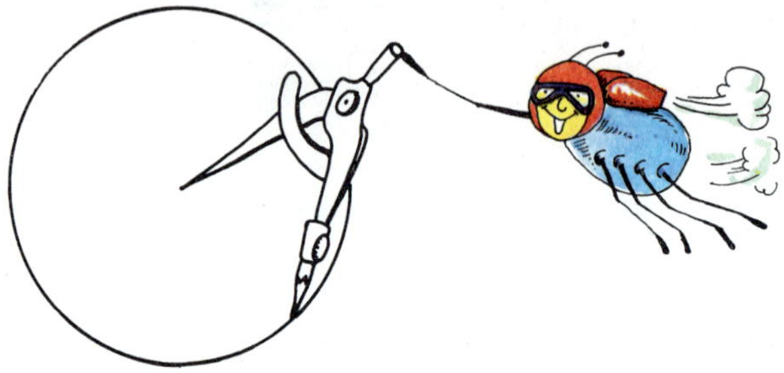

$\overline{AB}$ is a **chord** of this circle.

$\overline{RS}$ is a **diameter** of this circle.

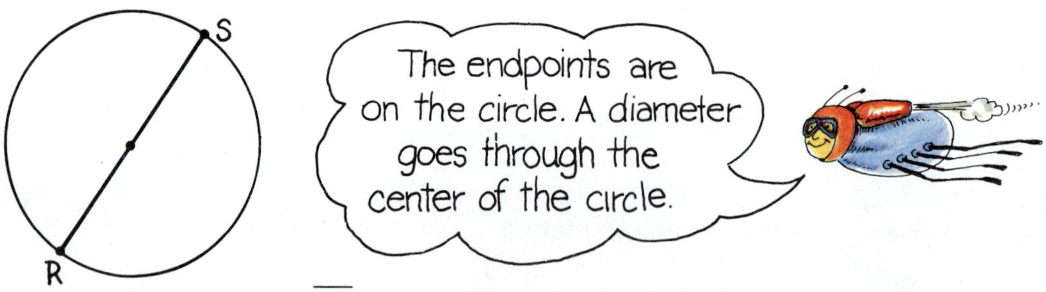


$\overline{CD}$ is a **radius** of this circle.

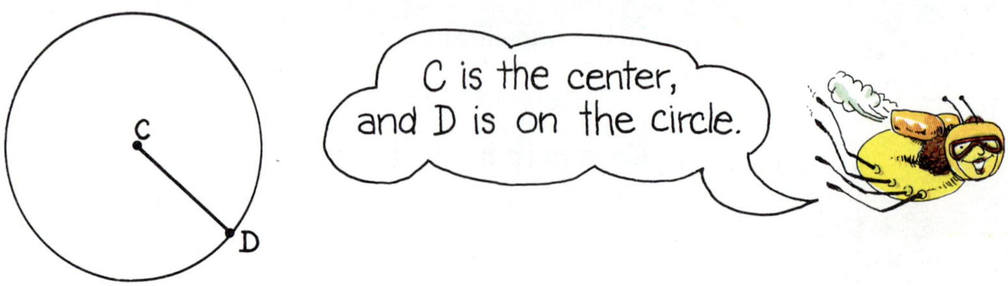

## EXERCISES

**Name the red part.**

1. 

2. 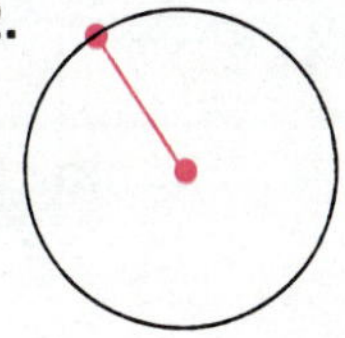

3. 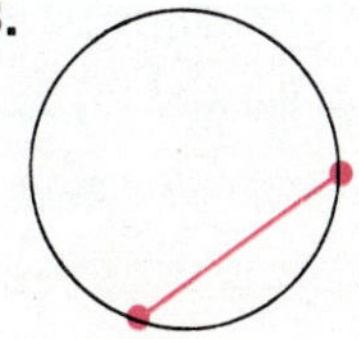

4. 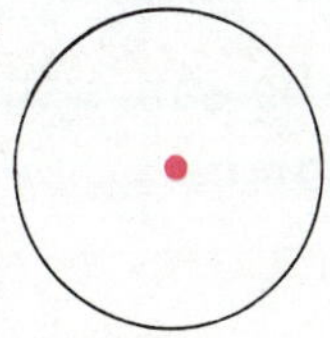

5. 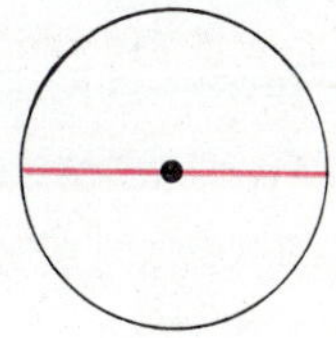

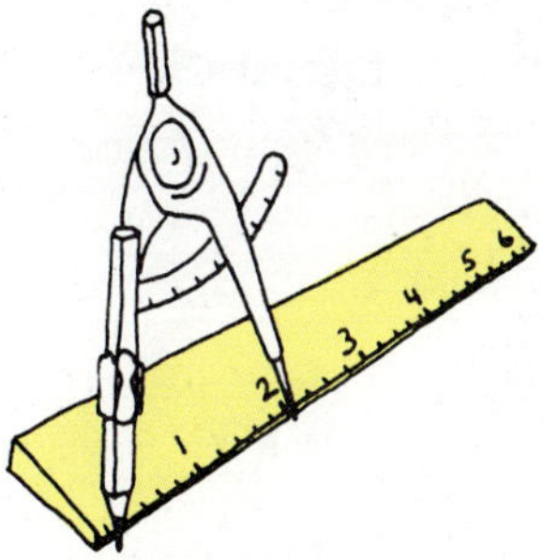


**Get a centimeter ruler and a compass.**
**Draw a circle with a radius of**

6. 2 cm
7. 3 cm
8. 4 cm

**Draw a circle with a diameter of**

9. 2 cm
10. 4 cm
11. 6 cm

The word **bisect** means "to cut into halves." These steps show how to bisect a segment.

**Step 1.**

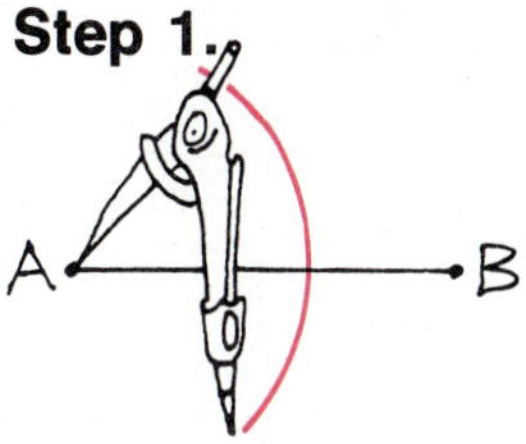


**Step 2.**

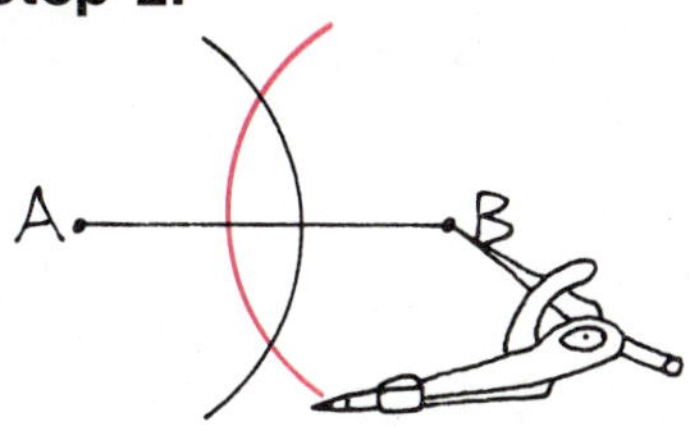


**Step 3.**

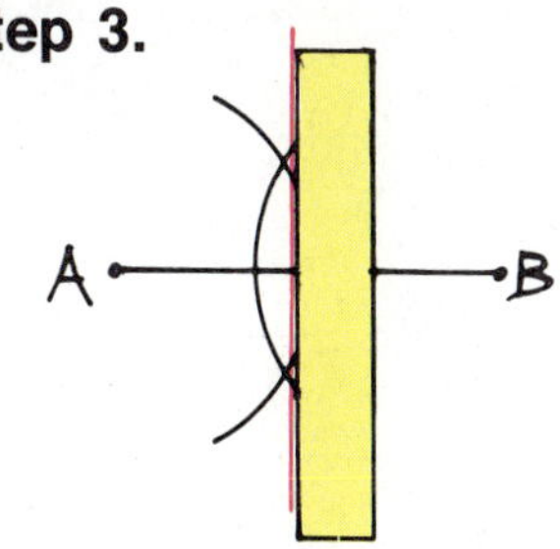


12. Draw a segment and bisect it.

Here is how to bisect an angle.

**Step 1.**

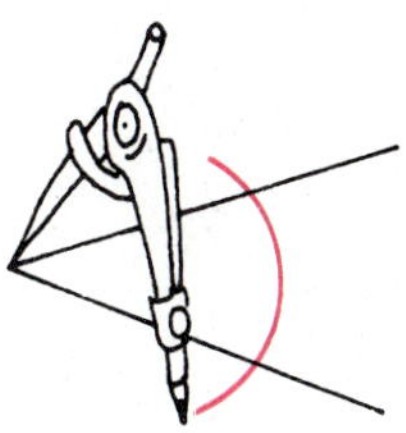

**Step 2.**

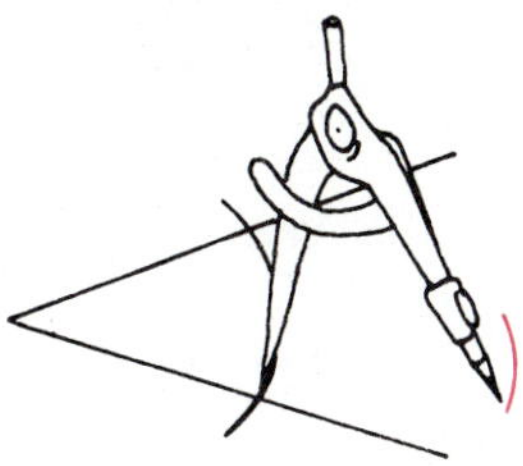

**Step 3.**

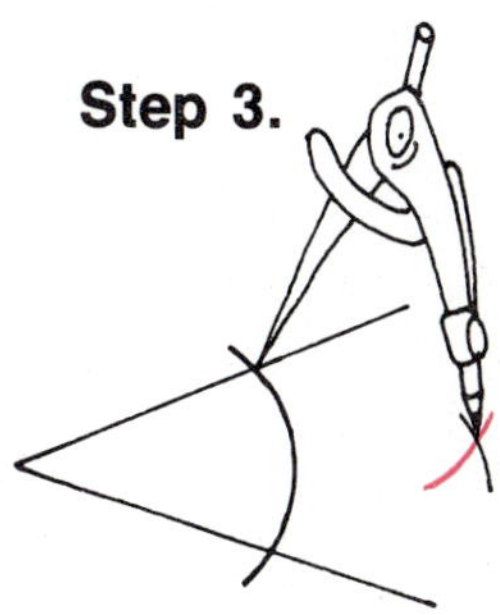

**Step. 4.**

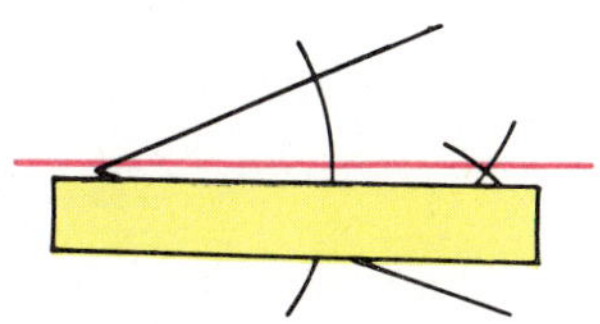

13. Draw an angle and bisect it.

# Similar figures and congruent figures

Figures that are the same shape are called **similar figures.**

Similar Rectangles

Figures that are the same shape and size are called **congruent figures.**

Congruent Triangles

**EXERCISES**

**Match similar figures.**

1. 

2. 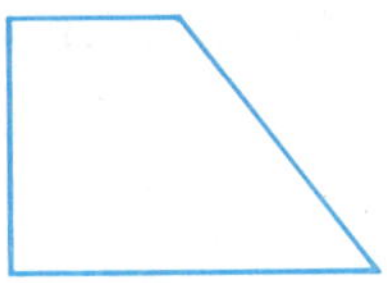

3. 

4. 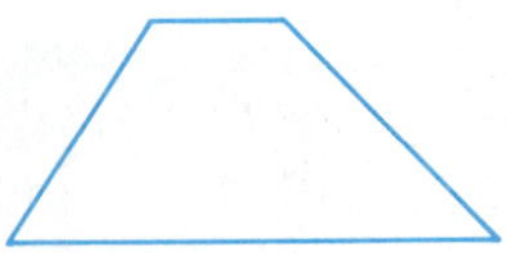

5. 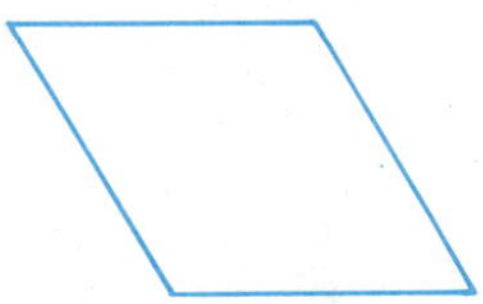

a. 

b. 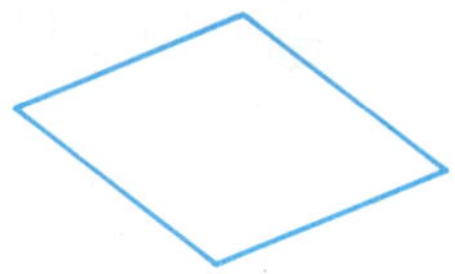

c. 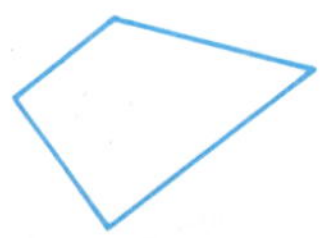

d. 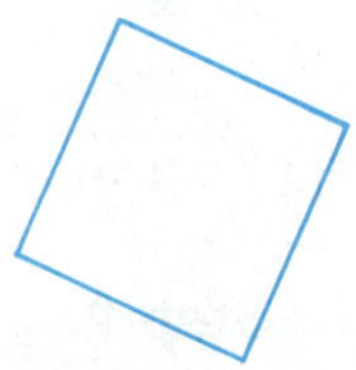

e. 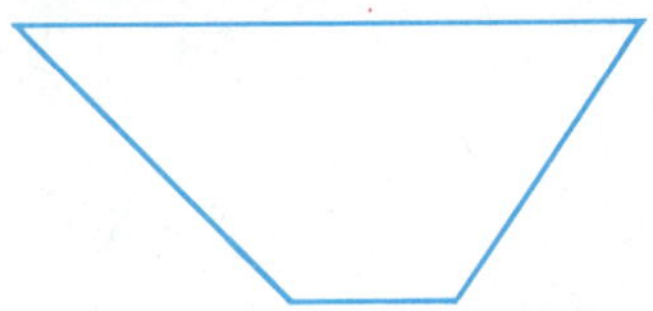

Three rubber stamps were used to make the red figures shown below. Which blue figures could be made with the same stamps? That is, which blue figures are congruent to a red figure?
(*Hint:* Try making tracings.)

6. 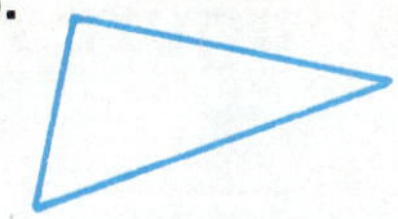

7. 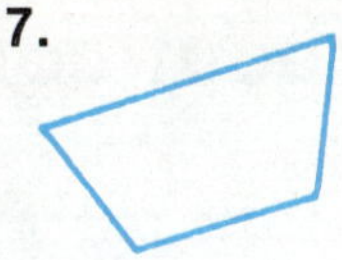

8. 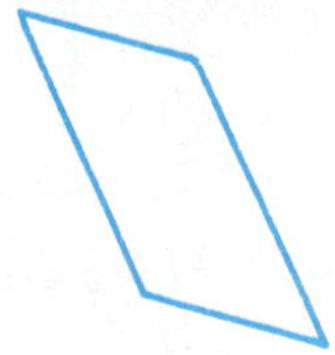

9. 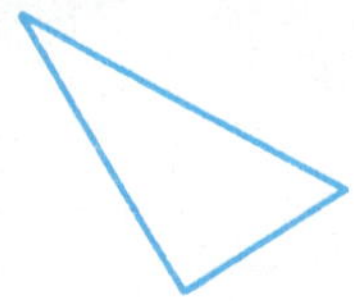

10. 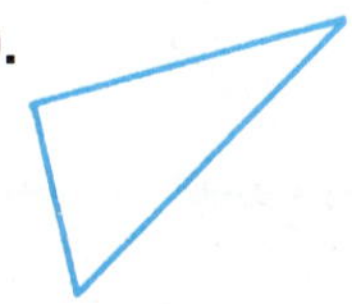

11. 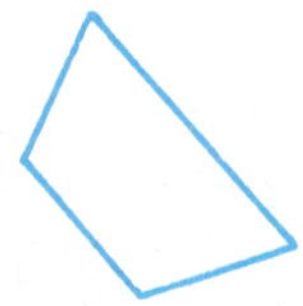

12. 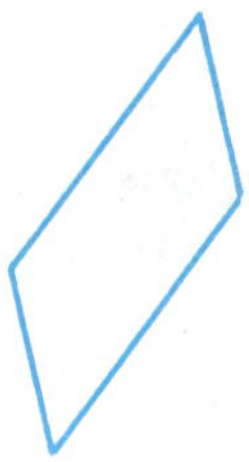

13. 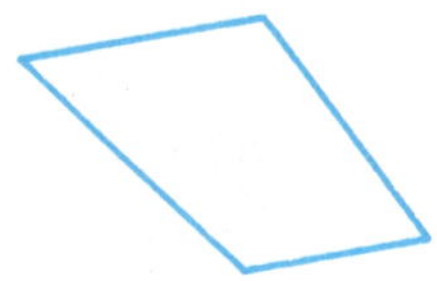

14. 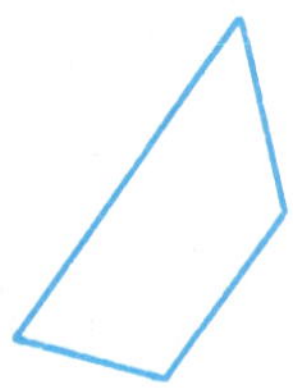

15. 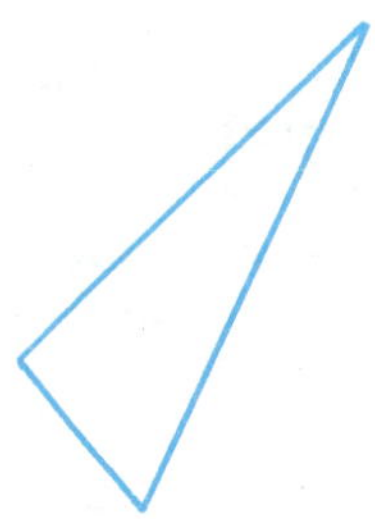

## KEEPING SKILLS SHARP

| | | | | |
|---|---|---|---|---|
| 1. 9)468 | 2. 7)658 | 3. 8)536 | 4. 5)7385 | 5. 3)1596 |
| 6. 14)784 | 7. 26)621 | 8. 37)1702 | 9. 52)1976 | 10. 81)5265 |
| 11. 27)596 | 12. 48)889 | 13. 63)1596 | 14. 95)3784 | 15. 74)6289 |

# More about congruent figures

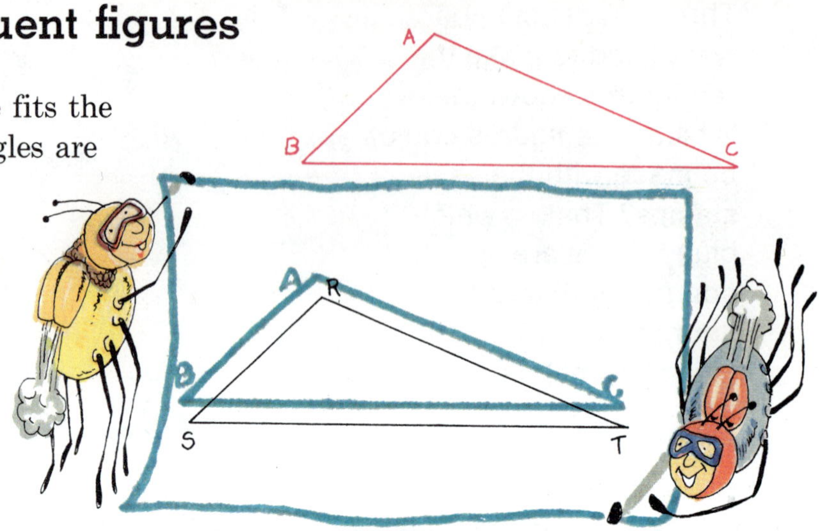


A tracing of the red triangle fits the black triangle. So, the triangles are congruent.

The parts that fit are called **corresponding parts.**

Corresponding angles

$\angle A \leftrightarrow \angle R$ Read as "$\angle A$ corresponds to $\angle R$."

$\angle B \leftrightarrow \angle S$

$\angle C \leftrightarrow \angle T$

Corresponding sides

$\overline{AB} \leftrightarrow \overline{RS}$

$\overline{BC} \leftrightarrow \overline{ST}$

$\overline{CA} \leftrightarrow \overline{TR}$

Notice that if two figures are congruent, then their corresponding parts are congruent. What angle is congruent to $\angle B$? What side is congruent to $\overline{CA}$?

**EXERCISES**

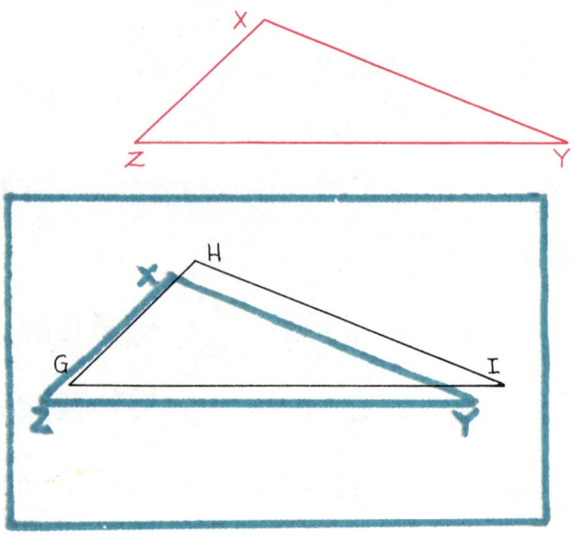


**Complete.**

Corresponding angles

1. $\angle X \leftrightarrow$ _?_
2. $\angle Y \leftrightarrow$ _?_
3. $\angle Z \leftrightarrow$ _?_

Corresponding sides

4. $\overline{XY} \leftrightarrow$ _?_
5. $\overline{YZ} \leftrightarrow$ _?_
6. $\overline{ZX} \leftrightarrow$ _?_

**The two triangles are congruent.**
**Complete the corresponding parts that fit.**

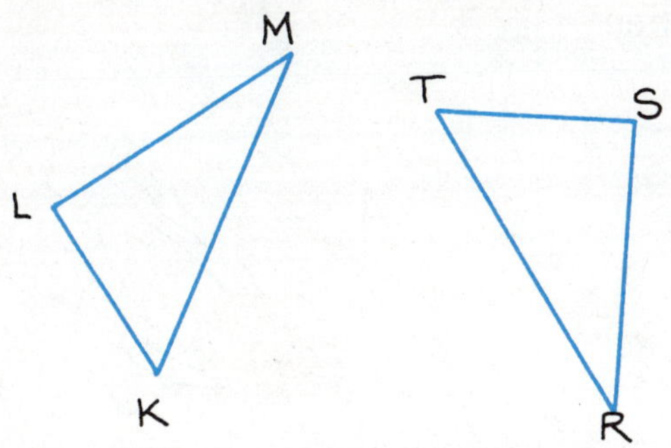


7. $\angle K \leftrightarrow$ ?
8. $\angle M \leftrightarrow$ ?
9. $\angle L \leftrightarrow$ ?
10. $\overline{KL} \leftrightarrow$ ?
11. $\overline{LM} \leftrightarrow$ ?
12. $\overline{MK} \leftrightarrow$ ?

**The two triangles are congruent.**
**Complete.**

13. $\angle A$ is congruent to ?
14. $\overline{FE}$ is congruent to ?
15. $\angle E$ is congruent to ?
16. $\overline{AB}$ is congruent to ?

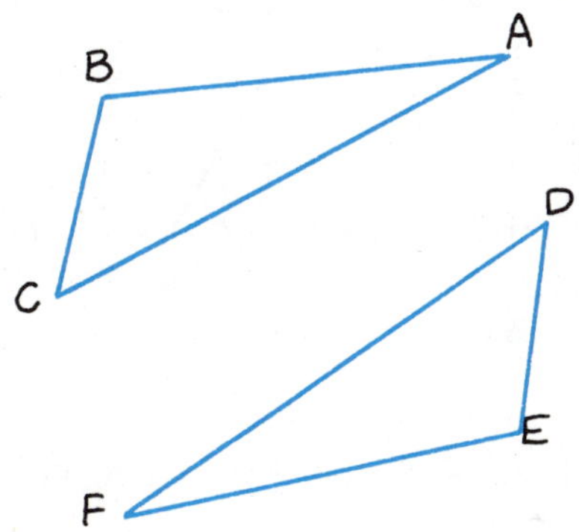


## Challenge!

**Part of each triangle below has been covered. Study the facts. Then decide whether the triangles are congruent.**

17. $\overline{AB}$ is congruent to $\overline{RS}$.
$\overline{BC}$ is congruent to $\overline{ST}$.
$\overline{CA}$ is congruent to $\overline{TR}$.

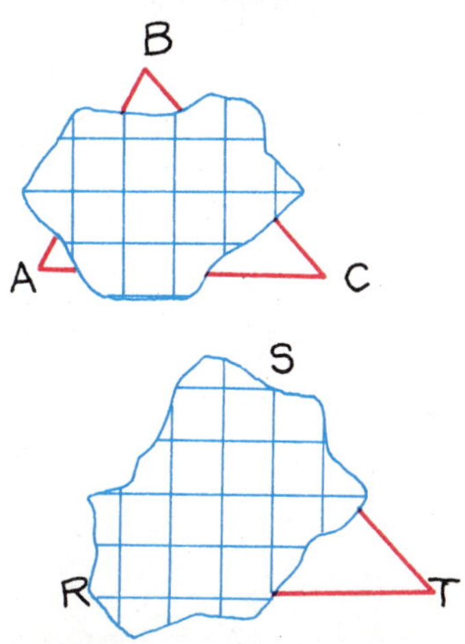


18. $\angle F$ is congruent to $\angle W$.
$\overline{FE}$ is congruent to $\overline{WV}$.
$\angle E$ is congruent to $\angle V$.

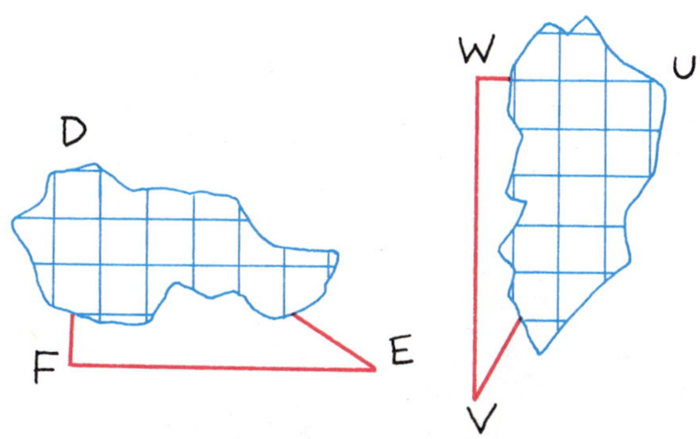

# Line of symmetry

Sylvester and Esther cut out a triangle and then folded it along the dotted line.

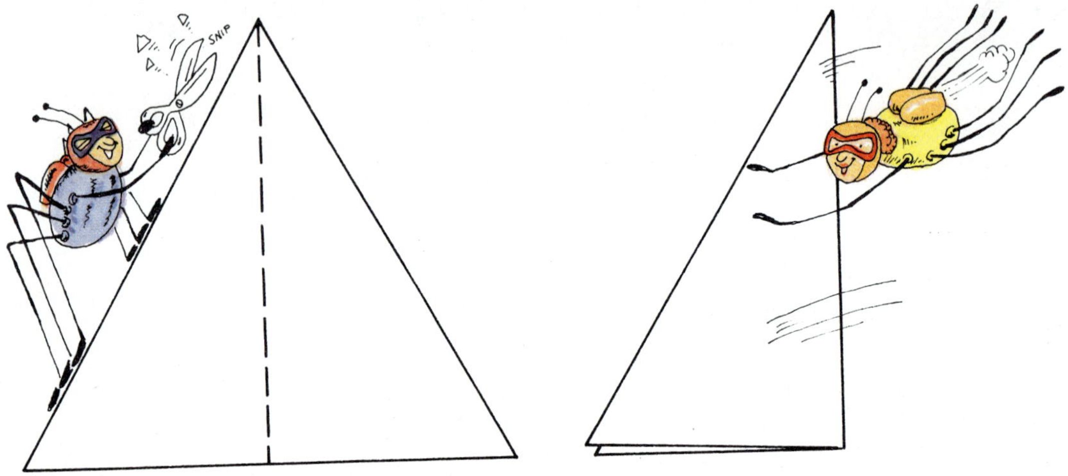


The two parts fit together. This means that the dotted line is a **line of symmetry** of the triangle.

## EXERCISES

**Is the red line a line of symmetry?**

**1.** 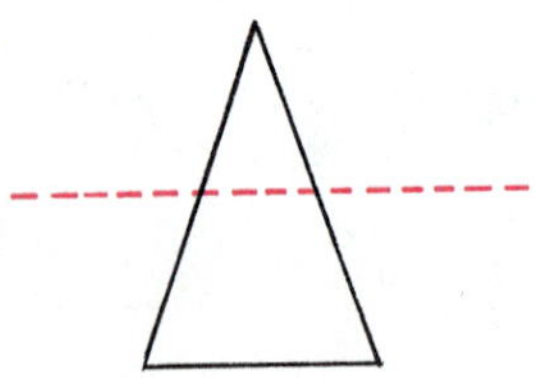

**2.** 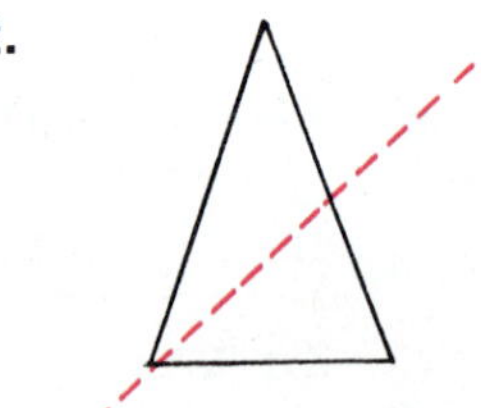

**3.** 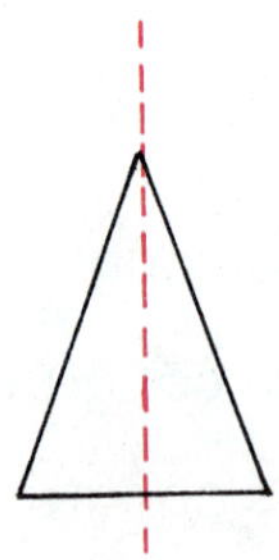

**4.** 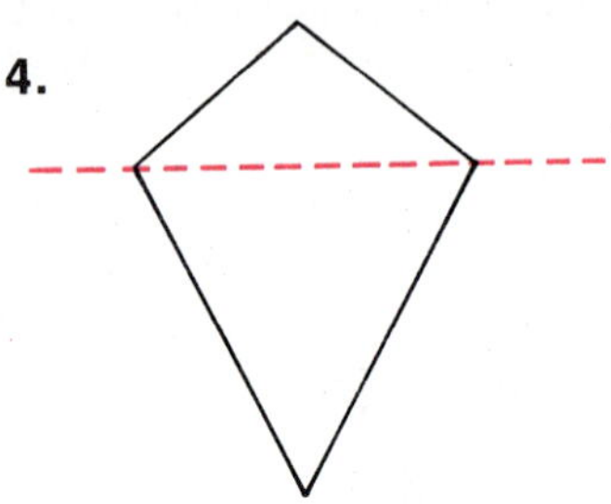

**5.** 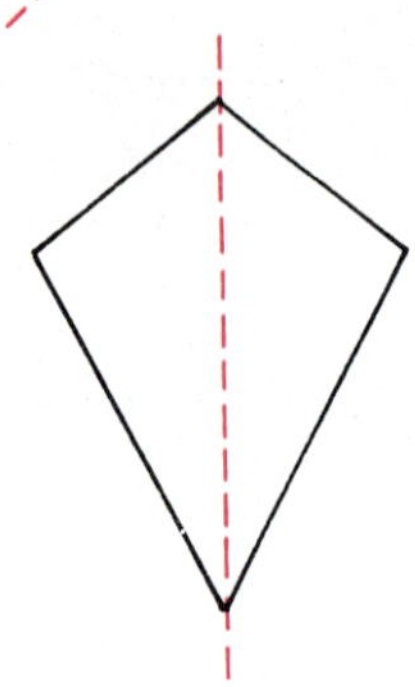

**6.** 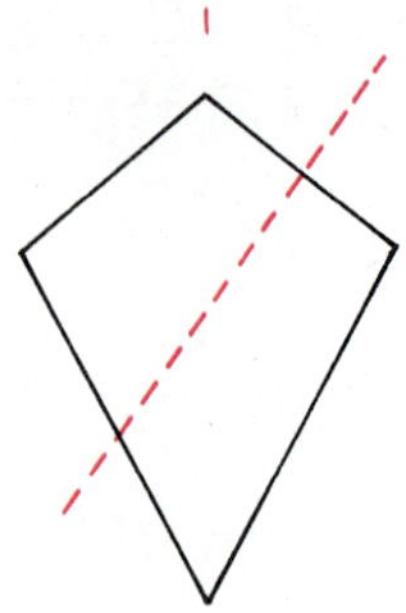

How many lines of symmetry? If you need to, trace each figure and draw all lines of symmetry.

7. 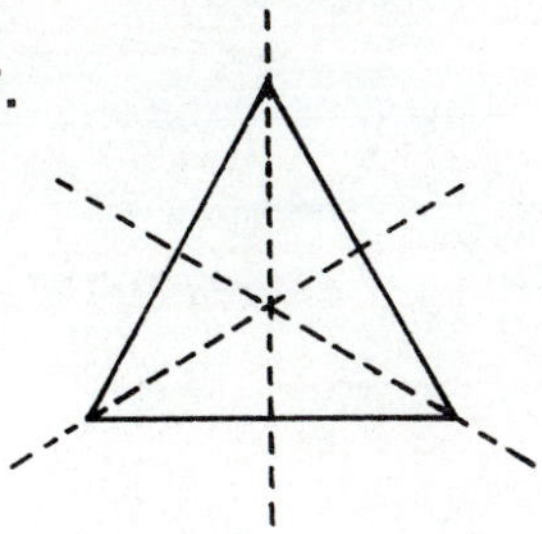

8. 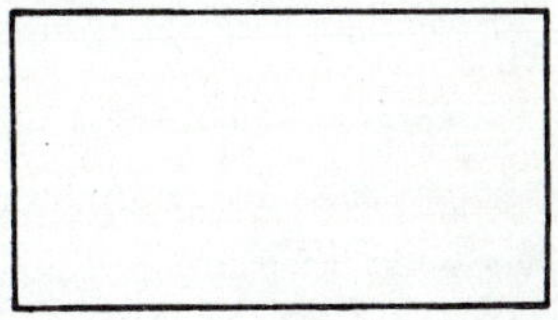

9. 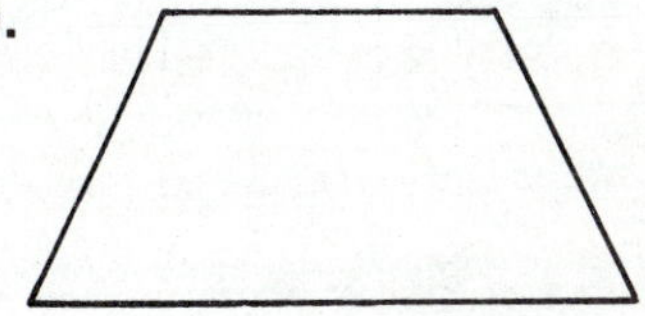

10. 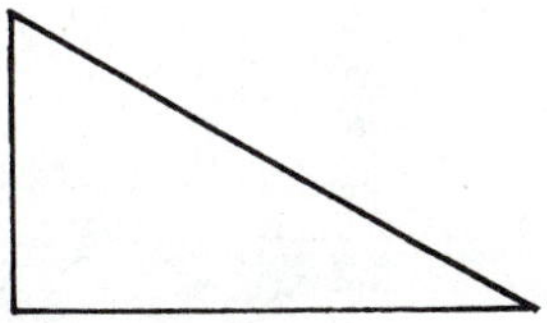

11. 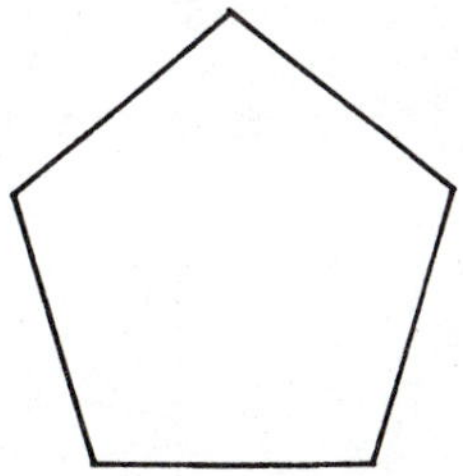

12. 

13. 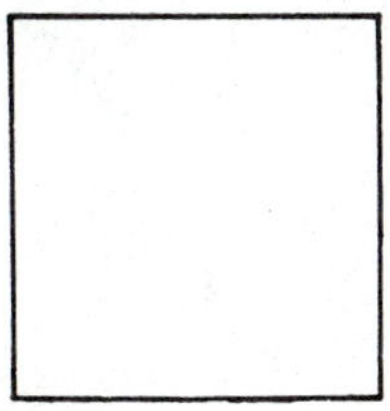

14. 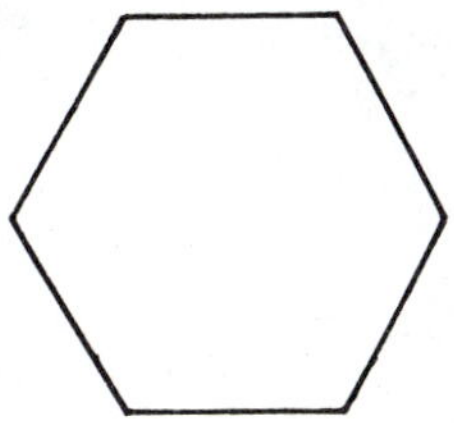

15. 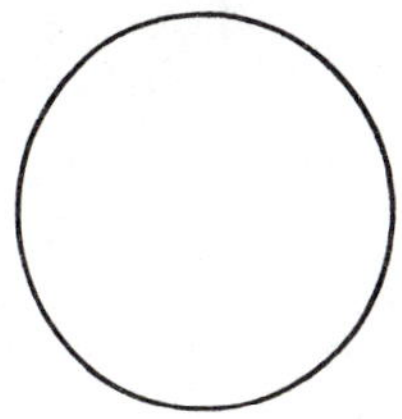

16. Which printed capital letters have a horizontal line of symmetry?

17. Which have a vertical line of symmetry?

18. Which have both a vertical and a horizontal line of symmetry?

★ 19. Can you find *words* that have a line of symmetry?

## Triangles and quadrilaterals

In this lesson you will study some triangles and quadrilaterals (4-sided figures).

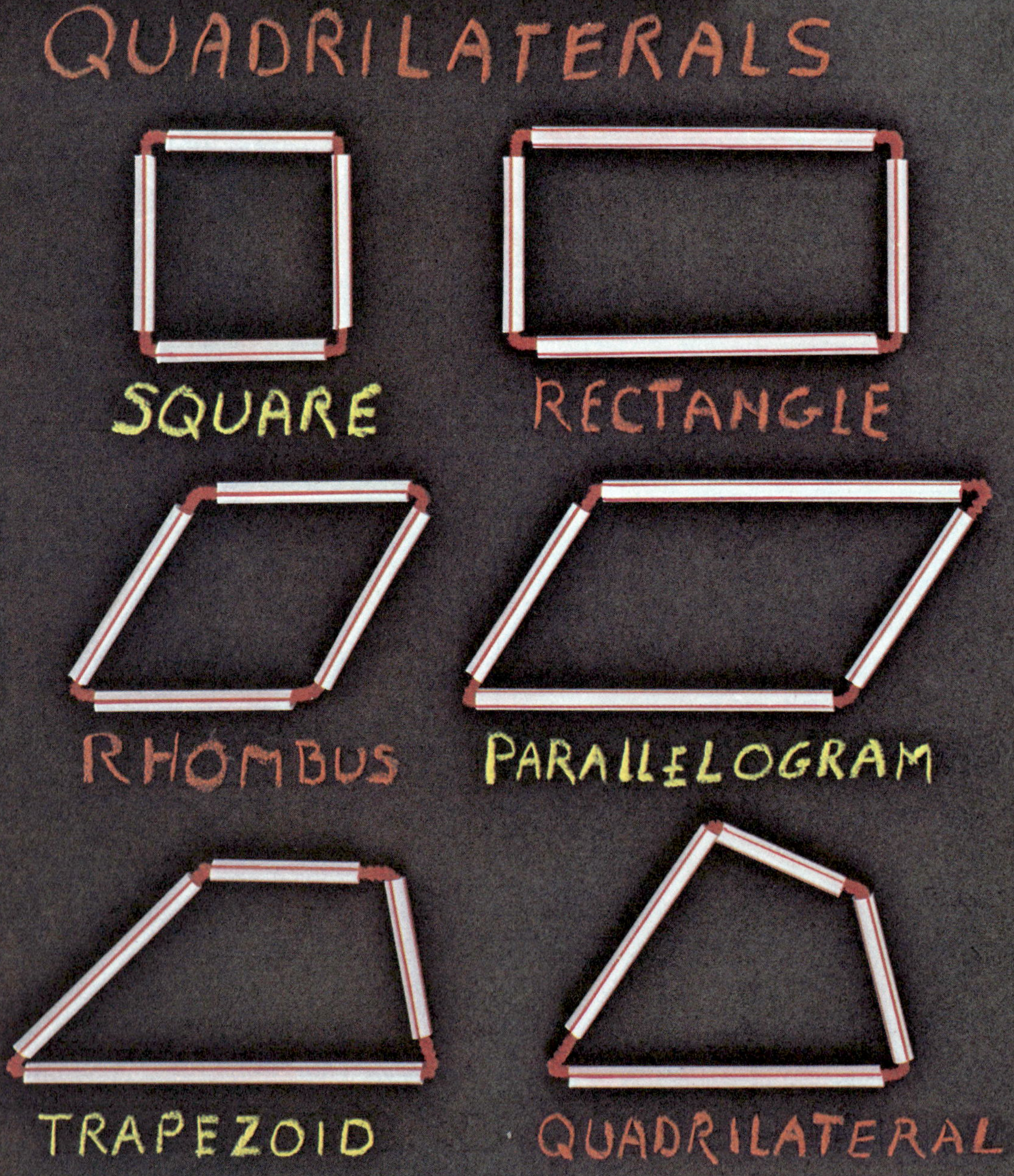

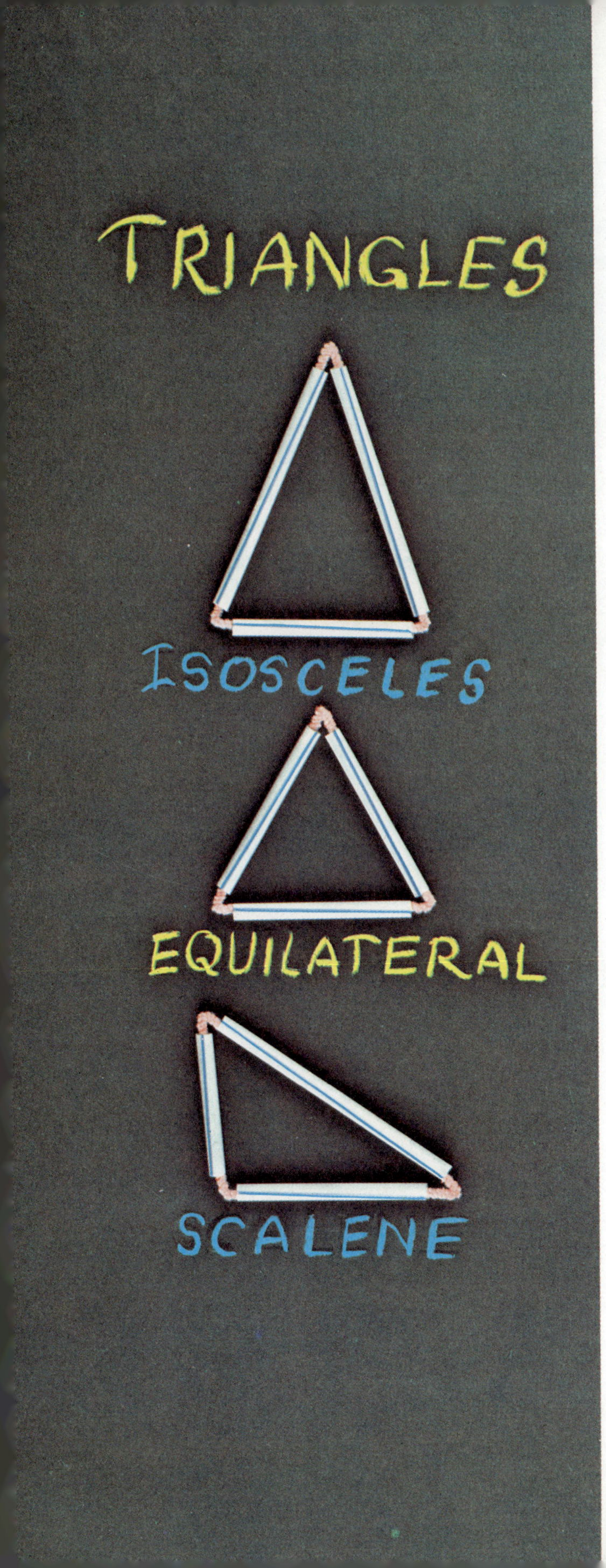


**EXERCISES**

**Which triangle has**

1. no two sides congruent?
2. all three sides congruent?
3. exactly two sides congruent?
4. all three angles congruent?
5. exactly two congruent angles?
6. no lines of symmetry?
7. exactly one line of symmetry?
8. exactly three lines of symmetry?

**Study the clues. Then name the quadrilateral.**

9. 4 sides congruent
   4 angles not congruent
10. 4 congruent sides
    4 congruent angles
11. 2 pairs of congruent sides
    4 congruent angles
12. 1 pair of parallel sides

***Challenge!***

13. How many of the quadrilaterals shown on page 152 have lines of symmetry? (*Hint:* The answer is **not** 4.)
14. Can you draw a quadrilateral that has just one line of symmetry?

# Graphing number pairs

A **number pair** is used to locate or graph a point on a grid. We use the number pair (5, 3) to give the location of point A. The first number tells us the number of units to the right, and the second number tells us the number of units up.

So to graph point A, start at 0, count 5 units to the right, and then count 3 units up.

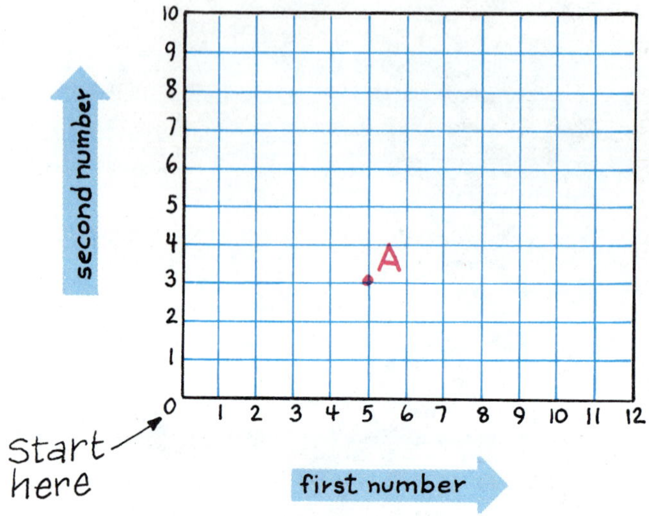


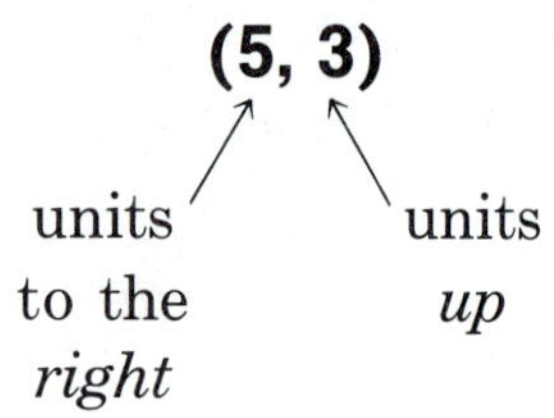


**EXERCISES**

**Give the number pair for each point.**

**1.** A **2.** E **3.** F
**4.** G **5.** B **6.** I
**7.** D **8.** H **9.** C

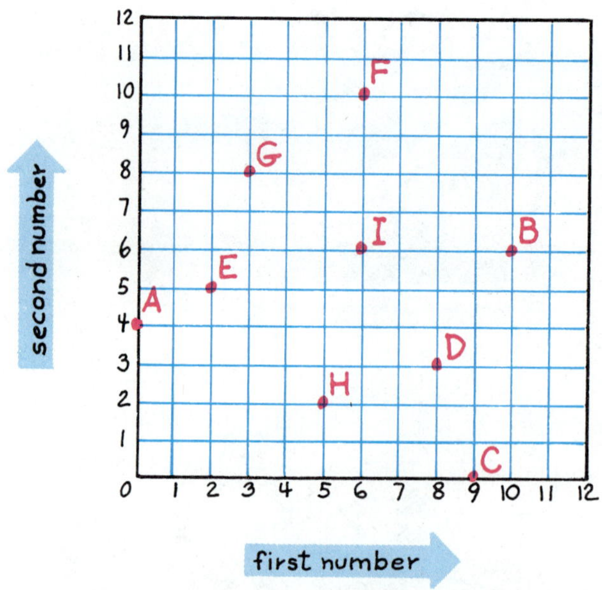

Make a grid like the one shown. Graph these points on your grid.

**10.** $A(6, 1)$
**11.** $B(7, 1)$
**12.** $C(1, 7)$
**13.** $D(2, 10)$
**14.** $E(5, 11)$
**15.** $F(7, 8)$
**16.** $G(9, 13)$
**17.** $H(10, 14)$
**18.** $I(12, 13)$
**19.** $J(11, 12)$
**20.** $K(12, 11)$
**21.** $L(11, 10)$
**22.** $M(12, 7)$
**23.** $N(9, 2)$
**24.** $O(10, 1)$

**25.** Draw these segments on your grid.

$\overline{AB}$ $\overline{BC}$ $\overline{CD}$ $\overline{DE}$ $\overline{EF}$

$\overline{FG}$ $\overline{GH}$ $\overline{HI}$ $\overline{IJ}$ $\overline{JK}$

$\overline{KL}$ $\overline{LM}$ $\overline{MN}$ $\overline{NO}$

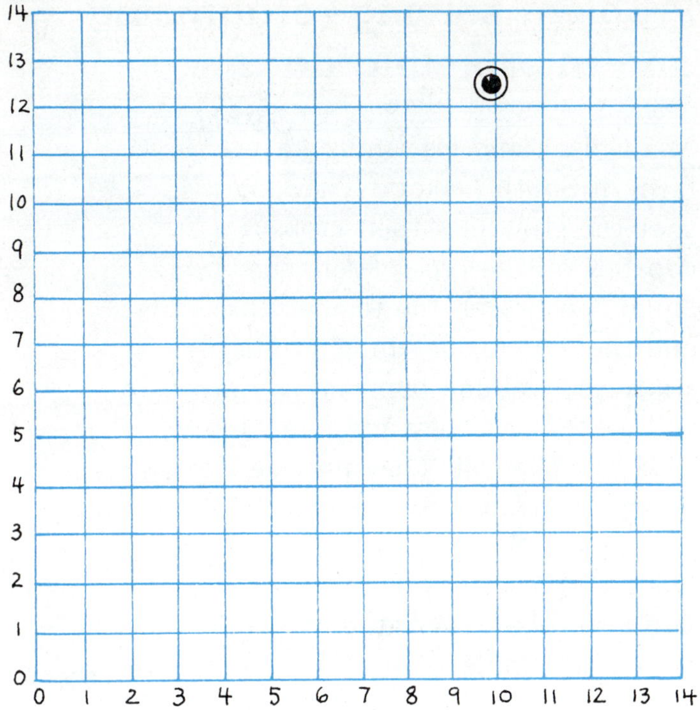


## Challenge!

**First, copy and complete the table of number pairs. Then graph the number pairs.**

**26.** Add 3.

| 0 | 3 |
|---|---|
| 1 | 4 |
| 2 | 5 |
| 3 | ? |
| 4 | ? |
| 5 | ? |
| 6 | ? |
| 7 | ? |

**27.** Subtract 2.

| 10 | 8 |
|---|---|
| 9 | 7 |
| 8 | 6 |
| 7 | ? |
| 6 | ? |
| 5 | ? |
| 4 | ? |
| 3 | ? |

**28.** Multiply by 3.

| 0 | 0 |
|---|---|
| 1 | 3 |
| 2 | 6 |
| 3 | ? |
| 4 | ? |
| 5 | ? |
| 6 | ? |

# Problem solving—organizing and interpreting data

Mr. Ellison and his family have a farm in South Dakota. They know that the growth of their crops depends on the type of soil, the amount of water, the temperature, and the amount of sunlight. They study the soil and plant crops that will grow best. They add plant food to enrich the soil. They provide water when there isn't enough rain.

Donna Ellison recorded the growth of a corn plant over an eight-week period in a table.

## CORN PLANT GROWTH

| Age in weeks | Height in centimeters |
|---|---|
| ( 2 , | 15 ) |
| ( 3 , | 30 ) |
| ( 4 , | 50 ) |
| ( 5 , | 65 ) |
| ( 6 , | 75 ) |
| ( 7 , | 90 ) |
| ( 8 , | 105 ) |

**EXERCISES**

1. Copy and complete the line graph of how the first corn plant grew.

## FIRST CORN PLANT

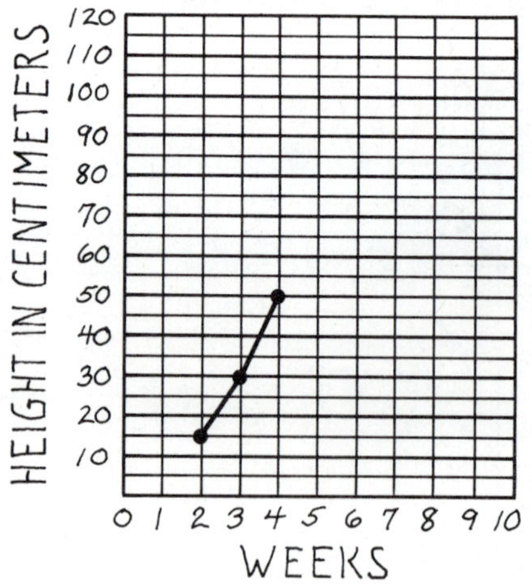

2. This graph shows the growth of a second corn plant that was given less water.

**SECOND CORN PLANT**

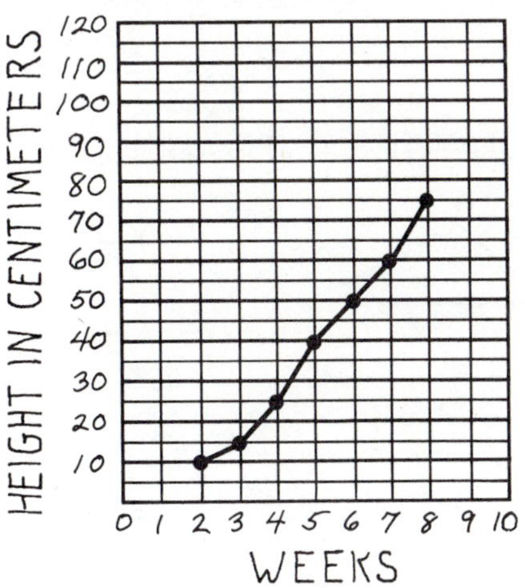


a. At what week was the plant 10 centimeters tall?

b. How tall was the plant at 4 weeks?

c. How much growth was there during the first 4 weeks?

d. How much growth was there during the second 4-week period?

e. How much growth was there between the third week and the seventh week?

f. How many days did it take for the plant to reach a height of 60 centimeters?

3. What was the difference in height between the two plants at the end of the fifth week?

4. You may wish to work with some classmates and do some plant research of your own. Keep a graph of your results.

# CHAPTER CHECKUP

**True or false?** [pages 134–143]

1. A segment has two endpoints.
2. A ray has two endpoints.

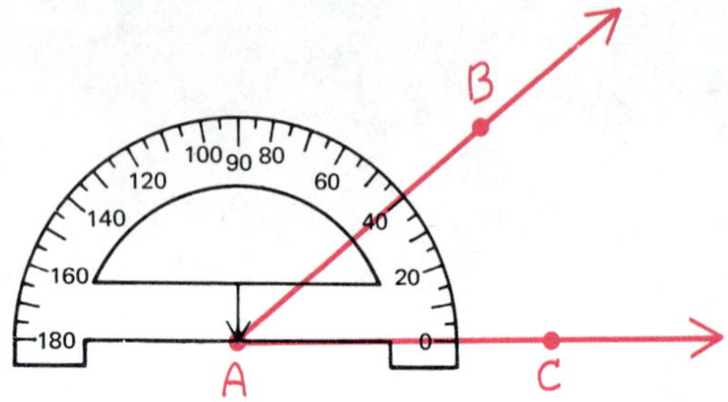


3. We can write $\angle ABC$ to name the angle pictured.
4. The measure of $\angle A$ is 40°.
5. The measure of a right angle is 90°.
6. Two lines in a plane that do not cross are called *perpendicular lines*.

**True or false?** [pages 144–145]

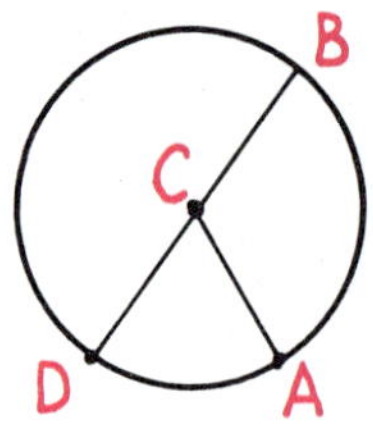


7. Point *C* is on the circle.
8. $\overline{BD}$ is a diameter of the circle.

**True or false?** [pages 146–153]

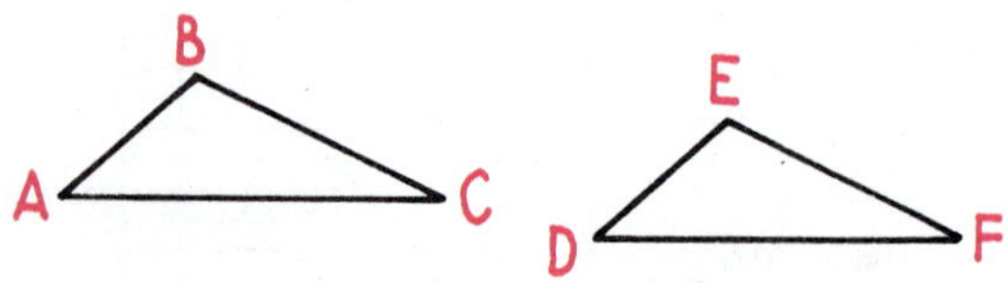


9. $\angle A$ and $\angle D$ are corresponding parts.
10. $\angle C$ is congruent to $\angle F$.
11. A rectangle has two lines of symmetry.
12. Some triangles have 3 lines of symmetry.

**Complete.** [pages 154–157]

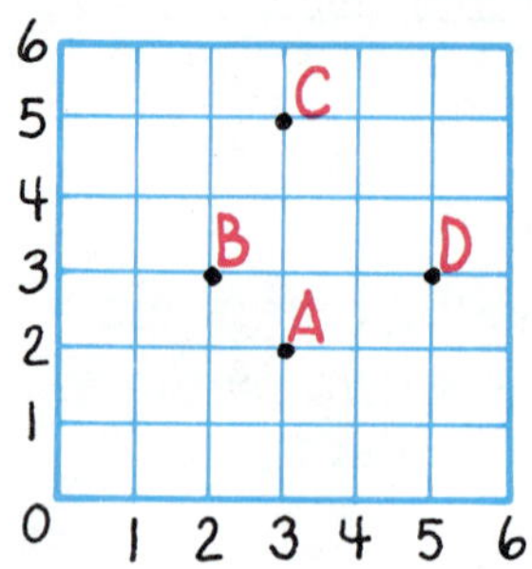


13. Give the number pair for point A.
14. What letter is at (3, 5)?

# CHAPTER PROJECT

**The art shown above was done by connecting points on the sides of angles. You can make your own segment art by following these steps.**

1. Draw an angle.

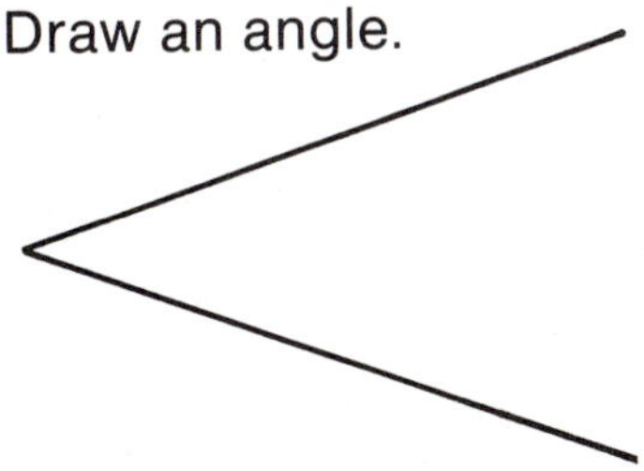

2. Use a ruler and mark off equally spaced dots as shown. Mark off the same number of dots on each side of the angle.

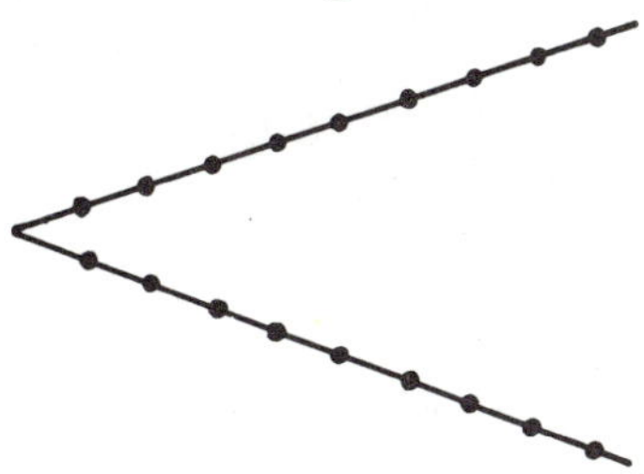

3. Use a sharp pencil and straightedge to connect the dots in the order shown.

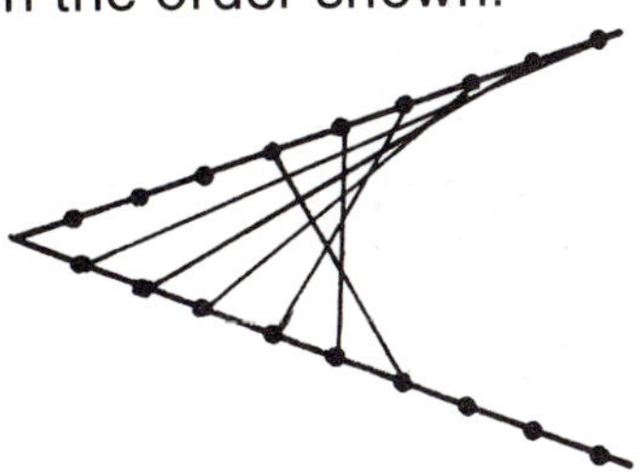

4. Design and draw your own segment art. You may wish to work with more than one angle.

# CHAPTER REVIEW

**Match.**

1. point
2. segment
3. ray
4. angle
5. line

a. 
b. 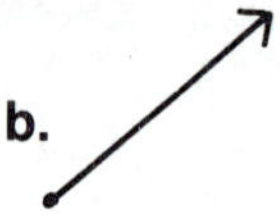
c. 
d. 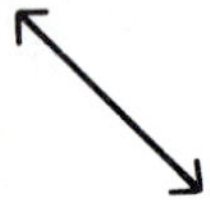
e. 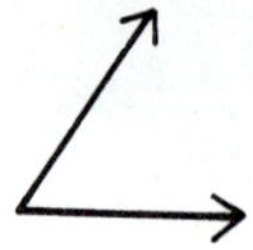

**Complete.**

6. In $\angle S$, point $S$ is called the ?.

7. The angle can be named in any of these three ways: $\angle S$, $\angle RST$, or ?.

8. The measure of $\angle RST$ is ? °.

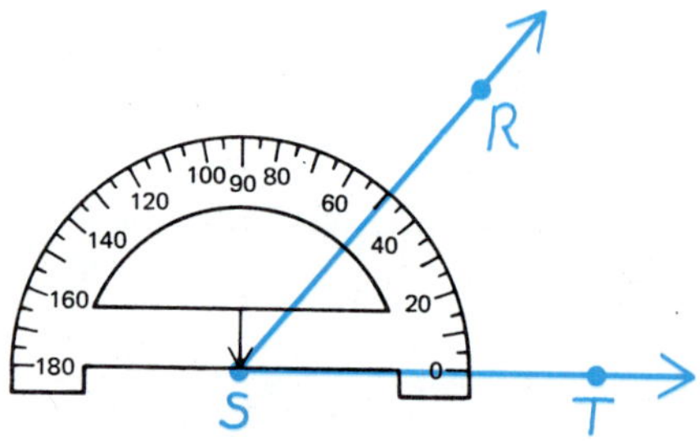


9. A right angle measures ?.

10. Two lines that intersect to form right angles are called ______?______ lines.
parallel / perpendicular

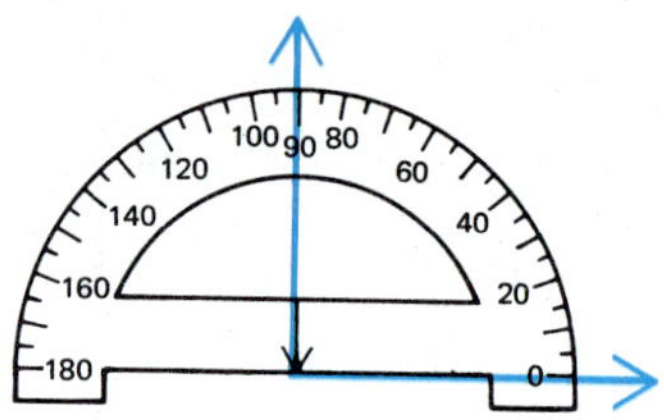

11. Point ? is the center of the circle.

12. $\overline{CD}$ is a radius and $\overline{AB}$ is a ?.

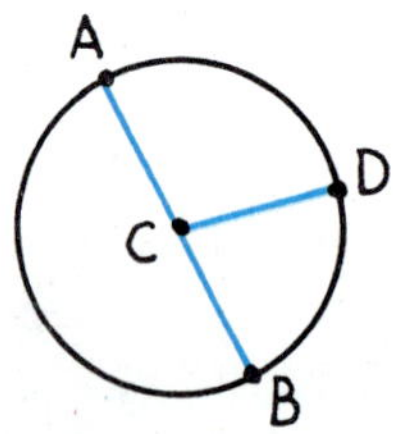


13. The congruent triangles are A and ?.

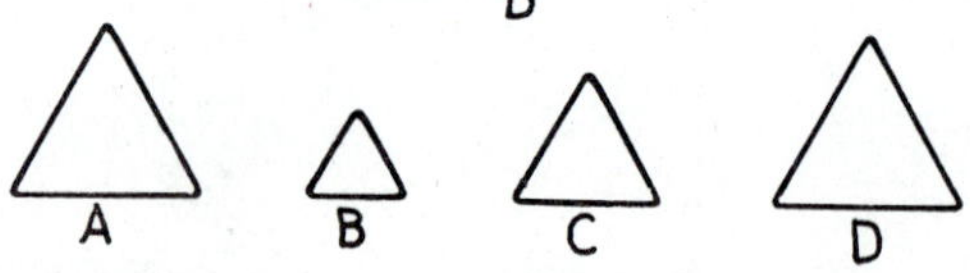


14. This quadrilateral (a 4-sided figure) has ? line(s) of symmetry.

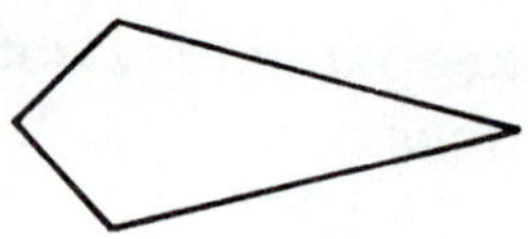

15. The number pair for point $A$ is ?.

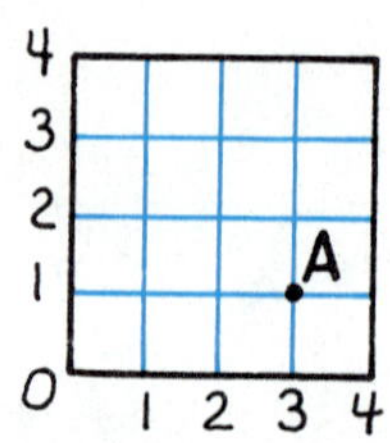

# CHAPTER CHALLENGE

The red segments in each figure are called **diagonals.**

**Complete the following table.**

| | Picture | Number of sides | Number of diagonals |
|---|---|---|---|
| 1. | 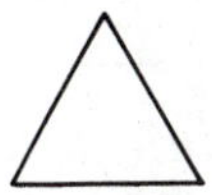 | | |
| 2. | 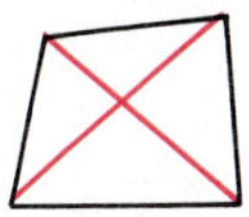 | | |
| 3. |  | | |
| 4. | 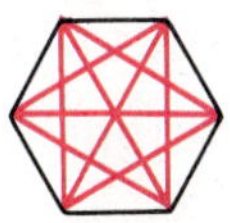 | | |
| 5. |  | | |

6. Study the numbers in your diagonals column. Do you see a pattern? (*Hint:* What is the difference between the first and second numbers? the second and third? the third and fourth?)

7. How many diagonals do you think an 8-sided figure has? Check your answer with a drawing.

8. How many diagonals does a 9-sided figure have?

# MAJOR CHECKUP
STANDARDIZED FORMAT

**Choose the correct letter.**

**1.** 578,500 rounded to the nearest thousand is

a. 579,000
b. 578,000
c. 580,000
d. none of these

**2.** Add.

$$\begin{array}{r} 297 \\ 4283 \\ 197 \\ +65 \\ \hline \end{array}$$

a. 4832
b. 4522
c. 4732
d. none of these

**3.** Subtract.

$$\begin{array}{r} 4902 \\ -578 \\ \hline \end{array}$$

a. 4476
b. 4434
c. 4324
d. none of these

**4.** Which number is the least common multiple of 6 and 8?

a. 14
b. 48
c. 24
d. none of these

**5.** Multiply.

$$\begin{array}{r} 328 \\ \times 59 \\ \hline \end{array}$$

a. 19,352
b. 18,352
c. 4592
d. none of these

**6.** Divide.

$3\overline{)744}$

a. 211
b. 248
c. 244
d. none of these

**7.** What is the average price?

a. 80¢
b. 75¢
c. 85¢
d. none of these

**8.** Divide.

$58\overline{)7946}$

a. 138
b. 127
c. 137
d. none of these

**9.** What kind of angle is $\angle T$?

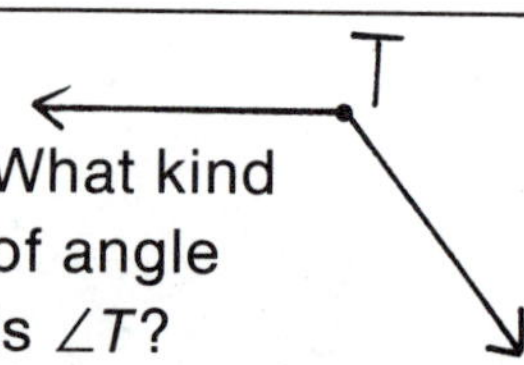


a. acute
b. obtuse
c. right
d. none of these

**10.** Which is a line of symmetry of the red figure?

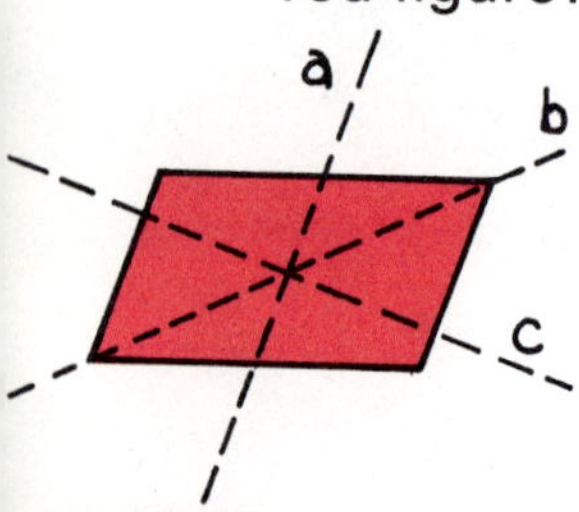


a. line a
b. line b
c. line c
d. none of these

**11.** Amy had 80¢. She bought 4 balloons for 19¢ each. How much did she spend?

a. 76¢
b. 4¢
c. 61¢
d. none of these

**12.** Lin bought 3 stamps for 18¢ each and some envelopes for 29¢. How much did she spend?

a. 47¢
b. 83¢
c. $1.41
d. none of these

# Fractions— Addition and Subtraction

# 6

# READY OR NOT?

**What fraction is colored?**

1. 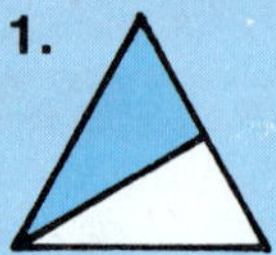

2. 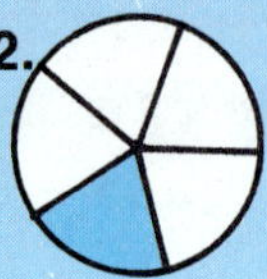

3. 

4. 

5. 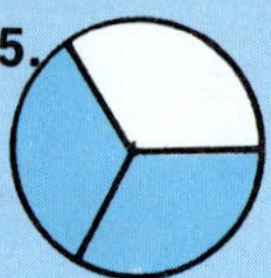

6. 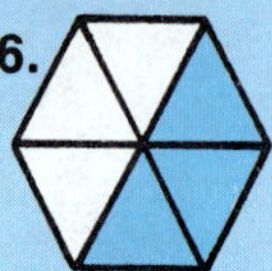

7. 

8. 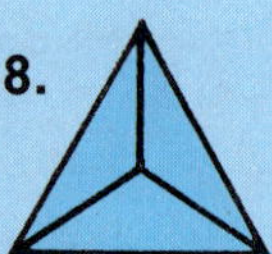

9. 

10. 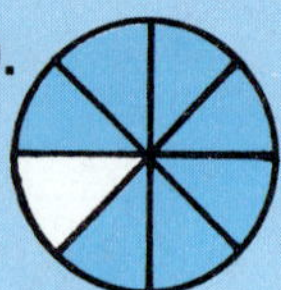



# Using fractions

3 green balls

5 balls in all

numerator → 3
denominator → 5

$\frac{3}{5}$ of the balls are green

Read as "three fifths."

The numerator and denominator of a fraction are sometimes called the **terms** of the fraction.

**EXERCISES**

**What fraction of the balls are**

1. blue?
2. yellow?
3. red?
4. not yellow?

**Give the fraction.**

5. 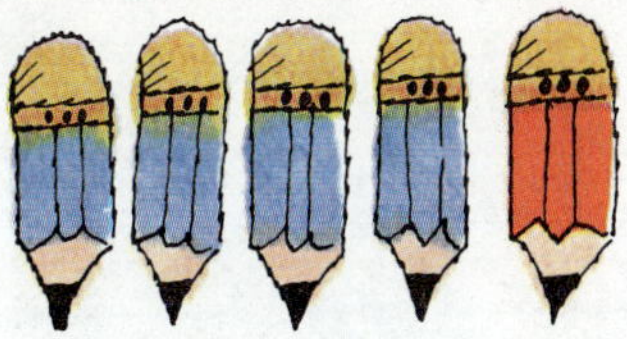

a. red pencils

b. blue pencils

6. 

a. empty bottles

b. full bottles

7. What fraction of the figures are

a. red?

b. blue?

c. yellow?

d. squares?

e. large circles?

f. not triangles?

**Give the fraction.**

8. 

? of a cup

9. 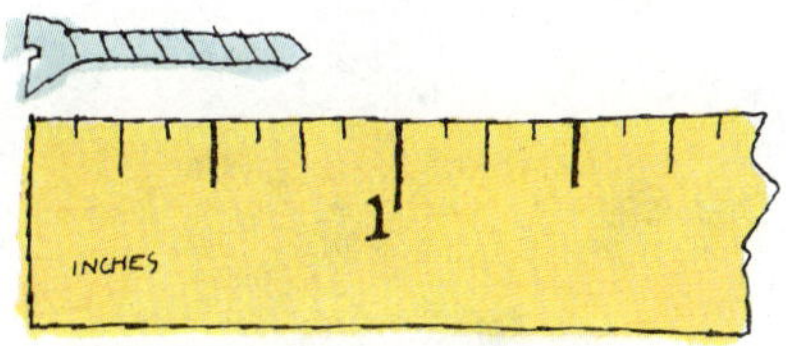


? of an inch long

10. 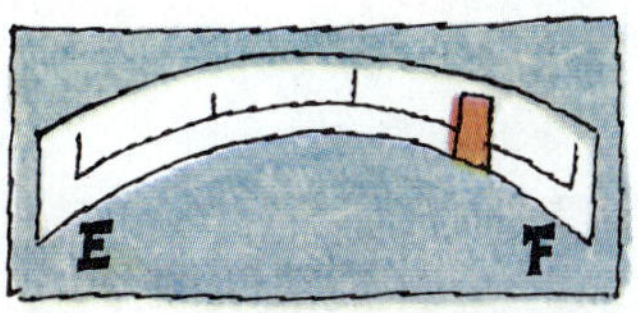


? of a tank of gasoline

11. 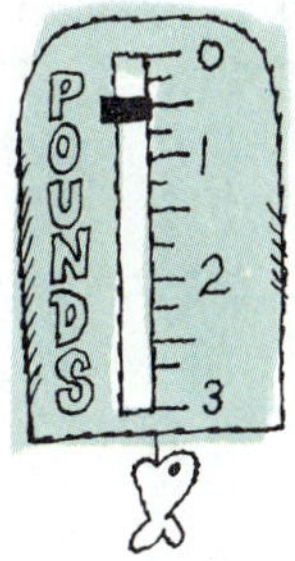


? of a pound

**What fraction of the students in your class**

12. are boys?

13. are girls?

14. wear glasses?

15. do not wear glasses?

16. have first names beginning with the letter D?

17. have last names beginning with a vowel?

# Equivalent fractions

When you multiply both the numerator and the denominator of a fraction by the same number (not zero), you get an equivalent fraction.

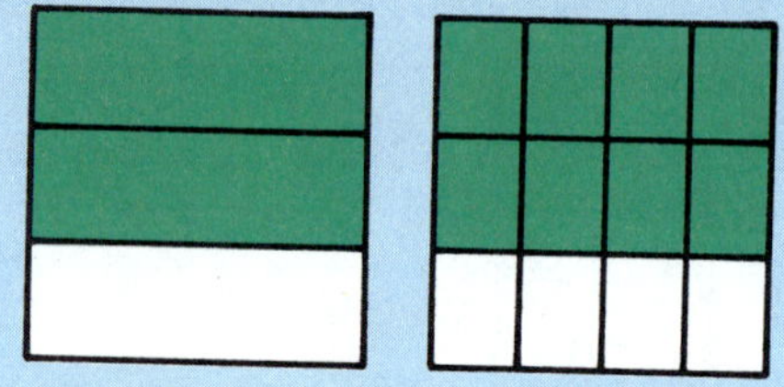

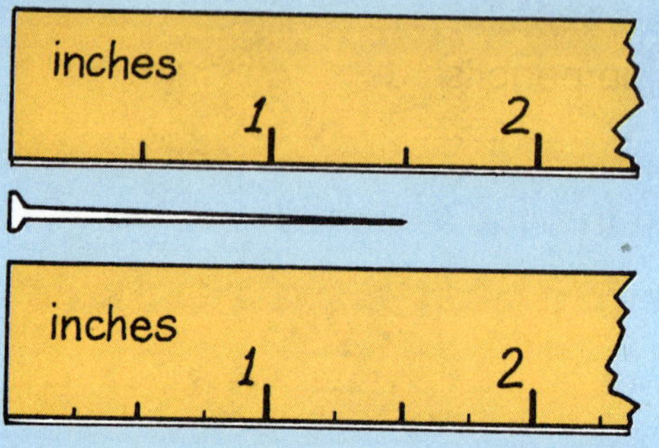


Multiply numerator and denominator by 4.

$$\frac{2}{3} = \frac{8}{12} \quad (\times 4)$$

Multiply numerator and denominator by 2.

$$\frac{3}{2} = \frac{6}{4} \quad (\times 2)$$

**EXERCISES**

**Complete.**

1. 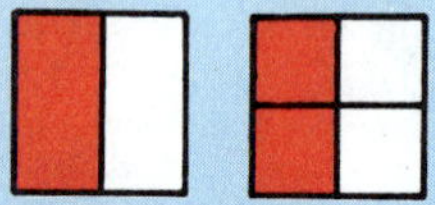

$\frac{1}{2} = ?$

2. 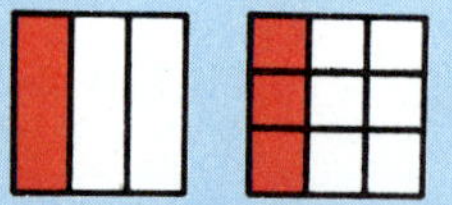

$\frac{1}{3} = ?$

3. 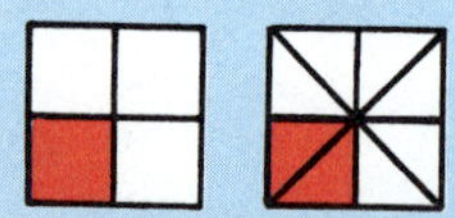

$\frac{1}{4} = ?$

4. 

$\frac{1}{2} = ?$

5. 

$\frac{2}{3} = ?$

6. 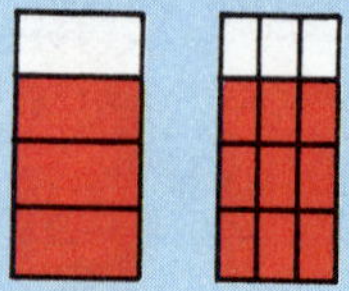

$\frac{3}{4} = ?$

7. 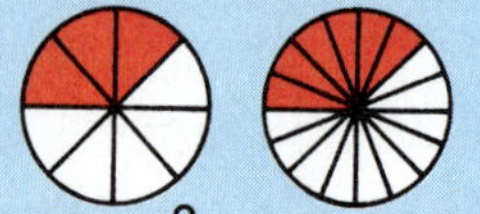

$\frac{3}{8} = ?$

8. 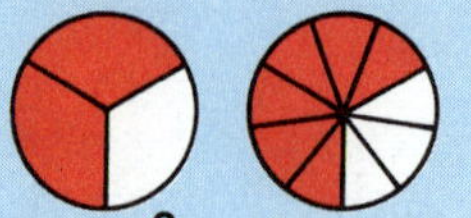

$\frac{2}{3} = ?$

9. 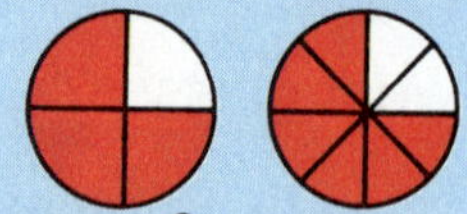

$\frac{3}{4} = ?$

**Complete.**

10. $\frac{1}{2} = ?$ (×2, ×2)
11. $\frac{1}{4} = ?$ (×3, ×3)
12. $\frac{1}{8} = ?$ (×4, ×4)
13. $\frac{1}{5} = ?$ (×2, ×2)

14. $\frac{2}{3} = ?$ (×4, ×4)
15. $\frac{1}{3} = ?$ (×2, ×2)
16. $\frac{5}{3} = ?$ (×4, ×4)
17. $\frac{7}{8} = ?$ (×2, ×2)

18. $\frac{3}{2} = ?$ (×3, ×3)
19. $\frac{2}{5} = ?$ (×4, ×4)
20. $\frac{3}{4} = ?$ (×2, ×2)
21. $\frac{4}{5} = ?$ (×3, ×3)

22. $\frac{3}{5} = ?$ (×4, ×4)
23. $\frac{1}{2} = ?$ (×6, ×6)
24. $\frac{5}{8} = ?$ (×3, ×3)
25. $\frac{4}{3} = ?$ (×2, ×2)

**Look for a pattern.**
**Give the next three equivalent fractions.**

26. $\frac{1}{2}, \frac{2}{4}, \frac{3}{6},$ ?, ?, ?

27. $\frac{1}{3}, \frac{2}{6}, \frac{3}{9},$ ?, ?, ?

28. $\frac{1}{4}, \frac{2}{8}, \frac{3}{12},$ ?, ?, ?

29. $\frac{2}{3}, \frac{4}{6},$ ?, ?, ?

30. $\frac{3}{4},$ ?, ?, ?

31. $\frac{3}{2},$ ?, ?, ?

32. $\frac{4}{3},$ ?, ?, ?

**Solve.**

★ 33.

What fraction of the butter has been used? Give two fractions.

★ 34.

What fraction of the eggs have been used? Give three fractions.

# Fractions in lowest terms

When you divide the terms of a fraction by the same number (called a **common factor**), you get an equivalent fraction in **lower terms.**

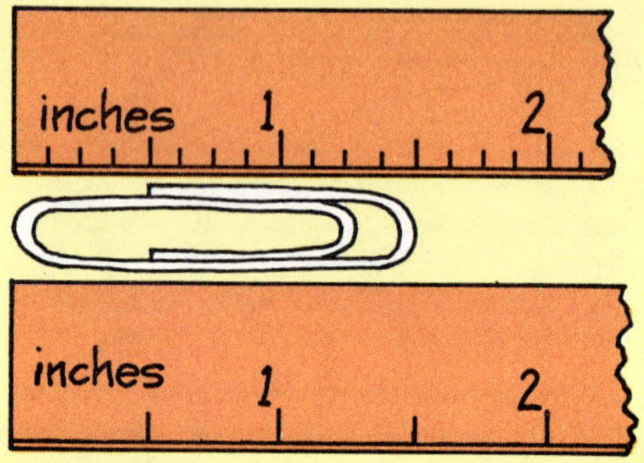


Divide numerator and denominator by 2.

÷ 2

$$\frac{6}{8} = \frac{3}{4}$$

÷ 2

Divide numerator and denominator by 4.

÷ 4

$$\frac{12}{8} = \frac{3}{2}$$

÷ 4

When you divide numerator and denominator by the **greatest common factor,** you get the fraction in **lowest terms.**

**EXERCISES**

**Complete.**

**1.** 

$\frac{2}{4} = \underline{?}$

**2.** 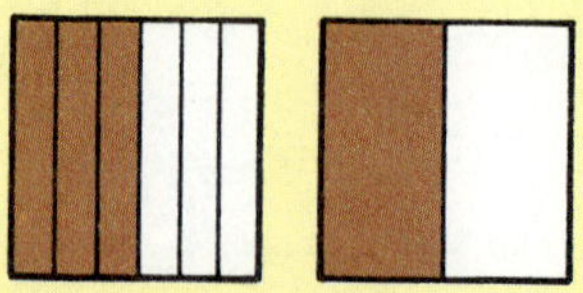

$\frac{3}{6} = \underline{?}$

**3.** 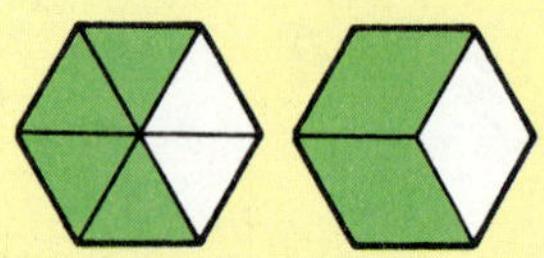

$\frac{4}{6} = \underline{?}$

**4.** 

$\frac{3}{9} = \underline{?}$

**5.** 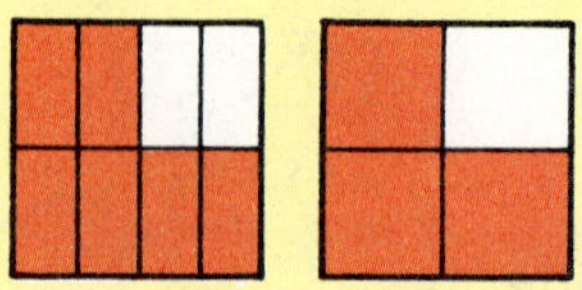

$\frac{6}{8} = \underline{?}$

**6.** 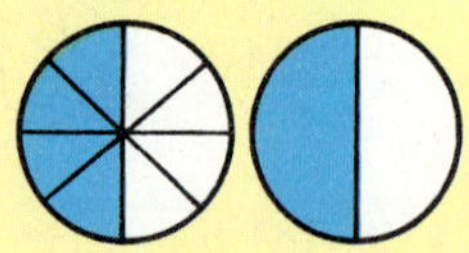

$\frac{4}{8} = \underline{?}$

Complete.

7. ÷2 $\frac{6}{8} = \frac{?}{?}$ ÷2
8. ÷4 $\frac{4}{8} = \frac{?}{?}$ ÷4
9. ÷3 $\frac{6}{9} = \frac{?}{?}$ ÷3
10. ÷2 $\frac{6}{4} = \frac{?}{?}$ ÷2

11. ÷3 $\frac{9}{6} = \frac{?}{?}$ ÷3
12. ÷4 $\frac{12}{8} = \frac{?}{?}$ ÷4
13. ÷4 $\frac{12}{16} = \frac{?}{?}$ ÷4
14. ÷5 $\frac{10}{15} = \frac{?}{?}$ ÷5

15. ÷8 $\frac{8}{16} = \frac{?}{?}$ ÷8
16. ÷4 $\frac{8}{12} = \frac{?}{?}$ ÷4
17. ÷8 $\frac{24}{16} = \frac{?}{?}$ ÷8
18. ÷6 $\frac{18}{24} = \frac{?}{?}$ ÷6

Write each fraction in lowest terms.

19. $\frac{3}{6}$ 20. $\frac{2}{6}$ 21. $\frac{4}{8}$ 22. $\frac{4}{12}$ 23. $\frac{5}{10}$ 24. $\frac{6}{8}$

25. $\frac{3}{12}$ 26. $\frac{4}{6}$ 27. $\frac{6}{9}$ 28. $\frac{6}{4}$ 29. $\frac{12}{16}$ 30. $\frac{3}{9}$

31. $\frac{12}{15}$ 32. $\frac{9}{6}$ 33. $\frac{12}{18}$ 34. $\frac{16}{12}$ 35. $\frac{18}{12}$ 36. $\frac{16}{20}$

37. $\frac{5}{15}$ 38. $\frac{2}{8}$ 39. $\frac{20}{16}$ 40. $\frac{15}{18}$ 41. $\frac{2}{12}$ 42. $\frac{10}{30}$

43. $\frac{2}{10}$ 44. $\frac{12}{9}$ 45. $\frac{8}{12}$ 46. $\frac{14}{16}$ 47. $\frac{7}{14}$ 48. $\frac{15}{20}$

49. $\frac{4}{10}$ 50. $\frac{8}{16}$ 51. $\frac{8}{24}$ 52. $\frac{8}{18}$ 53. $\frac{4}{20}$ 54. $\frac{10}{8}$

55. $\frac{4}{16}$ 56. $\frac{24}{9}$ 57. $\frac{6}{18}$ 58. $\frac{10}{6}$ 59. $\frac{5}{20}$ 60. $\frac{10}{20}$

## *Challenge!*

**Guess my number.**

61. My number is equivalent to $\frac{18}{24}$. The denominator is 8.

★62. My number is equivalent to $\frac{2}{3}$. The numerator is 6.

# More on fractions in lowest terms

When you divide the numerator and the denominator by the greatest common factor, you get the fraction in lowest terms.

$$\frac{6}{8} = \frac{3}{4}$$

$$\frac{\overset{3}{\cancel{6}}}{\underset{4}{\cancel{8}}}$$

One way to decide whether two fractions are equivalent is to write each fraction in lowest terms.

$$\frac{\overset{3}{\cancel{9}}}{\underset{2}{\cancel{6}}} = \frac{\overset{3}{\cancel{12}}}{\underset{2}{\cancel{8}}}$$

So, $\frac{9}{6}$ and $\frac{12}{8}$ are equivalent fractions.

**EXERCISES**

**Write each fraction in lowest terms.**

| | | | | | |
|---|---|---|---|---|---|
| **1.** $\frac{5}{10}$ | **2.** $\frac{6}{8}$ | **3.** $\frac{4}{8}$ | **4.** $\frac{6}{15}$ | **5.** $\frac{4}{6}$ | **6.** $\frac{15}{10}$ |
| **7.** $\frac{3}{9}$ | **8.** $\frac{15}{18}$ | **9.** $\frac{16}{12}$ | **10.** $\frac{8}{24}$ | **11.** $\frac{6}{4}$ | **12.** $\frac{6}{24}$ |
| **13.** $\frac{2}{10}$ | **14.** $\frac{4}{16}$ | **15.** $\frac{10}{20}$ | **16.** $\frac{10}{15}$ | **17.** $\frac{3}{12}$ | **18.** $\frac{12}{9}$ |

**Are the two fractions equivalent?**
**(*Hint:* Write each fraction in lowest terms.)**

**19.** $\frac{3}{12}, \frac{4}{16}$ **20.** $\frac{6}{12}, \frac{7}{14}$ **21.** $\frac{10}{12}, \frac{8}{10}$ **22.** $\frac{5}{15}, \frac{6}{18}$

Use the digits to build as many fractions as you can that are equivalent to the "red" fraction.

1. $\frac{1}{2}$ 4 6 1 2

2. $\frac{3}{4}$ 9 2 1 0

3. $\frac{6}{9}$ 3 6 2 4

4. $\frac{6}{8}$ 3 2 4 2

**Play the game.**

1. Make a set of digit cards.
2. Choose a game leader and divide the class into two teams, Team A and Team B.
3. The leader writes a fraction on the chalkboard.
4. The leader shows Team A four digits. Team A gets one point for each equivalent fraction that it can build with the digits.
5. The leader replaces the cards, and step 4 is repeated with Team B.
6. Next, the leader writes a different fraction on the chalkboard and it is Team A's turn.
7. When both teams have had ten turns, the team with more points wins!

# Comparing fractions

Both of these fractions have 5 as their denominator. We call 5 a **common denominator.** To compare two fractions that have a common denominator, compare the numerators.

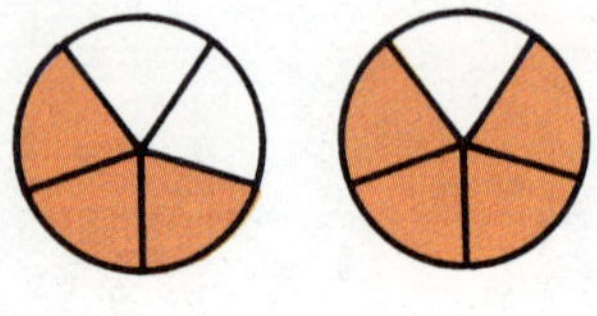

$\frac{3}{5} < \frac{4}{5}$

is less than

To compare two fractions with different denominators, you can use pictures or a number line.

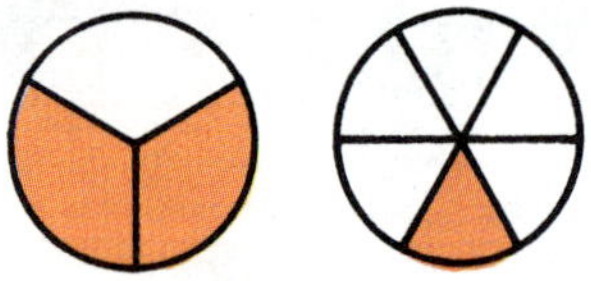

$\frac{2}{3} > \frac{1}{6}$

is greater than

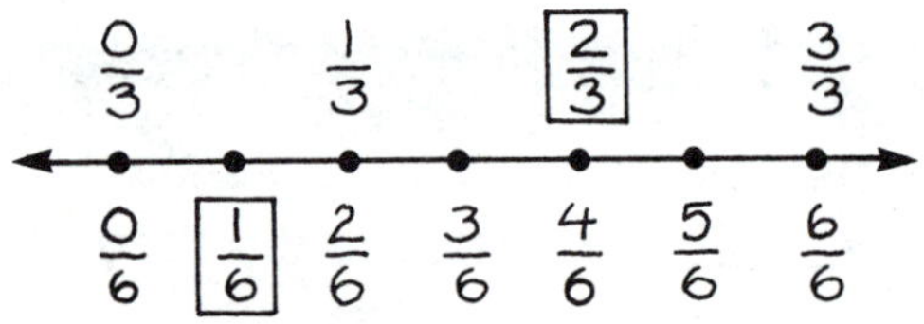


$\frac{2}{3} > \frac{1}{6}$

is greater than

To compare two fractions with different denominators, you can first change to equivalent fractions with a common denominator. Then compare the numerators.

$\frac{2}{3}$ ? $\frac{3}{4}$

$\frac{2}{3} < \frac{3}{4}$

**EXERCISES**

**< or >?**

**1.** 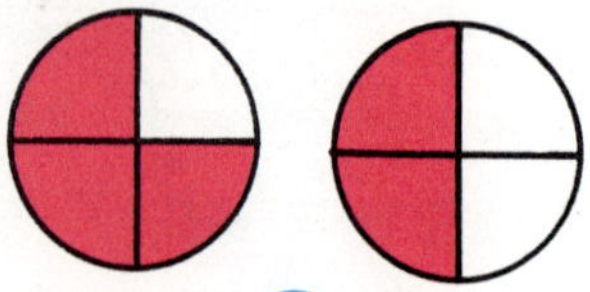

$\frac{3}{4}$ ● $\frac{2}{4}$

**2.** 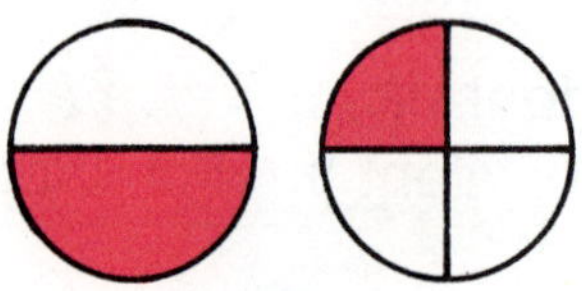

$\frac{1}{2}$ ● $\frac{1}{4}$

**3.** 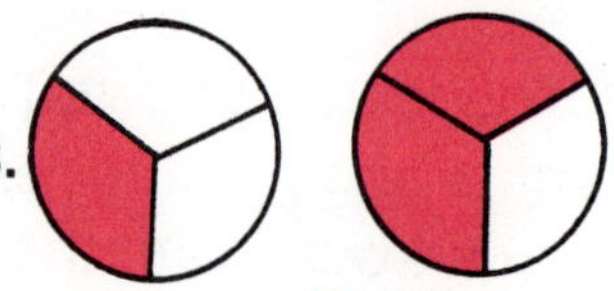

$\frac{1}{3}$ ● $\frac{2}{3}$

**4.** 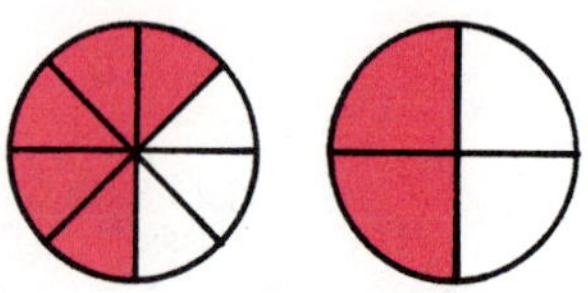

$\frac{5}{8}$ ● $\frac{2}{4}$

**5.** 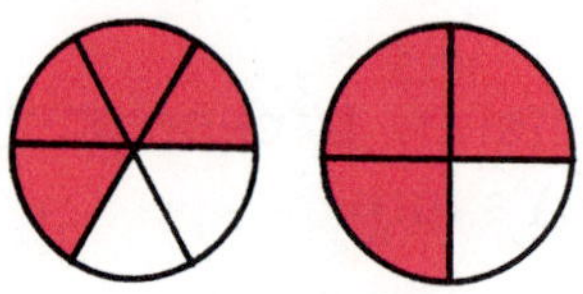

$\frac{4}{6}$ ● $\frac{3}{4}$

**6.** 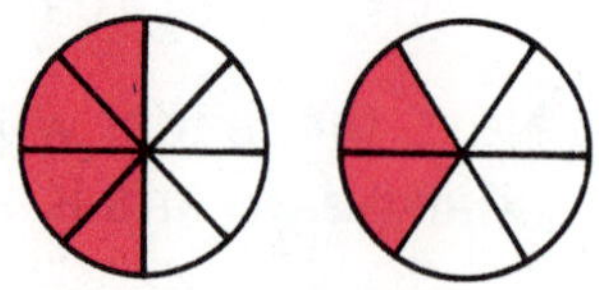

$\frac{4}{8}$ ● $\frac{2}{6}$

< or >? (If you need to, use the number line.)

7. $\frac{3}{8}$ ● $\frac{5}{8}$ 8. $\frac{3}{4}$ ● $\frac{1}{4}$ 9. $\frac{1}{2}$ ● $\frac{2}{2}$

10. $\frac{4}{4}$ ● $\frac{7}{8}$ 11. $\frac{3}{8}$ ● $\frac{2}{2}$ 12. $\frac{2}{4}$ ● $\frac{2}{2}$

13. $\frac{1}{8}$ ● $\frac{1}{2}$ 14. $\frac{5}{8}$ ● $\frac{3}{4}$ 15. $\frac{1}{2}$ ● $\frac{5}{8}$

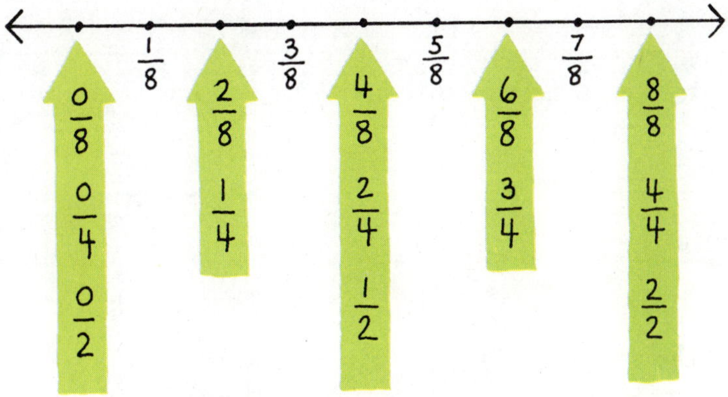


<, >, or =?

16. $\frac{1}{2}$ ● $\frac{1}{3}$ 17. $\frac{2}{3}$ ● $\frac{5}{6}$ 18. $\frac{2}{3}$ ● $\frac{4}{6}$

19. $\frac{5}{9}$ ● $\frac{2}{3}$ 20. $\frac{3}{5}$ ● $\frac{2}{3}$ 21. $\frac{3}{4}$ ● $\frac{2}{3}$

22. $\frac{3}{5}$ ● $\frac{1}{2}$ 23. $\frac{2}{5}$ ● $\frac{4}{10}$ 24. $\frac{5}{6}$ ● $\frac{5}{8}$

**Which is more?**

25. $\frac{1}{2}$ of an hour
$\frac{3}{4}$ of an hour

26. $\frac{1}{4}$ of a pound
$\frac{3}{8}$ of a pound

27. $\frac{3}{4}$ of a yard
$\frac{2}{3}$ of a yard

28. $\frac{1}{2}$ of a foot
$\frac{1}{3}$ of a foot

29. $\frac{2}{3}$ of a minute
$\frac{1}{2}$ of a minute

30. $\frac{5}{16}$ of a mile
$\frac{1}{4}$ of a mile

## KEEPING SKILLS SHARP

1. $591 \times 82$
2. $826 \times 91$
3. $4150 \times 36$
4. $3942 \times 54$
5. $6077 \times 45$
6. $375 \times 23$
7. $296 \times 46$
8. $840 \times 93$
9. $911 \times 108$
10. $783 \times 743$

# Adding fractions

$\frac{2}{4} + \frac{1}{4} = \frac{3}{4}$

$$\begin{array}{r} \frac{3}{6} \\ +\frac{2}{6} \\ \hline \frac{5}{6} \end{array}$$

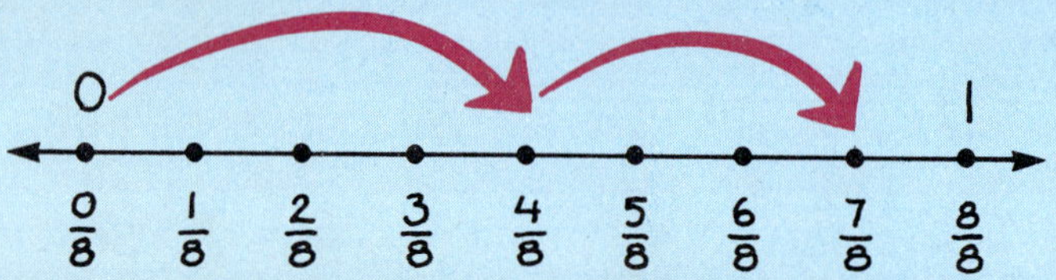


$\frac{4}{8} + \frac{3}{8} = \frac{7}{8}$

To add two fractions that have the same denominator:

Add the numerators to get the numerator of the sum.

Use the common denominator for the denominator of the sum.

**EXERCISES**

**Give each sum.**

**1.** 

$\frac{1}{6} + \frac{4}{6}$

**2.** 

$\frac{1}{4} + \frac{2}{4}$

**3.** 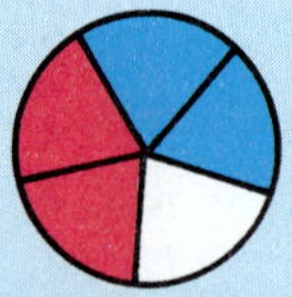

$\frac{2}{5} + \frac{2}{5}$

**4.** 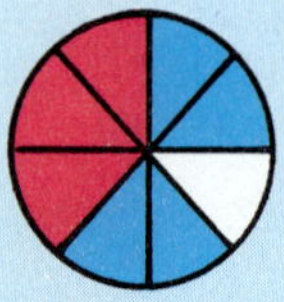

$\frac{3}{8} + \frac{4}{8}$

**5.** 

$\frac{2}{6} + \frac{3}{6}$

**6.** 

$\frac{3}{9} + \frac{2}{9}$

**Add.**

7. $\frac{3}{5} + \frac{1}{5}$
8. $\frac{2}{7} + \frac{2}{7}$
9. $\frac{1}{4} + \frac{2}{4}$
10. $\frac{2}{6} + \frac{3}{6}$
11. $\frac{1}{5} + \frac{2}{5}$
12. $\frac{3}{8} + \frac{4}{8}$
13. $\frac{1}{4} + \frac{1}{4}$
14. $\frac{1}{3} + \frac{1}{3}$
15. $\frac{1}{8} + \frac{1}{8}$
16. $\frac{1}{6} + \frac{4}{6}$
17. $\frac{3}{7} + \frac{2}{7}$
18. $\frac{3}{5} + \frac{1}{5}$

**Add. Write each sum in lowest terms.**

19. $\frac{3}{8} + \frac{3}{8}$
20. $\frac{4}{6} + \frac{1}{6}$
21. $\frac{5}{8} + \frac{1}{8}$
22. $\frac{1}{4} + \frac{1}{4}$
23. $\frac{2}{5} + \frac{2}{5}$
24. $\frac{3}{6} + \frac{1}{6}$
25. $\frac{1}{4} + \frac{2}{4}$
26. $\frac{3}{6} + \frac{2}{6}$
27. $\frac{2}{8} + \frac{4}{8}$
28. $\frac{3}{10} + \frac{3}{10}$
29. $\frac{1}{8} + \frac{2}{8}$
30. $\frac{4}{8} + \frac{2}{8}$
31. $\frac{2}{7} + \frac{4}{7}$
32. $\frac{1}{8} + \frac{5}{8}$
33. $\frac{2}{4} + \frac{1}{4}$
34. $\frac{1}{6} + \frac{3}{6}$
35. $\frac{2}{6} + \frac{2}{6}$
36. $\frac{1}{9} + \frac{2}{9}$
37. $\frac{4}{12} + \frac{5}{12}$
38. $\frac{2}{9} + \frac{4}{9}$
39. $\frac{1}{8} + \frac{3}{8}$
40. $\frac{1}{6} + \frac{1}{6}$
41. $\frac{3}{10} + \frac{5}{10}$
42. $\frac{1}{9} + \frac{7}{9}$

**Solve.**

43. Jonathan jogged for $\frac{1}{4}$ hour in the morning and $\frac{1}{4}$ hour in the afternoon. What fraction of an hour did he jog in all?

44. Rebecca read $\frac{3}{8}$ of the book one night and $\frac{3}{8}$ of the book the next night. What fraction of the book did she read on those two nights?

45. Hal is selling tickets for the school play. He sold $\frac{1}{5}$ of his tickets the first night and $\frac{2}{5}$ of his tickets the second night. What fraction of his tickets did he sell on the two nights?

46. Alice earned some money for raking leaves. She spent $\frac{1}{8}$ of it for a malt and $\frac{3}{8}$ of it for a movie ticket. What fraction of the money did she spend?

# Adding, unlike denominators

If fractions have the same denominator, adding them is easy. So, if the denominators are different, we change to equivalent fractions that have the same denominator.

different denominators $\frac{3}{8} + \frac{1}{4} = \frac{3}{8} + \frac{2}{8}$ same denominators

$= \frac{5}{8}$

*Remember:* 8 is called a **common denominator.**

$$\begin{array}{r} \frac{2}{3} \\ +\frac{1}{6} \\ \hline \end{array} = \begin{array}{r} \frac{4}{6} \\ +\frac{1}{6} \\ \hline \frac{5}{6} \end{array}$$

**EXERCISES**

**Add. Write each sum in lowest terms.**

1. $\frac{1}{4} + \frac{1}{2}$
2. $\frac{1}{6} + \frac{1}{2}$
3. $\frac{3}{8} + \frac{1}{4}$
4. $\frac{5}{9} + \frac{1}{3}$
5. $\frac{1}{3} + \frac{1}{6}$
6. $\frac{1}{8} + \frac{1}{2}$
7. $\frac{1}{8} + \frac{3}{4}$
8. $\frac{1}{6} + \frac{2}{3}$
9. $\frac{1}{3} + \frac{2}{9}$
10. $\frac{1}{4} + \frac{3}{8}$
11. $\frac{3}{8} + \frac{5}{16}$
12. $\frac{1}{5} + \frac{3}{10}$

**Add. Write each sum in lowest terms.**

13. $\frac{1}{6} + \frac{1}{3}$
14. $\frac{1}{4} + \frac{1}{8}$
15. $\frac{1}{9} + \frac{2}{3}$
16. $\frac{3}{8} + \frac{1}{2}$
17. $\frac{2}{3} + \frac{1}{6}$
18. $\frac{3}{10} + \frac{2}{5}$
19. $\frac{1}{6} + \frac{7}{12}$
20. $\frac{5}{16} + \frac{1}{2}$
21. $\frac{1}{8} + \frac{5}{16}$
22. $\frac{3}{4} + \frac{1}{8}$
23. $\frac{5}{12} + \frac{1}{4}$
24. $\frac{3}{10} + \frac{1}{2}$
25. $\frac{3}{8} + \frac{3}{16}$
26. $\frac{2}{9} + \frac{2}{3}$
27. $\frac{1}{2} + \frac{1}{4}$
28. $\frac{1}{3} + \frac{4}{9}$
29. $\frac{1}{2} + \frac{1}{6}$
30. $\frac{2}{3} + \frac{1}{9}$
31. $\frac{7}{16} + \frac{1}{2}$
32. $\frac{3}{8} + \frac{1}{4}$
33. $\frac{2}{7} + \frac{5}{14}$
34. $\frac{5}{12} + \frac{1}{3}$
35. $\frac{1}{8} + \frac{3}{4}$
36. $\frac{9}{16} + \frac{3}{8}$

**Solve.**

37. Jack spent $\frac{1}{2}$ of his allowance for a movie ticket and $\frac{1}{4}$ of his allowance for lunch. What fraction of his allowance did he spend?

38. Ralph painted $\frac{3}{8}$ of a fence the first day and $\frac{1}{4}$ of the fence the second day. How much of the fence did he paint during the two days?

## *Challenge!*

**Add across. Add down.**

39.

| $\frac{1}{9}$ | $\frac{2}{9}$ | ? |
|---|---|---|
| $\frac{1}{3}$ | $\frac{2}{9}$ | ? |
| ? | ? | ? |

40.

| $\frac{1}{10}$ | $\frac{1}{5}$ | ? |
|---|---|---|
| $\frac{3}{10}$ | $\frac{1}{5}$ | ? |
| ? | ? | ? |

41.

| $\frac{1}{12}$ | $\frac{1}{12}$ | ? |
|---|---|---|
| $\frac{1}{3}$ | $\frac{1}{6}$ | ? |
| ? | ? | ? |

# More about adding, unlike denominators

The least common multiple of the denominators is always the least common denominator.

**EXAMPLE 1.**

$$\frac{1}{2} + \frac{1}{3} = \frac{3}{6} + \frac{2}{6}$$

{2, 4, 6} {3, 6}

$$= \frac{5}{6}$$

**EXAMPLE 2.**

$$\frac{3}{4} + \frac{1}{6} = \frac{9}{12} + \frac{2}{12}$$

{12}

$$= \frac{11}{12}$$

$$\frac{3}{4} + \frac{1}{6} = \frac{18}{24} + \frac{4}{24}$$

$$= \frac{22}{24}$$

$$= \frac{11}{12}$$

Will the product of the denominators always be the least common denominator?

**EXERCISES**

**Add. Write each sum in lowest terms.**

1. $\frac{1}{3} + \frac{1}{4}$ {12}
2. $\frac{2}{3} + \frac{1}{5}$ {15}
3. $\frac{1}{2} + \frac{1}{3}$ {6}
4. $\frac{3}{4} + \frac{1}{5}$ {20}
5. $\frac{2}{3} + \frac{1}{8}$ {24}
6. $\frac{2}{3} + \frac{1}{4}$ {12}

7. $\frac{1}{4} + \frac{2}{5}$

8. $\frac{3}{5} + \frac{1}{6}$

9. $\frac{2}{5} + \frac{1}{3}$

10. $\frac{1}{5} + \frac{1}{2}$

11. $\frac{3}{8} + \frac{1}{6}$

12. $\frac{2}{9} + \frac{1}{3}$

13. $\frac{1}{6} + \frac{3}{4}$

14. $\frac{3}{4} + \frac{1}{8}$

15. $\frac{3}{8} + \frac{3}{5}$

**Add. Write each sum in lowest terms.**

16. $\frac{1}{3} + \frac{1}{4}$

17. $\frac{1}{4} + \frac{1}{6}$

18. $\frac{1}{2} + \frac{3}{8}$

19. $\frac{1}{4} + \frac{2}{3}$

20. $\frac{1}{4} + \frac{3}{10}$

21. $\frac{1}{3} + \frac{1}{5}$

22. $\frac{1}{8} + \frac{2}{3}$

23. $\frac{1}{6} + \frac{1}{2}$

24. $\frac{5}{12} + \frac{1}{3}$

25. $\frac{3}{10} + \frac{1}{5}$

26. $\frac{1}{5} + \frac{3}{5}$

27. $\frac{3}{8} + \frac{1}{3}$

28. $\frac{1}{4} + \frac{3}{8}$

29. $\frac{1}{6} + \frac{7}{10}$

30. $\frac{1}{5} + \frac{1}{3}$

31. $\frac{2}{3} + \frac{1}{4}$

32. $\frac{1}{10} + \frac{3}{4}$

33. $\frac{1}{9} + \frac{2}{9}$

34. $\frac{3}{10} + \frac{2}{10}$

35. $\frac{5}{9} + \frac{1}{6}$

36. $\frac{1}{3} + \frac{1}{6}$

37. $\frac{3}{4} + \frac{1}{8}$

38. $\frac{1}{5} + \frac{2}{7}$

39. $\frac{4}{5} + \frac{1}{10}$

**This circle graph shows how James spent his allowance. What fraction of his allowance did he spend for**

40. food?

41. record?

42. movie and food?

43. record and clothes?

## *Challenge!*

**Find the missing digit.**

44. $\frac{1}{3} + \frac{1}{\square} = \frac{5}{6}$

45. $\frac{1}{\square} + \frac{3}{4} = \frac{11}{12}$

46. $\frac{\square}{8} + \frac{1}{3} = \frac{17}{24}$

# Subtracting fractions

$$\frac{3}{4} - \frac{2}{4} = \frac{1}{4}$$

$$\begin{array}{r} \frac{4}{5} \\ -\frac{1}{5} \\ \hline \frac{3}{5} \end{array}$$

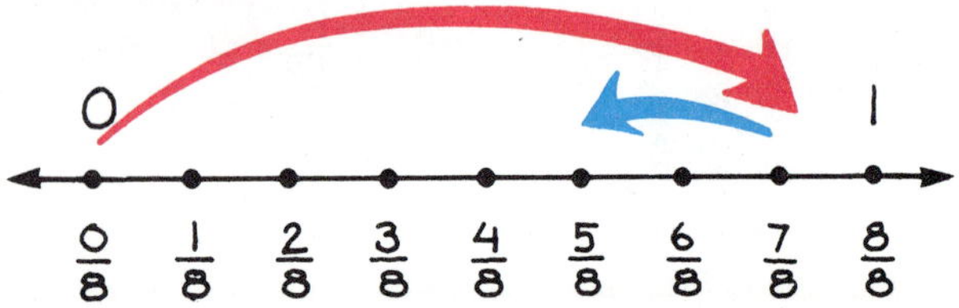


$$\frac{7}{8} - \frac{2}{8} = \frac{5}{8}$$

To subtract fractions that have the same denominator:

Subtract the numerators to get the numerator of the difference.

Use the common denominator for the denominator of the difference.

**EXERCISES**

**Give each difference.**

**1.** 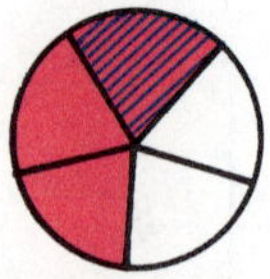

$\frac{3}{5} - \frac{1}{5}$

**2.** 

$\frac{2}{4} - \frac{1}{4}$

**3.** 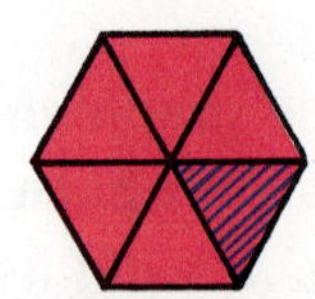

$\frac{6}{6} - \frac{1}{6}$

**4.** 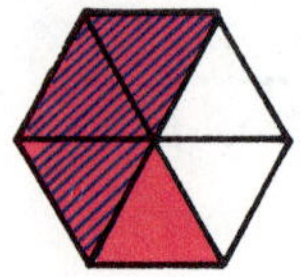

$\frac{4}{6} - \frac{3}{6}$

**5.** 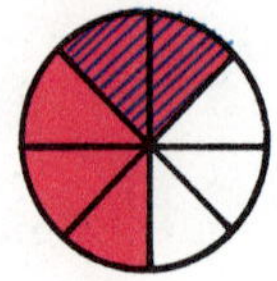

$\frac{5}{8} - \frac{2}{8}$

**6.** 

$\frac{7}{8} - \frac{6}{8}$

7. $\frac{5}{6} - \frac{4}{6}$

8. $\frac{3}{5} - \frac{2}{5}$

9. $\frac{4}{8} - \frac{1}{8}$

10. $\frac{5}{9} - \frac{4}{9}$

11. $\frac{6}{7} - \frac{1}{7}$

12. $\frac{2}{3} - \frac{1}{3}$

13. $\frac{9}{10} - \frac{6}{10}$

14. $\frac{1}{4} - \frac{1}{4}$

15. $\frac{3}{8} - \frac{3}{8}$

16. $\frac{7}{8} - \frac{2}{8}$

17. $\frac{7}{8} - \frac{6}{8}$

18. $\frac{5}{8} - \frac{2}{8}$

**Subtract. Write each difference in lowest terms.**

19. $\frac{5}{6} - \frac{1}{6}$

20. $\frac{3}{4} - \frac{2}{4}$

21. $\frac{3}{5} - \frac{1}{5}$

22. $\frac{4}{4} - \frac{1}{4}$

23. $\frac{7}{8} - \frac{5}{8}$

24. $\frac{5}{9} - \frac{2}{9}$

25. $\frac{7}{8} - \frac{6}{8}$

26. $\frac{9}{10} - \frac{5}{10}$

27. $\frac{2}{3} - \frac{2}{3}$

28. $\frac{9}{10} - \frac{1}{10}$

29. $\frac{9}{10} - \frac{8}{10}$

30. $\frac{5}{8} - \frac{1}{8}$

31. $\frac{5}{6} - \frac{2}{6}$

32. $\frac{5}{8} - \frac{1}{8}$

33. $\frac{5}{6} - \frac{3}{6}$

34. $\frac{5}{6} - \frac{5}{6}$

35. $\frac{8}{9} - \frac{4}{9}$

36. $\frac{7}{12} - \frac{5}{12}$

37. $\frac{7}{8} - \frac{3}{8}$

38. $\frac{11}{12} - \frac{7}{12}$

39. $\frac{7}{8} - \frac{2}{8}$

40. $\frac{7}{12} - \frac{3}{12}$

41. $\frac{13}{16} - \frac{7}{16}$

42. $\frac{3}{6} - \frac{1}{6}$

**Solve.**

43. Kevin bought $\frac{3}{4}$ dozen cookies. He ate $\frac{1}{4}$ dozen. What fraction of a dozen was left?

44. Gerry bought $\frac{7}{8}$ yard of yellow ribbon and $\frac{2}{8}$ yard of red ribbon. How much more yellow ribbon did he buy?

45. Marie jogged $\frac{9}{10}$ mile on Monday and $\frac{7}{10}$ mile on Tuesday. How much farther did she jog on Monday?

46. Jill bought $\frac{7}{8}$ pound of mixed nuts for a party. She ate $\frac{1}{8}$ pound. How much was left?

# Subtracting, unlike denominators

*Remember:* If one denominator is a multiple of the other, the greater number can be used as a common denominator.

Since 8 is a multiple of 4, 8 is a common denominator.

$$\frac{7}{8} - \frac{1}{4} = \frac{7}{8} - \frac{2}{8}$$
$$= \frac{5}{8}$$

6 is a common denominator.

$$\begin{array}{r} \frac{2}{3} = \frac{4}{6} \\ -\frac{1}{6} = -\frac{1}{6} \\ \hline \frac{3}{6} = \frac{1}{2} \end{array}$$

10 is a common denominator.

$$\begin{array}{r} \frac{1}{2} = \frac{5}{10} \\ -\frac{3}{10} = -\frac{3}{10} \\ \hline \frac{2}{10} = \frac{1}{5} \end{array}$$

## EXERCISES

**Subtract. Write each difference in lowest terms.**

| | | | |
|---|---|---|---|
| **1.** $\frac{1}{2} - \frac{1}{4}$ | **2.** $\frac{3}{4} - \frac{1}{2}$ | **3.** $\frac{3}{4} - \frac{3}{8}$ | **4.** $\frac{1}{2} - \frac{1}{6}$ |
| **5.** $\frac{1}{2} - \frac{1}{8}$ | **6.** $\frac{7}{10} - \frac{1}{2}$ | **7.** $\frac{5}{6} - \frac{2}{3}$ | **8.** $\frac{3}{4} - \frac{1}{8}$ |
| **9.** $\frac{2}{3} - \frac{4}{9}$ | **10.** $\frac{5}{8} - \frac{1}{2}$ | **11.** $\frac{5}{8} - \frac{1}{4}$ | **12.** $\frac{1}{2} - \frac{3}{8}$ |
| **13.** $\frac{3}{4} - \frac{5}{8}$ | **14.** $\frac{9}{10} - \frac{3}{5}$ | **15.** $\frac{7}{8} - \frac{3}{4}$ | **16.** $\frac{4}{5} - \frac{4}{10}$ |

**Subtract. Write each difference in lowest terms.**

**17.** $\frac{1}{2} - \frac{1}{4}$ **18.** $\frac{5}{8} - \frac{1}{4}$ **19.** $\frac{1}{3} - \frac{1}{6}$ **20.** $\frac{1}{2} - \frac{1}{8}$ **21.** $\frac{5}{6} - \frac{1}{3}$ **22.** $\frac{3}{4} - \frac{1}{8}$

**23.** $\frac{1}{2} - \frac{1}{8}$ **24.** $\frac{3}{4} - \frac{1}{2}$ **25.** $\frac{5}{9} - \frac{1}{3}$ **26.** $\frac{8}{9} - \frac{2}{3}$ **27.** $\frac{5}{6} - \frac{2}{3}$ **28.** $\frac{5}{8} - \frac{1}{2}$

**29.** $\frac{3}{4} - \frac{5}{8}$ **30.** $\frac{7}{8} - \frac{1}{2}$ **31.** $\frac{2}{3} - \frac{1}{6}$ **32.** $\frac{3}{4} - \frac{3}{8}$ **33.** $\frac{4}{4} - \frac{1}{2}$ **34.** $\frac{7}{10} - \frac{1}{5}$

**35.** $\frac{3}{4} - \frac{5}{16}$ **36.** $\frac{9}{16} - \frac{1}{4}$ **37.** $\frac{3}{4} - \frac{9}{16}$ **38.** $\frac{5}{6} - \frac{1}{2}$ **39.** $\frac{1}{2} - \frac{3}{16}$ **40.** $\frac{9}{10} - \frac{1}{2}$

**Solve.**

**41.**

**a.** How many pounds of nuts?
**b.** How much more in the large box?

**42.** Terry swam $\frac{1}{2}$ mile in the morning and $\frac{3}{10}$ mile in the afternoon.

**a.** How far did she swim that day?
**b.** How much farther did she swim in the morning?

**Who am I?**

★ **43.** I am $\frac{2}{3}$ greater than $\frac{1}{6}$.

★ **44.** I am $\frac{1}{2}$ less than the sum of $\frac{3}{4}$ and $\frac{1}{8}$.

# More about subtracting, unlike denominators

*Remember:* The least common multiple of the denominators is the least common denominator.

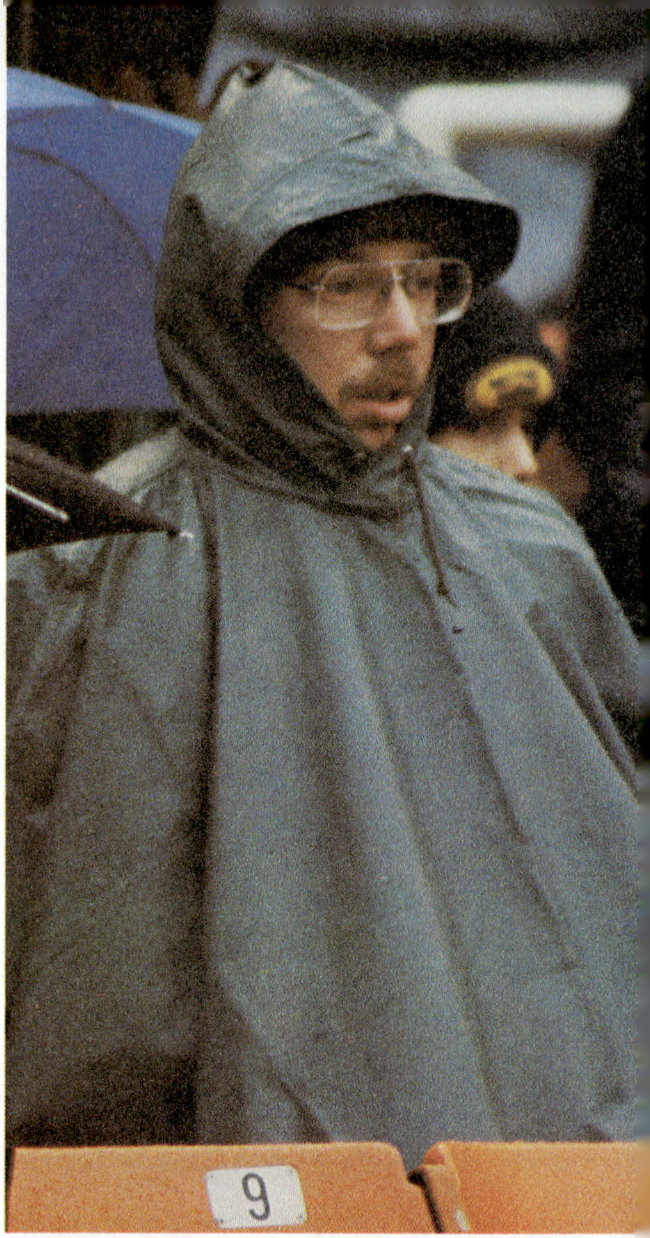


**EXAMPLE 1.**

$$\frac{1}{2} - \frac{1}{3} = \frac{3}{6} - \frac{2}{6}$$

{2,4,6} {3,6}

$$= \frac{1}{6}$$

**EXAMPLE 2.**

$$\frac{3}{4} - \frac{1}{6} = \frac{9}{12} - \frac{2}{12}$$

{4,8,12} {6,12}

$$= \frac{7}{12}$$

**EXERCISES**

**Subtract. Write each difference in lowest terms.**

1. $\frac{1}{2} - \frac{1}{3} = \frac{3}{6} - \frac{2}{6}$ {2,4,6} {3,6}
2. $\frac{3}{4} - \frac{1}{2} = \frac{3}{4} - \frac{2}{4}$ {4} {2,4}
3. $\frac{3}{4} - \frac{2}{3} = \frac{9}{12} - \frac{8}{12}$ {4,8,12} {3,6,9,12}
4. $\frac{2}{3} - \frac{1}{5}$
5. $\frac{5}{6} - \frac{3}{4}$
6. $\frac{7}{8} - \frac{1}{4}$
7. $\frac{2}{3} - \frac{1}{2}$
8. $\frac{1}{2} - \frac{1}{6}$
9. $\frac{3}{5} - \frac{1}{2}$
10. $\frac{3}{4} - \frac{1}{3}$
11. $\frac{5}{6} - \frac{2}{9}$
12. $\frac{5}{6} - \frac{1}{2}$
13. $\frac{7}{10} - \frac{2}{5}$
14. $\frac{2}{3} - \frac{1}{4}$
15. $\frac{3}{8} - \frac{1}{3}$

16. $\frac{5}{8} - \frac{1}{4}$

17. $\frac{6}{8} - \frac{1}{4}$

18. $\frac{5}{8} - \frac{3}{8}$

19. $\frac{8}{9} - \frac{1}{6}$

20. $\frac{3}{5} - \frac{1}{5}$

21. $\frac{2}{3} - \frac{4}{9}$

22. $\frac{7}{10} - \frac{1}{2}$

23. $\frac{5}{6} - \frac{2}{3}$

24. $\frac{3}{8} - \frac{1}{3}$

25. $\frac{5}{12} - \frac{2}{12}$

26. $\frac{2}{3} - \frac{1}{4}$

27. $\frac{5}{8} - \frac{1}{2}$

28. $\frac{7}{9} - \frac{1}{3}$

29. $\frac{3}{4} - \frac{1}{10}$

30. $\frac{5}{6} - \frac{3}{4}$

31. $\frac{2}{3} - \frac{3}{5}$

32. $\frac{3}{4} - \frac{1}{2}$

33. $\frac{9}{10} - \frac{1}{2}$

The table shows the rainfall each day for a week.

| *Day* | *Rainfall in inches* |
|---|---|
| Sun. | $\frac{3}{10}$ |
| Mon. | $\frac{1}{2}$ |
| Tues. | $\frac{1}{4}$ |
| Wed. | none |
| Thurs. | none |
| Fri. | $\frac{3}{4}$ |
| Sat. | none |

34. How many days did it rain?

35. On which day did it rain the most?

36. What was the Sunday–Monday total?

37. How much more did it rain on Friday than on Sunday?

## *Challenge!*

**Add across. Subtract down.**

38.

| $\frac{1}{2}$ | $\frac{1}{4}$ | ? |
|---|---|---|
| $\frac{1}{8}$ | $\frac{1}{8}$ | ? |
| ? | ? | ? |

39.

| $\frac{1}{3}$ | $\frac{1}{2}$ | ? |
|---|---|---|
| $\frac{1}{4}$ | $\frac{1}{6}$ | ? |
| ? | ? | ? |

40.

| $\frac{1}{2}$ | $\frac{1}{3}$ | ? |
|---|---|---|
| $\frac{3}{8}$ | $\frac{1}{4}$ | ? |
| ? | ? | ? |

## Problem solving

Pam Johnson is a bicycle mechanic. She replaces broken or worn parts, adjusts gears, changes tires, and services all moving parts on a bicycle.

She has a variety of tools. Many of her wrenches are marked in fractions of an inch.

The size is given as a fraction in lowest terms.

**Solve.**

1. Suppose that you are working with a set of wrenches that is marked in sixteenths. What wrench would you use if you wanted the next larger wrench? [*Remember:* the size is given in lowest terms.]

   a. $\frac{1}{4}$ $\left(\frac{1}{4} + \frac{1}{16} = \frac{4}{16} + \frac{1}{16} = \frac{5}{16}\right)$

   b. $\frac{3}{8}$ c. $\frac{9}{16}$

   d. $\frac{3}{4}$ e. $\frac{1}{2}$

   f. $\frac{7}{8}$ g. $\frac{7}{16}$

2. Suppose that you are using a set of wrenches that is marked in thirty-seconds. Give the next smaller wrench.

   a. $\frac{1}{2}$ $\left(\frac{1}{2} - \frac{1}{32}\right)$ b. $\frac{9}{16}$

   c. $\frac{3}{4}$ d. $\frac{3}{8}$

3. Which wrench is larger?

   a. $\frac{9}{16}, \frac{5}{8}$ b. $\frac{1}{2}, \frac{15}{32}$

   c. $\frac{13}{32}, \frac{7}{16}$ d. $\frac{3}{4}, \frac{7}{8}$

   e. $\frac{5}{32}, \frac{1}{4}$ f. $\frac{5}{8}, \frac{19}{32}$

★ 4. What size wrench is halfway between these wrenches?

   $\left(\frac{4}{8}\right)$ $\left(\frac{6}{8}\right)$

   a. $\frac{1}{2}, \frac{3}{4}$ b. $\frac{5}{8}, \frac{3}{4}$

   c. $\frac{3}{8}, \frac{1}{2}$ d. $\frac{3}{4}, \frac{7}{8}$

   e. $\frac{5}{8}, \frac{11}{16}$ f. $\frac{1}{2}, \frac{9}{16}$

# Practice

**Give each sum or difference in lowest terms.**

1. $\frac{1}{8} + \frac{1}{8} + \frac{1}{8}$
2. $\frac{1}{4} + \frac{1}{4} + \frac{1}{4}$
3. $\frac{1}{5} + \frac{1}{5} + \frac{1}{5}$
4. $\frac{1}{6} + \frac{1}{6} + \frac{1}{6}$
5. $\frac{1}{4} + \frac{1}{4} + \frac{1}{8}$
6. $\frac{1}{6} + \frac{1}{3} + \frac{1}{6}$
7. $\frac{1}{8} + \frac{1}{4} + \frac{1}{8}$
8. $\frac{1}{6} + \frac{1}{2} + \frac{1}{6}$
9. $\frac{1}{2} + \frac{1}{4} + \frac{3}{8}$
10. $\frac{1}{3} + \frac{1}{6} + \frac{5}{12}$
11. $\frac{3}{10} + \frac{2}{5} + \frac{1}{10}$
12. $\frac{5}{12} + \frac{1}{4} + \frac{1}{8}$
13. $\frac{1}{3} + \frac{1}{12} + \frac{3}{8}$
14. $\frac{3}{8} + \frac{1}{16} + \frac{1}{2}$
15. $\frac{3}{8} + \frac{1}{4} + \frac{3}{16}$
16. $\frac{5}{16} + \frac{1}{8} + \frac{1}{2}$
17. $\frac{5}{32} + \frac{5}{16} + \frac{1}{4}$
18. $\frac{3}{4} + \frac{1}{8} + \frac{3}{32}$
19. $\frac{3}{8} + \frac{7}{32} + \frac{5}{16}$
20. $\frac{1}{3} + \frac{5}{12} + \frac{1}{8}$

21. $\begin{array}{r} \frac{3}{5} \\ -\frac{1}{5} \\ \hline \end{array}$
22. $\begin{array}{r} \frac{5}{8} \\ +\frac{1}{8} \\ \hline \end{array}$
23. $\begin{array}{r} \frac{2}{3} \\ -\frac{1}{6} \\ \hline \end{array}$
24. $\begin{array}{r} \frac{3}{4} \\ +\frac{1}{6} \\ \hline \end{array}$
25. $\begin{array}{r} \frac{5}{8} \\ -\frac{1}{3} \\ \hline \end{array}$
26. $\begin{array}{r} \frac{7}{8} \\ -\frac{5}{6} \\ \hline \end{array}$
27. $\begin{array}{r} \frac{7}{10} \\ -\frac{3}{5} \\ \hline \end{array}$
28. $\begin{array}{r} \frac{3}{10} \\ +\frac{1}{4} \\ \hline \end{array}$
29. $\begin{array}{r} \frac{5}{8} \\ -\frac{2}{5} \\ \hline \end{array}$
30. $\begin{array}{r} \frac{3}{7} \\ +\frac{2}{5} \\ \hline \end{array}$
31. $\begin{array}{r} \frac{6}{7} \\ -\frac{3}{4} \\ \hline \end{array}$
32. $\begin{array}{r} \frac{5}{9} \\ -\frac{2}{9} \\ \hline \end{array}$
33. $\begin{array}{r} \frac{3}{10} \\ -\frac{1}{4} \\ \hline \end{array}$
34. $\begin{array}{r} \frac{2}{3} \\ +\frac{1}{4} \\ \hline \end{array}$
35. $\begin{array}{r} \frac{9}{10} \\ -\frac{1}{5} \\ \hline \end{array}$
36. $\begin{array}{r} \frac{3}{8} \\ +\frac{1}{2} \\ \hline \end{array}$
37. $\begin{array}{r} \frac{2}{3} \\ -\frac{1}{4} \\ \hline \end{array}$
38. $\begin{array}{r} \frac{7}{8} \\ -\frac{1}{3} \\ \hline \end{array}$

## *Challenge!*

**Guess my number.**

39. If you add $\frac{3}{10}$ to my number, you get $\frac{4}{5}$. What is my number?

40. If you subtract $\frac{1}{4}$ from my number, you get $\frac{2}{5}$. What is my number?

## Problem solving

**Solve. (One of these problems does not have enough information.)**

1. On the average, each American throws away about 2000 pounds of trash each year. How many pounds would a family of 4 throw away?

2. Of the 2000 pounds of trash, about 1200 pounds are paper. What fraction of the trash is paper?

3. If you throw away 2000 pounds of trash in a year, how much do you average a week? Give your answer as the nearest whole number.

4. It is estimated that $\frac{3}{5}$ of the trash is paper and $\frac{1}{6}$ is cans. What fraction is either paper or cans?

5. About $\frac{1}{6}$ of the trash is cans and $\frac{1}{16}$ is bottles. How much greater is the fraction for cans?

6. One-eighth of the trash is bottles and plastic products. What fraction of the trash is not bottles or plastic?

7. A merchant will pay 7¢ each for returnable soft-drink bottles. How much refund will there be for 2 cases of bottles?

8. An elementary school raised money by collecting 6000 pounds of aluminum cans. They were paid $34.75 a ton (2000 pounds). How much money did they get?

# Changing whole numbers and mixed numbers to fractions

There are 2 halves in 1.

$$1 = \frac{2}{2}$$

To write a whole number as a fraction, you can write the whole number over a denominator of 1. Then to change to an equivalent fraction, you multiply both numerator and denominator by the same whole number (not *0*).

$$2 = \frac{2}{1} \overset{\times 3}{=} \frac{6}{3} \qquad 5 = \frac{5}{1} \overset{\times 2}{=} \frac{10}{2}$$

A **mixed number** has a whole number part and a fraction part:

$$1 + \frac{1}{4}$$

$1\frac{1}{4}$ ← mixed number

You can change a mixed number to a fraction by adding.

$$\begin{aligned} 2\frac{1}{2} &= 2 + \frac{1}{2} \\ &= \frac{2}{1} + \frac{1}{2} \\ &= \frac{4}{2} + \frac{1}{2} \\ &= \frac{5}{2} \end{aligned}$$

## EXERCISES

**Give a mixed number for the red point.**

1. 

2. 

3. 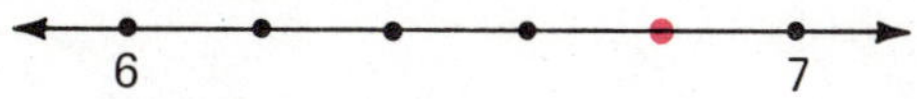


4. 

**Copy and complete.**

5. $1 = \frac{\bullet}{2}$  6. $2 = \frac{\bullet}{2}$  7. $1 = \frac{\bullet}{3}$  8. $2 = \frac{\bullet}{3}$

9. $3 = \frac{\bullet}{2}$  10. $2 = \frac{\bullet}{4}$  11. $1 = \frac{\bullet}{4}$  12. $4 = \frac{\bullet}{2}$

**Change to fractions.**

13. $1\frac{1}{2}$ ($1 + \frac{1}{2}$)  14. $2\frac{1}{2}$  15. $3\frac{1}{4}$  16. $1\frac{3}{4}$  17. $2\frac{2}{3}$

18. $1\frac{1}{5}$  19. $1\frac{2}{5}$  20. $1\frac{3}{5}$  21. $1\frac{4}{5}$  22. $3\frac{1}{2}$  23. $4\frac{1}{2}$

24. $2\frac{1}{3}$  25. $2\frac{2}{3}$  26. $1\frac{1}{8}$  27. $1\frac{3}{8}$  28. $1\frac{5}{8}$  29. $2\frac{1}{8}$

**Use the shortcut to change to fractions.**

30. $1\frac{1}{2}$  31. $3\frac{1}{3}$  32. $2\frac{2}{3}$  33. $2\frac{1}{2}$

34. $2\frac{2}{5}$  35. $3\frac{1}{8}$  36. $4\frac{2}{5}$  37. $3\frac{1}{2}$

38. $2\frac{3}{5}$  39. $5\frac{3}{8}$  40. $4\frac{5}{8}$  41. $3\frac{7}{8}$

### KEEPING SKILLS SHARP

1. $9\overline{)639}$  2. $8\overline{)736}$  3. $6\overline{)5658}$  4. $5\overline{)3640}$

5. $56\overline{)3210}$  6. $27\overline{)5963}$  7. $63\overline{)7000}$  8. $49\overline{)4297}$

# Changing fractions to whole numbers or mixed numbers

If the numerator is greater than or equal to the denominator, then the fraction is greater than or equal to 1. To change such a fraction to a whole number or a mixed number, we divide the numerator by the denominator.

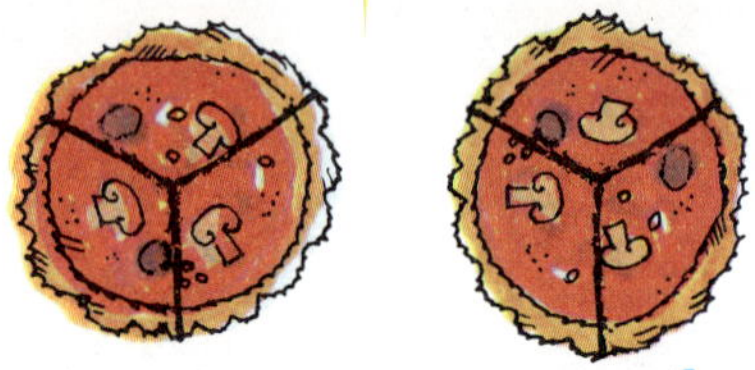

$\frac{6}{3} = ?$

number of thirds in a whole pizza → $3\overline{)6}$ ← number of thirds

quotient 2: number of whole pizzas

So, $\frac{6}{3} = 2$

$\frac{22}{8} = ?$

number of eighths in a whole pizza → $8\overline{)22}$ ← number of eighths

quotient 2: number of whole pizzas

$-16$

$6$ ← number of eighths left over

So, $\frac{22}{8} = 2\frac{6}{8} = 2\frac{3}{4}$

Write all fractions that are less than 1 in lowest terms.

$$\frac{6}{8} = \frac{3}{4}$$

Write all mixed numbers as a whole number and a fraction less than 1 in lowest terms.

$$2\frac{5}{10} = 2\frac{1}{2}$$

Write all fractions greater than or equal to 1 as either a whole number or a mixed number in simplest form.

$$\frac{12}{3} = 4 \qquad \frac{14}{4} = 3\frac{2}{4} = 3\frac{1}{2}$$

## EXERCISES

**Change each fraction to either a whole number or a mixed number.**

1. 

$\frac{9}{4} = ?$

2. 

$\frac{6}{2} = ?$

3. 

$\frac{8}{3} = ?$

4. 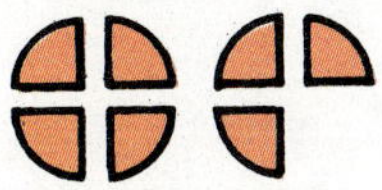

$\frac{7}{4} = ?$

5. 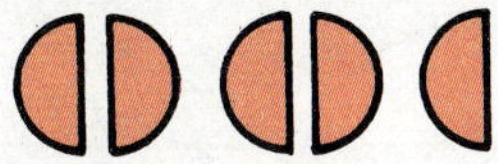

$\frac{5}{2} = ?$

6. 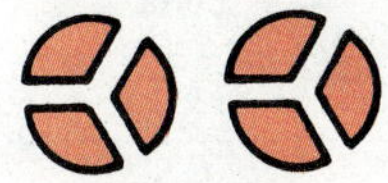

$\frac{6}{3} = ?$

**Write in simplest form.**

| | | | | | |
|---|---|---|---|---|---|
| 7. $\frac{6}{8}$ | 8. $\frac{4}{3}$ | 9. $\frac{4}{6}$ | 10. $\frac{16}{8}$ | 11. $\frac{12}{3}$ | 12. $\frac{8}{2}$ |
| 13. $\frac{11}{6}$ | 14. $\frac{8}{5}$ | 15. $3\frac{2}{4}$ | 16. $\frac{4}{2}$ | 17. $\frac{10}{4}$ | 18. $\frac{14}{6}$ |
| 19. $\frac{5}{10}$ | 20. $\frac{10}{2}$ | 21. $\frac{15}{6}$ | 22. $5\frac{8}{10}$ | 23. $\frac{5}{3}$ | 24. $\frac{6}{2}$ |
| 25. $\frac{8}{4}$ | 26. $\frac{7}{3}$ | 27. $\frac{4}{16}$ | 28. $\frac{11}{5}$ | 29. $\frac{6}{5}$ | 30. $4\frac{6}{8}$ |
| 31. $\frac{7}{2}$ | 32. $5\frac{4}{6}$ | 33. $\frac{20}{8}$ | 34. $\frac{2}{10}$ | 35. $\frac{9}{3}$ | 36. $\frac{10}{3}$ |

**Add. Write the sum in simplest form.**

| | | | |
|---|---|---|---|
| 37. $\frac{5}{6} + \frac{5}{6}$ | 38. $\frac{3}{5} + \frac{2}{5}$ | 39. $\frac{5}{8} + \frac{7}{8}$ | 40. $\frac{3}{10} + \frac{1}{10}$ |
| 41. $\frac{2}{3} + \frac{1}{2}$ | 42. $\frac{3}{4} + \frac{5}{8}$ | 43. $\frac{1}{2} + \frac{4}{5}$ | 44. $\frac{5}{6} + \frac{1}{2}$ |

★ 45. $\frac{3}{4} + \frac{3}{4} + \frac{3}{4}$ ★ 46. $\frac{3}{5} + \frac{4}{5} + \frac{3}{5}$ ★ 47. $\frac{9}{10} + \frac{7}{10} + \frac{9}{10}$

**Subtract. Write the difference in simplest form.**

| | | | |
|---|---|---|---|
| 48. $\frac{5}{6} - \frac{1}{6}$ | 49. $\frac{5}{8} - \frac{1}{8}$ | 50. $\frac{7}{10} - \frac{3}{10}$ | 51. $\frac{7}{9} - \frac{1}{9}$ |
| 52. $\frac{1}{2} - \frac{1}{6}$ | 53. $\frac{5}{6} - \frac{1}{3}$ | 54. $\frac{4}{5} - \frac{3}{10}$ | 55. $\frac{7}{10} - \frac{1}{2}$ |

# Ratios

There are 5 puppies in all. The **ratio** of white puppies to spotted puppies is 3 to 2. Here are some ways that you can write the ratio of white puppies to spotted puppies:

$$3 \text{ to } 2 \qquad \frac{3}{2}$$

Read each ratio as "3 to 2." What is the ratio of spotted to white puppies?

**EXERCISES**

**Give the ratio.**

1. green pencils to yellow pencils
2. yellow pencils to green pencils
3. small drinks to large drinks
4. large drinks to small drinks

5. red buttons to blue buttons

6. blue buttons to red buttons

**Give each ratio for your class.**

7. girls to boys

8. boys to girls

9. last names beginning with vowels to last names beginning with consonants

10. students wearing glasses to students not wearing glasses

11. students not wearing blue jeans to students wearing blue jeans

12. students who walk to school to students who ride to school

Think about mixing these 3 cans of paint together. The ratio of yellow to red in the mixture would be $\frac{1}{2}$.

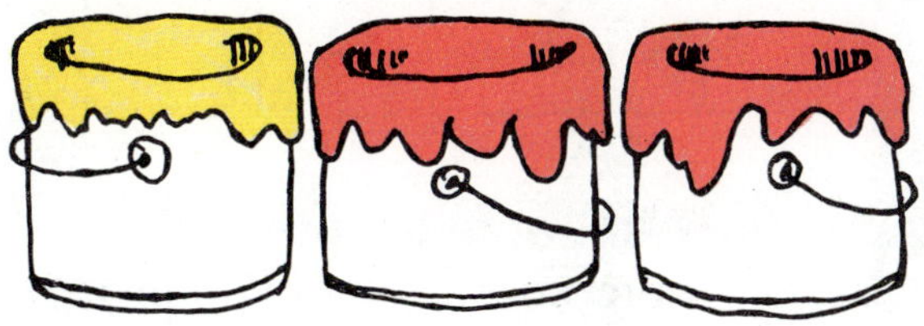

**Complete this table of ratios to get the same mixture.**

| | | | 13. | 14. | 15. | 16. |
|---|---|---|---|---|---|---|
| Liters of yellow | 1 | 2 | 3 | 4 | 5 | 6 |
| Liters of red | 2 | 4 | | | | |

17. The ratio of girls to boys in a certain class is 3 to 2. Are there more girls or boys in the class?

★ 18. The ratio of red marbles to blue marbles is $\frac{1}{3}$. There are 24 marbles in all. How many red marbles and how many blue marbles are there?

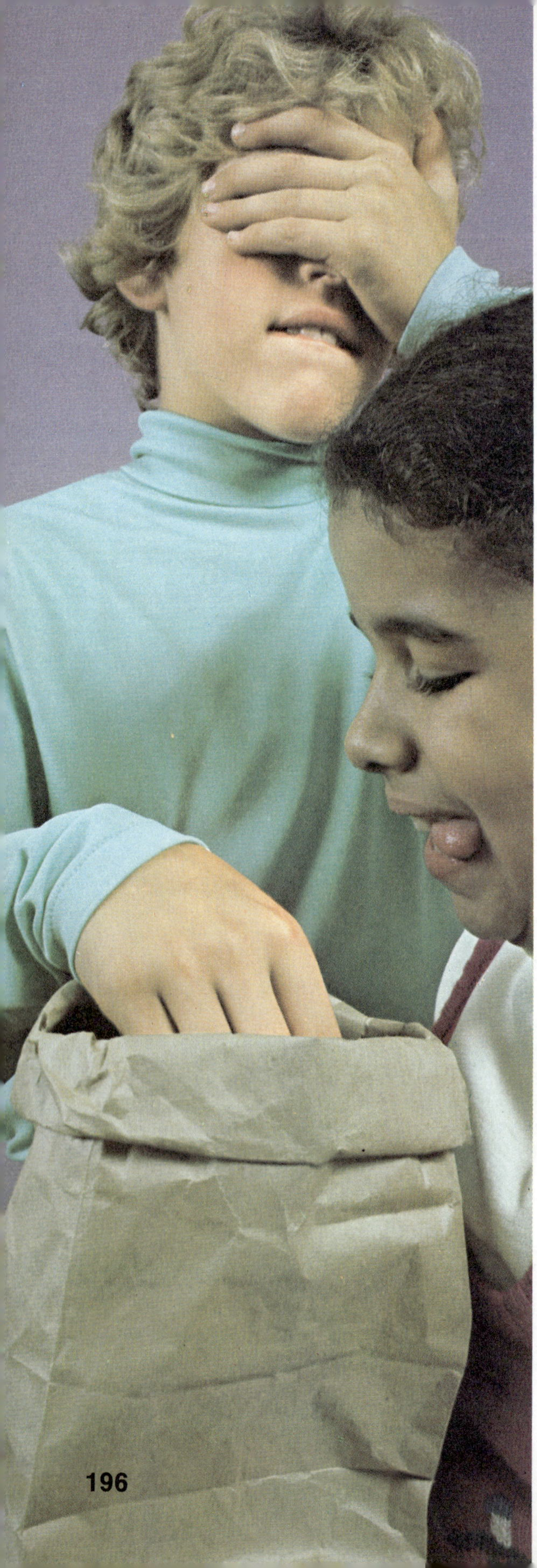

# Probability

Two students put some red, yellow, and blue marbles in the bag. One student mixed up the marbles and the other, without looking in the bag, took out a marble and then recorded its color. The marble was replaced in the bag and the experiment repeated. Here is the record they kept of their experiment.

| EXPERIMENT: Picking a Marble from a Bag | | | |
|---|---|---|---|
| Possible outcomes | 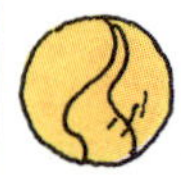 | 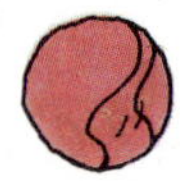 | 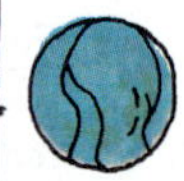 |
| Tally of outcomes | 𝍸 \|\|\| | \|\|\|\| | 𝍸 𝍸 \|\| |
| Fraction of all outcomes | $\frac{8}{24}$ | $\frac{4}{24}$ | $\frac{12}{24}$ |

1. What were the possible outcomes of the experiment?
2. How many times was a yellow marble picked?
3. How many times was a blue marble picked?
4. How many times was the experiment done?
5. What fraction of the outcomes were red marbles?
6. If you had to guess, would you guess that there were more blue marbles in the bag or more yellow marbles?

| EXPERIMENT: Tossing a Thumbtack | | |
|---|---|---|
| Possible outcomes |  |  |
| Tally of outcomes | 卌 卌 卌 \|\|\| | 卌 \|\| |
| Fraction of all outcomes | ? | ? |

7. How many possible outcomes?

8. How many times did the thumbtack land "point down"? "point up"?

9. What fraction of the outcomes were "point up"?

10. If you were to toss the thumbtack once, how do you think it would land?

11. Toss a penny 40 times and complete this record of your experiment.

| Possible outcomes |  Heads |  Tails |
|---|---|---|
| Tally of outcomes | | |
| Fraction of all outcomes | | |

Andrew tossed a paper cup and recorded the outcomes on a bar graph.

12. How many possible outcomes?

13. How many times did he toss the cup?

14. What fraction of the outcomes were right side up?

15. If you were to toss the cup one time, how do you think it would land?

# Problem solving—using graphs and maps

The circle graph shows the results of a school survey. For example, it shows that baseball was the favorite sport of $\frac{1}{4}$ of the students surveyed.

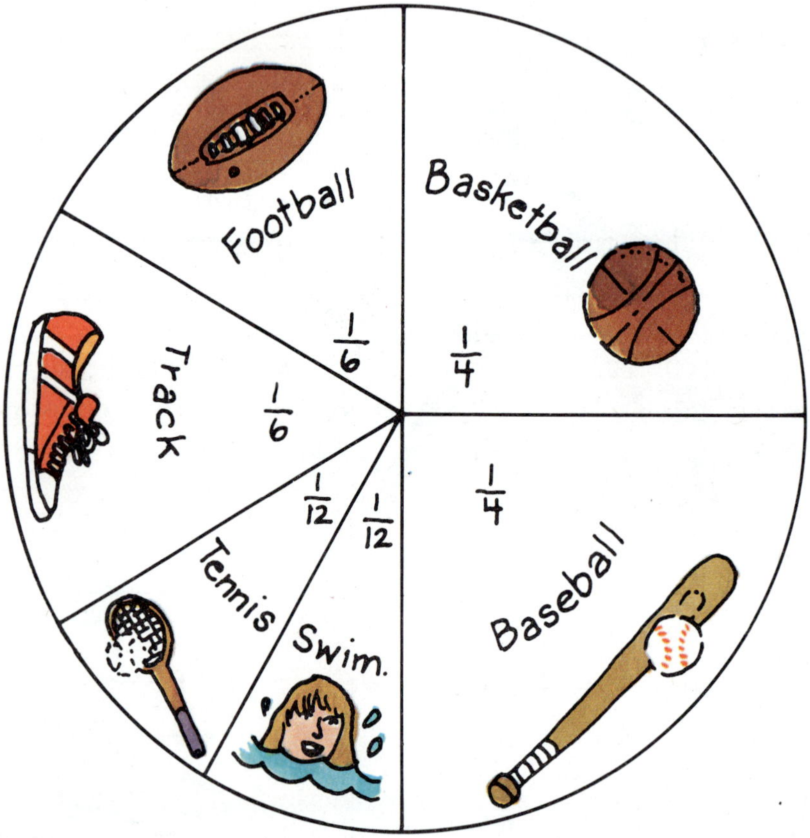


**EXERCISES**

**Solve.**

1. Did more students choose baseball or swimming?
2. Which two sports were chosen by half of the students?
3. What fraction liked either tennis or track?
4. How much greater was the fraction of students that chose baseball than the fraction that chose tennis?
5. What fraction of the students liked either football or baseball?
6. What fraction of the students did not choose baseball?

Yellowstone Park is the oldest national park in the United States. This map of the park helps visitors plan their travel.

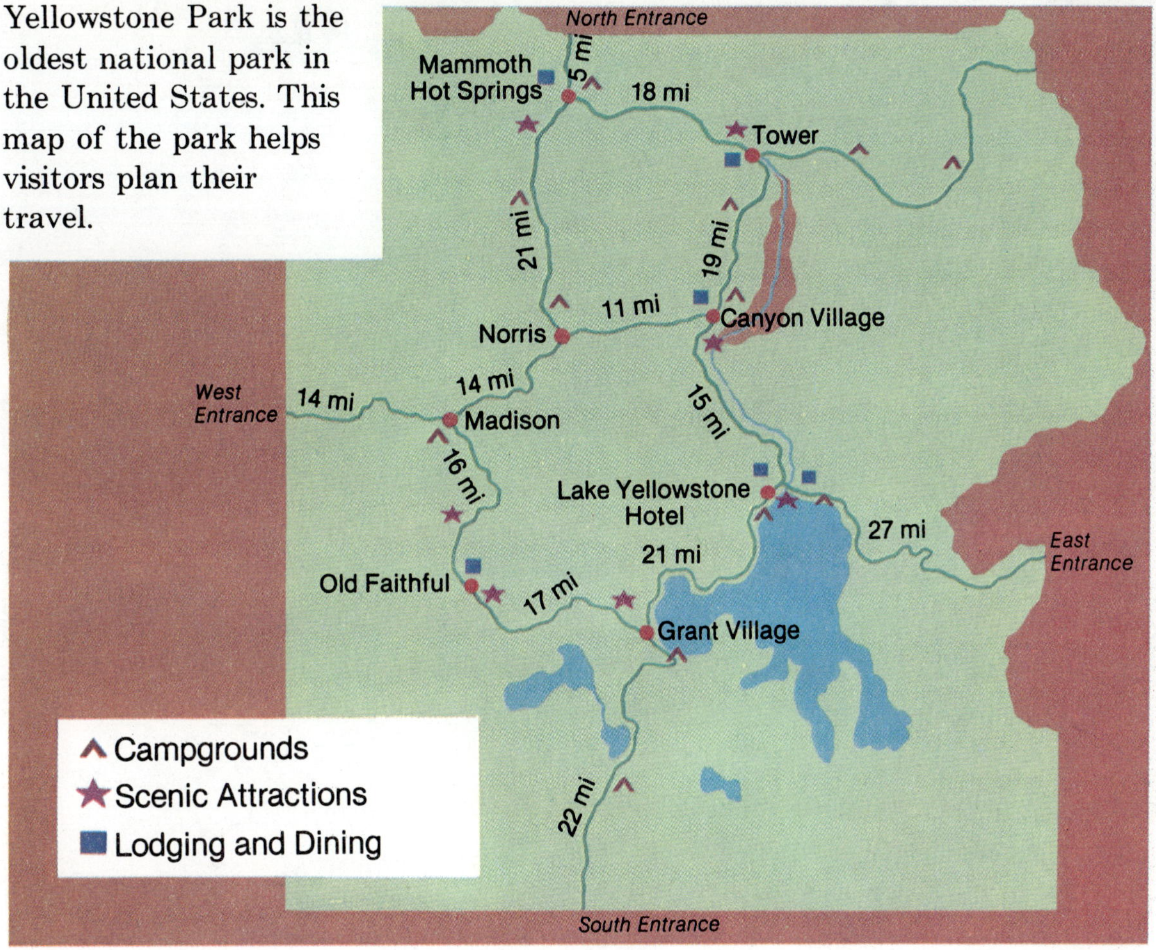


**Solve.**

7. How many campgrounds are shown on the map? How many scenic attractions? How many lodging and dining places?

8. If you enter the park at the South Entrance, how many miles will you drive to get to Old Faithful?

9. If you left Old Faithful and drove to Mammoth Hot Springs and on to Tower, how far would you drive?

10. Which is the shorter route from Old Faithful to Canyon Village?
    a. through Madison and Norris
    b. through Grant Village and Lake Yellowstone Hotel

11. If you traveled at 45 miles per hour, could you drive from Old Faithful to Canyon Village in less than 2 hours?

★ 12. Plan a two-day visit to the park. Where would you camp? What would you visit each day? How far would you travel each day?

# CHAPTER CHECKUP

**Write each fraction in lowest terms.** [pages 164–171]

1. $\frac{6}{8}$  2. $\frac{8}{10}$  3. $\frac{6}{9}$  4. $\frac{10}{16}$  5. $\frac{8}{12}$  6. $\frac{10}{15}$

$<$ **or** $>$**?** [pages 172–173]

7. $\frac{5}{8}$ ● $\frac{3}{4}$  8. $\frac{1}{2}$ ● $\frac{5}{6}$  9. $\frac{2}{3}$ ● $\frac{3}{5}$  10. $\frac{1}{4}$ ● $\frac{2}{3}$

**Add. Write each sum in lowest terms.** [pages 174–179, 188]

11. $\frac{1}{5} + \frac{3}{5}$  12. $\frac{3}{8} + \frac{1}{8}$  13. $\frac{1}{2} + \frac{3}{8}$  14. $\frac{1}{2} + \frac{2}{5}$  15. $\frac{2}{5} + \frac{1}{3}$  16. $\frac{2}{3} + \frac{2}{9}$

**Subtract. Write each difference in lowest terms.** [pages 180–185, 188]

17. $\frac{5}{9} - \frac{1}{9}$  18. $\frac{7}{8} - \frac{3}{8}$  19. $\frac{5}{8} - \frac{1}{4}$  20. $\frac{5}{6} - \frac{2}{3}$  21. $\frac{3}{5} - \frac{1}{3}$  22. $\frac{2}{3} - \frac{1}{4}$

**Change to a fraction.** [pages 190–191]

23. $1\frac{1}{2}$  24. $1\frac{2}{3}$  25. $2\frac{1}{2}$  26. $2\frac{3}{4}$  27. $3\frac{1}{8}$  28. $3\frac{5}{8}$

**Write in simplest form.** [pages 192–193]

29. $\frac{8}{2}$  30. $\frac{5}{2}$  31. $3\frac{8}{10}$  32. $\frac{12}{3}$  33. $\frac{7}{4}$  34. $\frac{14}{4}$

**Solve.** [pages 186–187, 189]

35. Sarah hiked $\frac{3}{10}$ of a mile the first hour and $\frac{1}{10}$ of a mile the second hour.

a. How far did she hike during the two hours?

b. How much farther did she hike the first hour?

# CHAPTER PROJECT

**You may wish to work on this experiment in small groups.**

1. Carry out the following experiment 50 times. For each toss, count the number of heads (H) and the number of tails (T). Record the results in a record like this:

| Experiment: Tossing 10 Pennies at Once | | | | | | | | | | | |
|---|---|---|---|---|---|---|---|---|---|---|---|
| Possible outcomes (H: heads; T: tails) | 10 H 0 T | 9 H 1 T | 8 H 2 T | 7 H 3 T | 6 H 4 T | 5 H 5 T | 4 H 6 T | 3 H 7 T | 2 H 8 T | 1 H 9 T | 0 H 10 T |
| Tally of outcomes | | | | | | | | | | | |
| Fraction of all outcomes | | | | | | | | | | | |

2. Show your results on a bar graph.

3. Which outcome occurred most often? least often?

4. If you tossed 10 coins, would you have a better chance of getting the outcome [10H 0T] or [5H 5T]?

# CHAPTER REVIEW

**Complete.**

1. 

(×3)
$\frac{2}{3} = ?$
(×3)

2. 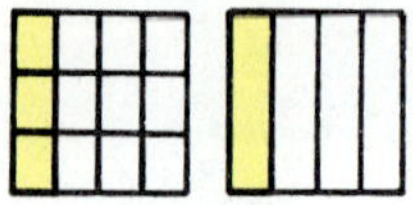

(÷3)
$\frac{3}{12} = ?$
(÷3)

3. 

(÷4)
$\frac{4}{8} = ?$
(÷4)

**Write each fraction in lowest terms.**

4. (÷2) $\frac{4}{6} = ?$ (÷2)
5. (÷3) $\frac{6}{9} = ?$ (÷3)
6. $\frac{4}{12} = ?$
7. $\frac{6}{8} = ?$
8. $\frac{5}{10} = ?$
9. $\frac{8}{12} = ?$

**< or >?**

10. $\frac{1}{4}$ ● $\frac{1}{2}$
11. $\frac{1}{2}$ ● $\frac{3}{8}$
12. $\frac{1}{2}$ ● $\frac{1}{3}$
13. $\frac{5}{8}$ ● $\frac{3}{4}$

**Add or subtract. Write each answer in lowest terms.**

14. $\frac{1}{2} = \frac{2}{4}$ $\quad + \frac{1}{4} = \frac{1}{4}$
15. $\frac{2}{3} + \frac{1}{6}$
16. $\frac{1}{2} + \frac{1}{3}$
17. $\frac{1}{2} - \frac{1}{3}$
18. $\frac{5}{6} - \frac{1}{2}$
19. $\frac{4}{5} - \frac{2}{3}$

**Change to a fraction.**

20. 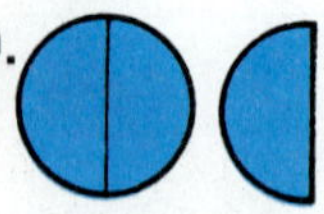

$1\frac{1}{2} = ?$

21. 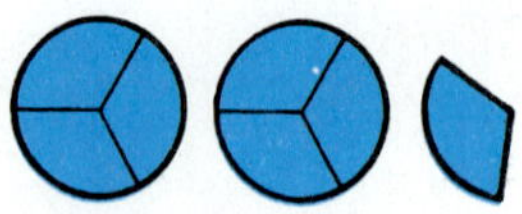

$2\frac{1}{3} = ?$

22. 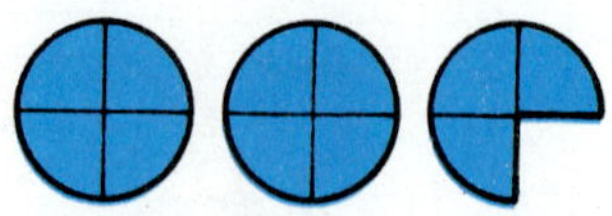

$2\frac{3}{4} = ?$

**Write in simplest form.**

23. 

$\frac{4}{2} = ?$

24. 

$\frac{7}{4} = ?$

25. 

$\frac{8}{3} = ?$

# CHAPTER CHALLENGE

1. Find the sum of the numbers in each row, column, and diagonal.

| $1$ | $\frac{1}{8}$ | $\frac{3}{4}$ |
|---|---|---|
| $\frac{3}{8}$ | $\frac{5}{8}$ | $\frac{7}{8}$ |
| $\frac{1}{2}$ | $\frac{9}{8}$ | $\frac{1}{4}$ |

2. If all the sums are the same, the square is a magic square. Is this one magic?

**Complete these magic squares.**

3.

|  | $\frac{1}{6}$ |  |
|---|---|---|
|  |  |  |
| $\frac{2}{3}$ | $\frac{3}{2}$ | $\frac{1}{3}$ |

4.

| $\frac{2}{3}$ |  |  |
|---|---|---|
|  | $\frac{5}{12}$ |  |
| $\frac{1}{3}$ |  | $\frac{1}{6}$ |

b c d 2. a b c d a b c d a b c d 5. a b c d 6. a b c

b c d 8. a b c d 9. a b c d 10. a b c d 11. a b c d 12. a b c

# MAJOR CHECKUP

## STANDARDIZED FORMAT

**Choose the correct letter.**

1. 783,463 rounded to the nearest hundred is
   - a. 783,000
   - b. 783,460
   - c. 783,500
   - d. none of these

2. Which number is least?
   - a. 6,983,425
   - b. 6,894,715
   - c. 6,099,098
   - d. 6,243,527

3. Add.
   $$\begin{array}{r} 3689 \\ +5747 \\ \hline \end{array}$$
   - a. 8326
   - b. 8436
   - c. 9436
   - d. none of these

4. Subtract.
   $$\begin{array}{r} 8026 \\ -594 \\ \hline \end{array}$$
   - a. 8572
   - b. 8532
   - c. 7532
   - d. none of these

5. Multiply.
   $$\begin{array}{r} 579 \\ \times 6 \\ \hline \end{array}$$
   - a. 3024
   - b. 3074
   - c. 3474
   - d. none of these

6. Divide.
   $39\overline{)4914}$
   - a. 126
   - b. 136
   - c. 125
   - d. none of these

7. The measure of a right angle is
   - a. 30°
   - b. 60°
   - c. 150°
   - d. none of these

8. These triangles are congruent.

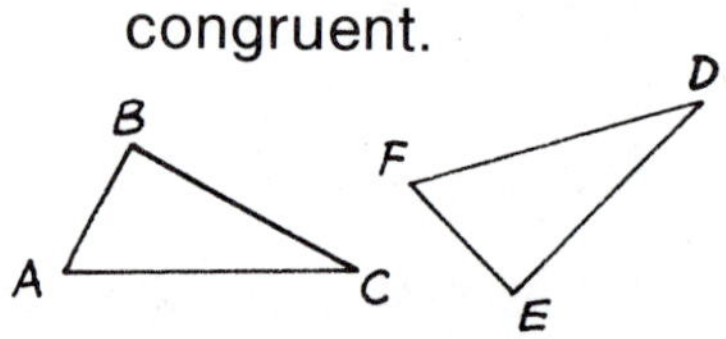


   $\angle A$ corresponds to
   - a. $\angle D$
   - b. $\angle E$
   - c. $\angle F$
   - d. none of these

9. $\frac{1}{2} + \frac{1}{3} = \underline{?}$
   - a. $\frac{2}{5}$
   - b. $\frac{2}{6}$
   - c. $\frac{5}{6}$
   - d. none of these

10. $\frac{5}{8} - \frac{1}{2} = \underline{?}$
    - a. $\frac{2}{3}$
    - b. $\frac{1}{8}$
    - c. $\frac{1}{16}$
    - d. none of these

11. Guy spent $1.75 for a grape drink and 2 hot dogs. The grape drink cost 45¢. How much was 1 hot dog?
    - a. $1.35
    - b. $1.30
    - c. $.60
    - d. none of these

12. A can of tennis balls costs $2.69. If you buy 3 cans, how much change will you get from a $10 bill?
    - a. $7.31
    - b. $1.93
    - c. $8.07
    - d. none of these

# Fractions — Multiplication and Division

# 7

## READY OR NOT?

**What fraction is colored?**

**1.** 

**2.** 

**3.** 

**Give the fraction of marbles that are**

**4.** 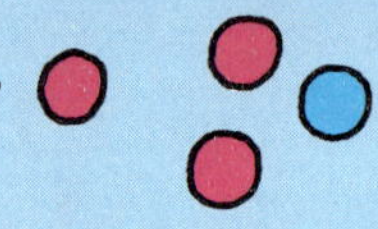

**a.** red
**b.** blue

**5.** 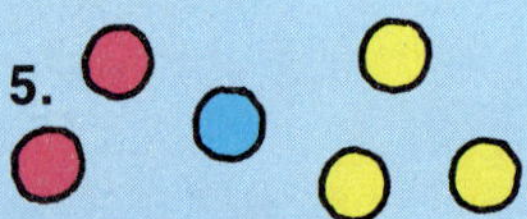

**a.** red
**b.** blue
**c.** yellow

**6.** 

**a.** red
**b.** green
**c.** blue
**d.** not blue
**e.** not green



# Fraction of a number

To find $\frac{1}{2}$ of a number, you can divide by 2.

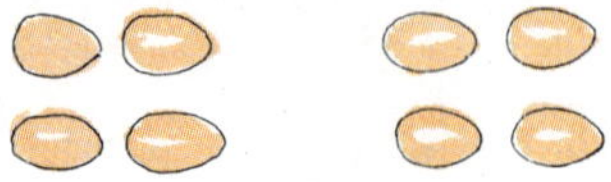

$\frac{1}{2}$ of 8 = 4

To find $\frac{1}{3}$ of a number, you can divide by 3.

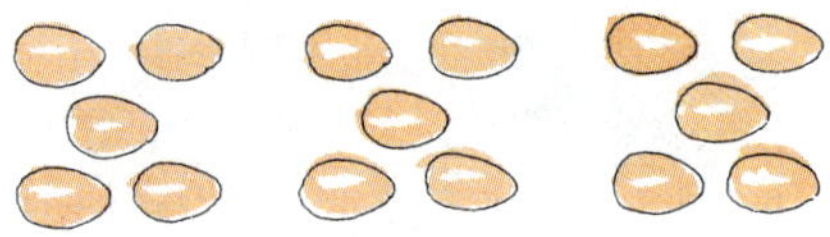

$\frac{1}{3}$ of 15 = 5

To find $\frac{1}{4}$ of a number, you can divide by 4.

$\frac{1}{4}$ of 12 = 3

What would you divide by to find $\frac{1}{5}$ of a number? $\frac{1}{8}$ of a number?

## EXERCISES

**Complete.**

1. $\frac{1}{2}$ of 12 = ?
2. $\frac{1}{3}$ of 12 = ?
3. $\frac{1}{4}$ of 8 = ?
4. $\frac{1}{5}$ of 10 = ?
5. $\frac{1}{8}$ of 8 = ?
6. $\frac{1}{6}$ of 12 = ?
7. $\frac{1}{3}$ of 18 = ?
8. $\frac{1}{2}$ of 18 = ?
9. $\frac{1}{3}$ of 30 = ?
10. $\frac{1}{8}$ of 24 = ?
11. $\frac{1}{4}$ of 16 = ?
12. $\frac{1}{3}$ of 15 = ?
13. $\frac{1}{2}$ of 20 = ?
14. $\frac{1}{5}$ of 15 = ?
15. $\frac{1}{4}$ of 36 = ?
16. $\frac{1}{6}$ of 18 = ?
17. $\frac{1}{6}$ of 36 = ?
18. $\frac{1}{4}$ of 32 = ?

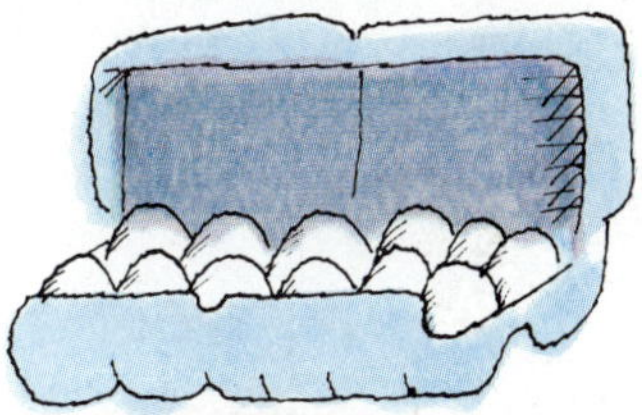

1 dozen = 12

**Complete.**

19. $\frac{1}{2}$ of a dozen = ?
20. $\frac{1}{3}$ of a dozen = ?
21. $\frac{1}{4}$ of a dozen = ?
22. $\frac{1}{6}$ of a dozen = ?

1 hour = 60 minutes

**How many minutes in**

23. $\frac{1}{2}$ of an hour?
24. $\frac{1}{3}$ of an hour?
25. $\frac{1}{4}$ of an hour?
26. $\frac{1}{5}$ of an hour?
27. $\frac{1}{6}$ of an hour?
28. $\frac{1}{10}$ of an hour?

**Solve.**

29. Jerry had 48¢. He spent $\frac{1}{3}$ of it for a candy bar. How much did he spend? How much did he have left?

30. Donna has 436 stamps in her collection. $\frac{1}{4}$ of them are Canadian stamps. How many Canadian stamps does she have? How many non-Canadian stamps?

# More about a fraction of a number

Here is how to find

$$\frac{2}{3} \text{ of } 12.$$

First, divide by 3 to find $\frac{1}{3}$ of 12.

$$\frac{1}{3} \text{ of } 12 = 4$$

Then, multiply by 2 to find $\frac{2}{3}$ of 12.

$$\frac{2}{3} \text{ of } 12 = 8$$

To find a fraction of a number:

**Step 1.** Divide by the denominator.

**Step 2.** Multiply that answer by the numerator.

**EXERCISES**

**Complete.**

**1.** $\frac{1}{6}$ of 24 = ?  **2.** $\frac{2}{6}$ of 24 = ?

**3.** $\frac{3}{6}$ of 24 = ?  **4.** $\frac{4}{6}$ of 24 = ?

**5.** $\frac{5}{6}$ of 24 = ?  **6.** $\frac{6}{6}$ of 24 = ?

**Complete.**

**7.** $\frac{2}{3}$ of 18 = ?

**8.** $\frac{5}{6}$ of 12 = ?

**9.** $\frac{3}{4}$ of 16 = ?

**10.** $\frac{5}{8}$ of 24 = ?

**11.** $\frac{1}{2}$ of 20 = ?

**12.** $\frac{3}{2}$ of 12 = ?

**13.** $\frac{3}{8}$ of 16 = ?

**14.** $\frac{4}{4}$ of 20 = ?

**15.** $\frac{4}{3}$ of 24 = ?

**16.** $\frac{6}{5}$ of 25 = ?

**17.** $\frac{3}{4}$ of 12 = ?

**18.** $\frac{5}{6}$ of 18 = ?

**19.** $\frac{1}{4}$ of 32 = ?

**20.** $\frac{5}{2}$ of 14 = ?

**21.** $\frac{2}{3}$ of 9 = ?

**22.** $\frac{5}{3}$ of 21 = ?

**23.** $\frac{1}{2}$ of 30 = ?

**24.** $\frac{7}{8}$ of 24 = ?

**25.** $\frac{4}{3}$ of 15 = ?

**26.** $\frac{3}{3}$ of 15 = ?

**27.** $\frac{3}{8}$ of 40 = ?

**Complete.**

**28.** $\frac{2}{3}$ of a dozen = ?

**29.** $\frac{3}{4}$ of a dozen = ?

**30.** $\frac{4}{3}$ of a dozen = ?

**How many minutes?**

**31.** $\frac{3}{4}$ of an hour

**32.** $\frac{5}{6}$ of an hour

**33.** $\frac{1}{10}$ of an hour

**How many inches?**

**34.** $\frac{2}{3}$ of a yard

**35.** $\frac{3}{4}$ of a yard

**36.** $\frac{5}{2}$ of a yard

**37.** $\frac{4}{9}$ of a yard

1 yard = 36 inches

**How many ounces?**

**38.** $\frac{1}{2}$ of a pound

**39.** $\frac{3}{4}$ of a pound

**40.** $\frac{3}{8}$ of a pound

**41.** $\frac{5}{8}$ of a pound

1 pound = 16 ounces

**How many inches?**

**42.** $\frac{1}{2}$ of a foot

**43.** $\frac{2}{3}$ of a foot

**44.** $\frac{3}{4}$ of a foot

**45.** $\frac{5}{6}$ of a foot

1 foot = 12 inches

# Using mixed numbers

The Math Club needed $2\frac{1}{2}$ dozen doughnuts for their meeting.

To find the cost of $2\frac{1}{2}$ dozen doughnuts:

(1) Find the cost of 2 dozen. $2 \times \$1.20$

(2) Find the cost of $\frac{1}{2}$ dozen. $\frac{1}{2}$ of $1.20

(3) Add.

$$\$2.40 + \$.60 = \$3.00.$$

The total cost is $3.00.

**EXERCISES**

**What is the cost?**

1. a. 2 dozen

   b. $\frac{1}{4}$ dozen

   c. $2\frac{1}{4}$ dozen

2. a. 1 dozen

   b. $\frac{3}{4}$ dozen

   c. $1\frac{3}{4}$ dozen

**How many minutes?**

3. a. 2 hours

   b. $\frac{1}{2}$ hour

   c. $2\frac{1}{2}$ hours

4. a. 1 hour

   b. $\frac{2}{3}$ hour

   c. $1\frac{2}{3}$ hours

5. a. 3 hours

   b. $\frac{3}{4}$ hour

   c. $3\frac{3}{4}$ hours

**How many ounces?**

6. a. 2 pounds

   b. $\frac{1}{4}$ pound

   c. $2\frac{1}{4}$ pounds

7. a. 1 pound

   b. $\frac{3}{8}$ pound

   c. $1\frac{3}{8}$ pounds

8. a. 3 pounds

   b. $\frac{5}{8}$ pound

   c. $3\frac{5}{8}$ pounds

**How many inches?**

9. a. 1 foot
   b. $\frac{1}{2}$ foot
   c. $1\frac{1}{2}$ feet

10. a. 2 feet
   b. $\frac{3}{4}$ foot
   c. $2\frac{3}{4}$ feet

11. a. 3 feet
   b. $\frac{2}{3}$ foot
   c. $3\frac{2}{3}$ feet

**How many inches?**

12. a. 1 yard
   b. $\frac{1}{2}$ yard
   c. $1\frac{1}{2}$ yards

13. a. 2 yards
   b. $\frac{5}{6}$ yard
   c. $2\frac{5}{6}$ yards

14. a. 4 yards
   b. $\frac{2}{3}$ yard
   c. $4\frac{2}{3}$ yards

**How many feet?**

15. a. 1 yard
   b. $\frac{2}{3}$ yard
   c. $1\frac{2}{3}$ yards

16. a. 2 yards
   b. $\frac{1}{3}$ yard
   c. $2\frac{1}{3}$ yards

17. a. 3 yards
   b. $\frac{2}{3}$ yard
   c. $3\frac{2}{3}$ yards

**Solve.**

18. There are 2 pints in a quart. How many pints are in $7\frac{1}{2}$ quarts?

19. There are 4 quarts in a gallon. How many quarts are in $3\frac{3}{4}$ gallons?

20. Linda earns $1.20 an hour for babysitting. How much does she earn for 2 hours and 15 minutes? *Hint:* 15 minutes = $\frac{1}{4}$ hour.

21. David rode his bicycle 8 miles in an hour. At that rate, how far could he ride his bicycle in $3\frac{3}{4}$ hours?

## KEEPING SKILLS SHARP

**Give each sum or difference in simplest form.**

1. $\frac{3}{8} + \frac{1}{8}$

2. $\frac{4}{5} - \frac{3}{10}$

3. $\frac{7}{8} - \frac{3}{4}$

4. $\frac{2}{3} + \frac{1}{5}$

5. $\frac{5}{6} - \frac{1}{4}$

6. $\frac{1}{6} + \frac{1}{2}$

# Multiplying fractions

You can find the product of 2 and 3 this way:

(1) Pick a unit square. 

(2) Color a rectangle that is as long as 3 squares and as wide as 2 squares.

(3) Count the total number of unit squares.

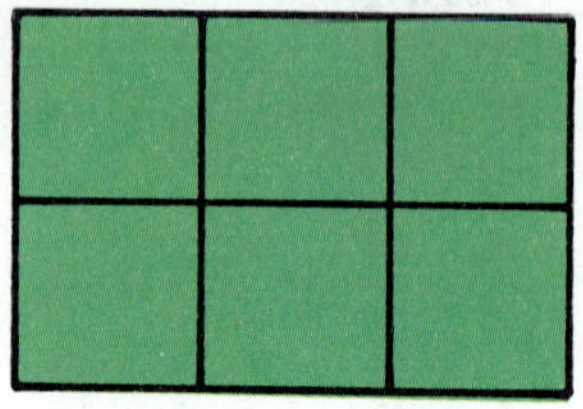

$$\mathbf{2 \times 3 = 6}$$

You can also use a unit square to find the product of $\frac{1}{3}$ and $\frac{1}{2}$.

(1) Pick a unit square. 

(2) Color a rectangle that is $\frac{1}{2}$ as long and $\frac{1}{3}$ as wide as the unit square.

(3) The colored rectangle is $\frac{1}{6}$ of the unit square.

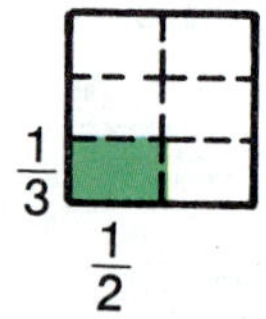


$$\frac{1}{3} \times \frac{1}{2} = \frac{1}{6}$$

Here are other examples.

Each small rectangle is $\frac{1}{15}$ of the unit square.

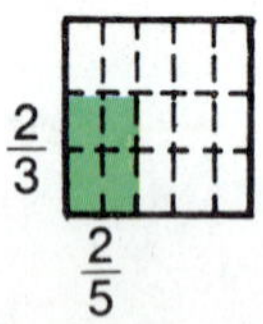


$$\frac{2}{3} \times \frac{2}{5} = \frac{4}{15}$$

Each small rectangle is $\frac{1}{3}$ of the unit square.

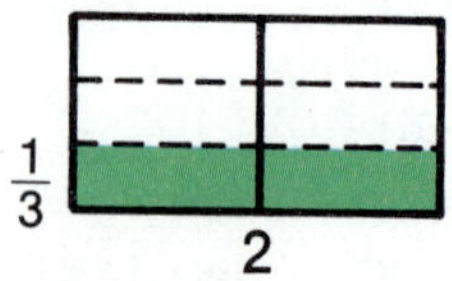


$$\frac{1}{3} \times 2 = \frac{2}{3}$$

**EXERCISES**

**Study the picture.**
**Then give the product.**

**1.** 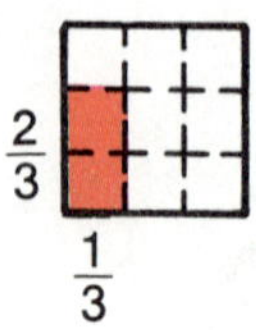

$\frac{2}{3} \times \frac{1}{3} = \underline{?}$

**2.** 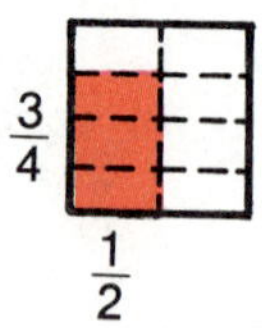

$\frac{3}{4} \times \frac{1}{2} = \underline{?}$

3. 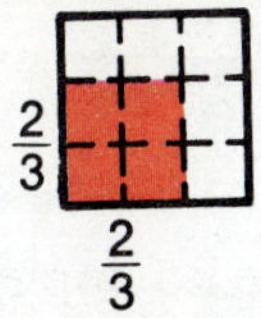


$\frac{2}{3} \times \frac{2}{3} = \underline{?}$

4. 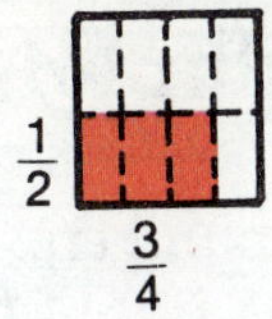


$\frac{1}{2} \times \frac{3}{4} = \underline{?}$

5. 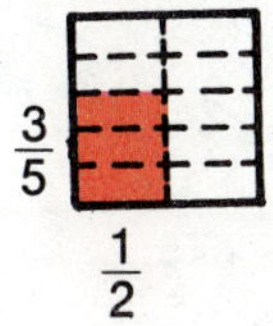


$\frac{3}{5} \times \frac{1}{2} = \underline{?}$

6. 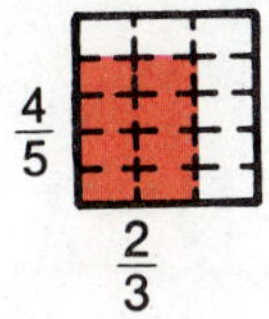


$\frac{4}{5} \times \frac{2}{3} = \underline{?}$

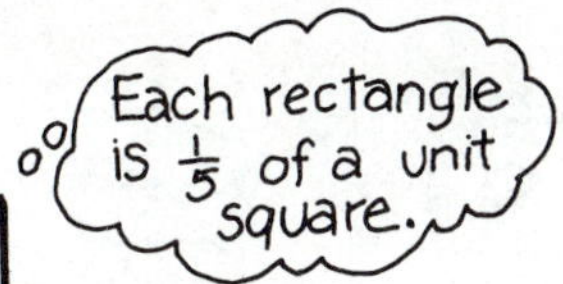


7. 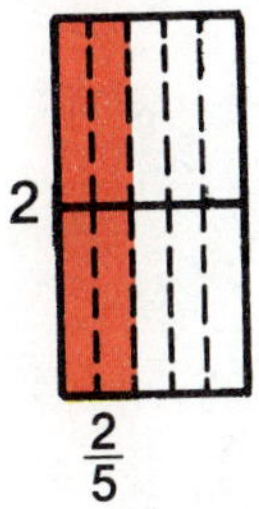


$2 \times \frac{2}{5} = \underline{?}$

8. 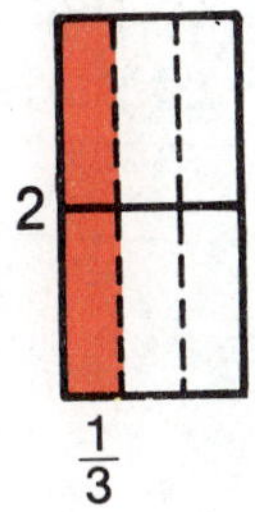


$2 \times \frac{1}{3} = \underline{?}$

9. 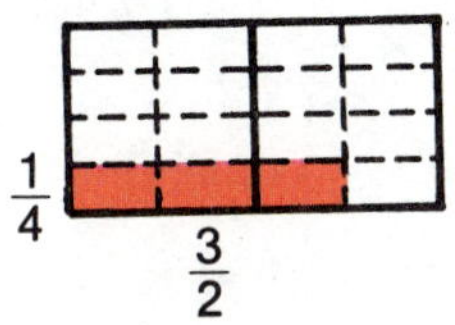


$\frac{1}{4} \times \frac{3}{2} = \underline{?}$

10. 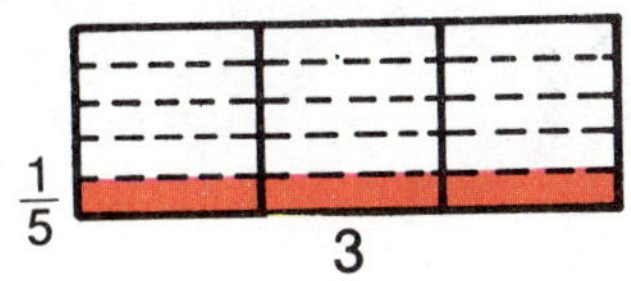


$\frac{1}{5} \times 3 = \underline{?}$

11. 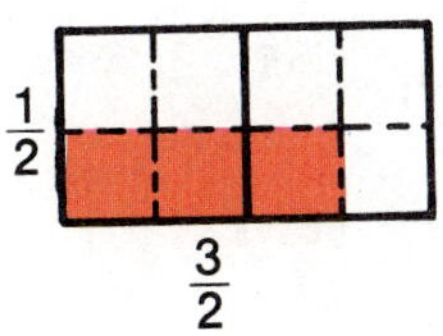


$\frac{1}{2} \times \frac{3}{2} = \underline{?}$

12. 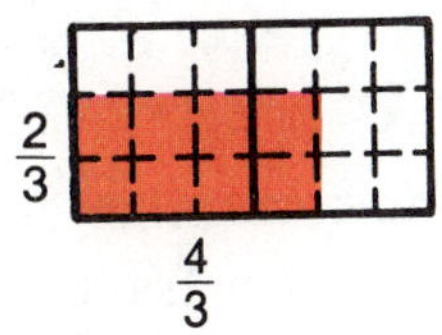


$\frac{2}{3} \times \frac{4}{3} = \underline{?}$

13. 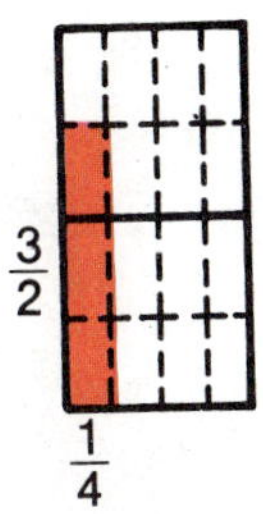


$\frac{3}{2} \times \frac{1}{4} = \underline{?}$

14. 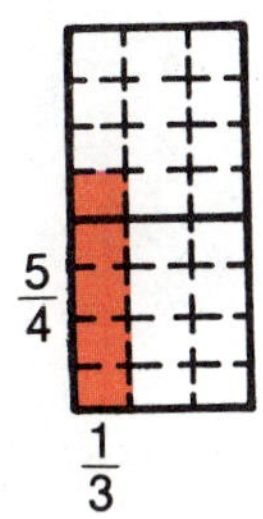


$\frac{5}{4} \times \frac{1}{3} = \underline{?}$

# More about multiplying fractions

To multiply fractional numbers:

**Step 1.** Multiply the numerators to get the numerator of the product.

**Step 2.** Multiply the denominators to get the denominator of the product.

**EXAMPLE 1.**

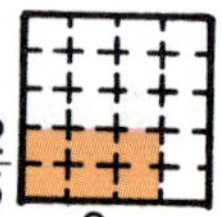


$\frac{2}{5} \times \frac{3}{4} = \frac{6}{20}$ ← parts colored / ← parts in unit square

$= \frac{3}{10}$

**EXAMPLE 2.**

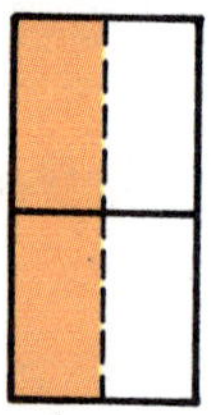


$2 \times \frac{1}{2} = \frac{2}{2}$

$= 1$

**EXAMPLE 3.**

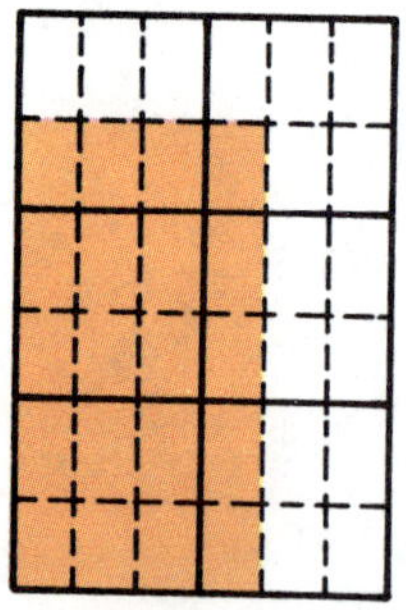


$\frac{5}{2} \times \frac{4}{3} = \frac{20}{6}$

$= \frac{10}{3}$

$= 3\frac{1}{3}$

Taking a fraction of a number is the same as multiplying by a fraction.

**EXAMPLE 4.**

$\frac{2}{3}$ of $15 = \frac{2}{3} \times 15$

$= \frac{30}{3}$

$= 10$

**EXAMPLE 5.**

$\frac{1}{2}$ of $\frac{1}{3} = \frac{1}{2} \times \frac{1}{3}$

$= \frac{1}{6}$

## EXERCISES

**Give each product in simplest form.**

1. $\frac{1}{2} \times \frac{2}{3}$
2. $\frac{2}{5} \times \frac{3}{2}$
3. $4 \times \frac{3}{4}$
4. $\frac{2}{3} \times 2$
5. $\frac{1}{2} \times \frac{1}{3}$
6. $\frac{3}{2} \times \frac{3}{4}$
7. $\frac{4}{3} \times \frac{2}{5}$
8. $\frac{1}{4} \times \frac{5}{8}$
9. $\frac{2}{3} \times \frac{3}{4}$
10. $\frac{3}{8} \times \frac{2}{3}$
11. $4 \times \frac{2}{3}$
12. $\frac{5}{6} \times \frac{1}{3}$
13. $3 \times \frac{3}{4}$
14. $\frac{3}{5} \times \frac{2}{5}$
15. $\frac{4}{5} \times 2$
16. $\frac{3}{2} \times \frac{1}{4}$
17. $\frac{3}{2} \times \frac{4}{3}$
18. $\frac{2}{5} \times \frac{5}{6}$
19. $3 \times \frac{6}{5}$
20. $\frac{3}{4} \times 1$
21. $\frac{3}{5} \times \frac{4}{3}$
22. $\frac{1}{10} \times 4$
23. $\frac{5}{8} \times \frac{4}{5}$
24. $\frac{5}{2} \times \frac{3}{2}$
25. $\frac{7}{10} \times \frac{5}{4}$
26. $\frac{3}{4} \times \frac{3}{4}$
27. $\frac{4}{3} \times \frac{4}{3}$
28. $\frac{6}{5} \times \frac{4}{3}$
29. $2 \times \frac{1}{2}$
30. $3 \times \frac{1}{3}$
31. $\frac{2}{3} \times \frac{3}{2}$
32. $\frac{3}{4} \times \frac{4}{3}$
33. $\frac{1}{4} \times 4$
34. $\frac{1}{6} \times 6$
35. $\frac{8}{5} \times \frac{5}{8}$
36. $\frac{3}{10} \times \frac{10}{3}$

**Solve.**

37. Larry baked 48 cookies. He ate $\frac{1}{6}$ of them. How many cookies did he eat?

38. Lucille lives $\frac{5}{12}$ of a mile from school. She walks $\frac{1}{10}$ of this distance to get to the bus stop. How far is it to the bus stop?

39. $\frac{1}{2}$ of the garden contains berries. $\frac{1}{3}$ of the berries are strawberries. How much of the garden contains strawberries?

40. $\frac{5}{8}$ of the class are girls. $\frac{1}{3}$ of the girls have blond hair. What fraction of the class are blond girls?

## *Challenge!*

**Copy and complete these multiplication boxes.**

41.

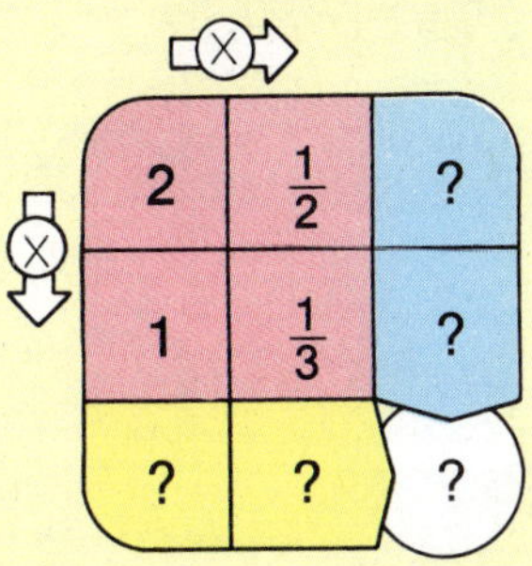


42.

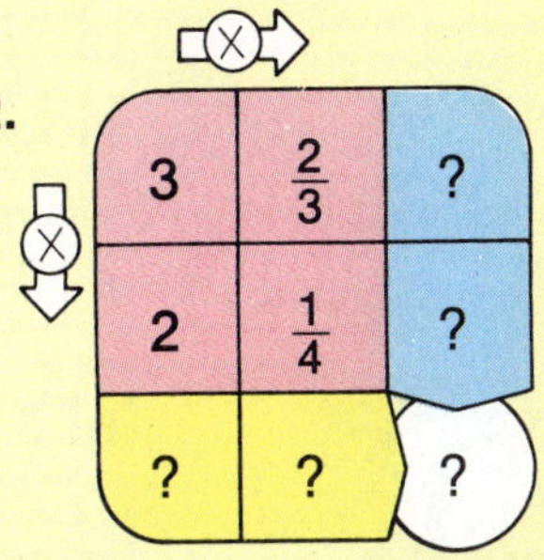


43.

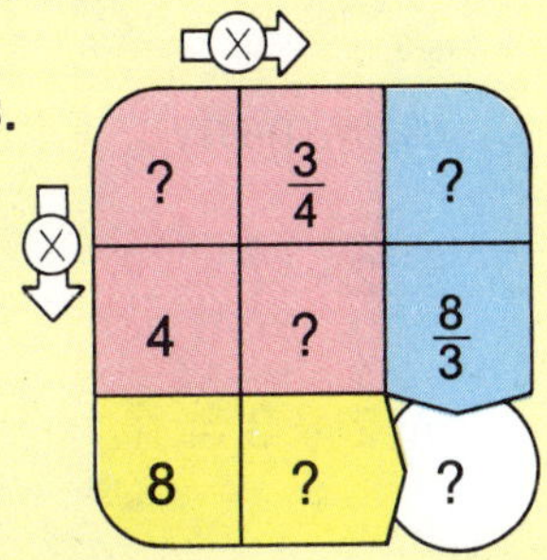

# Dividing fractions

How many $\frac{1}{2}$'s in 3?

$$3 \div \frac{1}{2} = 6$$

$$3 \times 2 = 6$$

**EXERCISES**

**First complete.**
**Then compare your answers in each exercise.**

1. 

   How many $\frac{1}{2}$'s in 2?

   a. $2 \div \frac{1}{2} = \underline{?}$

   b. $2 \times 2 = \underline{?}$

2. 

   How many $\frac{1}{4}$'s in 2?

   a. $2 \div \frac{1}{4} = \underline{?}$

   b. $2 \times 4 = \underline{?}$

3. 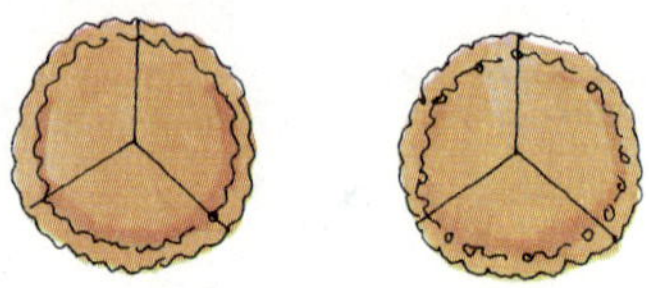

   How many $\frac{1}{3}$'s in 2?

   a. $2 \div \frac{1}{3} = \underline{?}$

   b. $2 \times 3 = \underline{?}$

4. 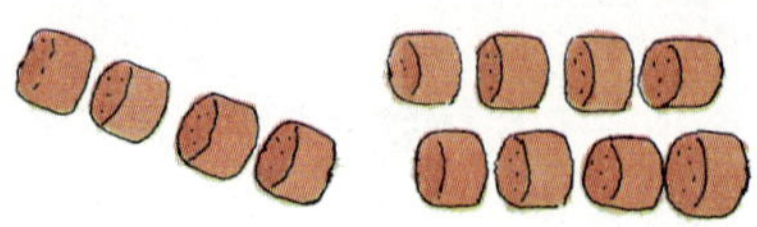

   How many $\frac{1}{4}$'s in 3?

   a. $3 \div \frac{1}{4} = \underline{?}$

   b. $3 \times 4 = \underline{?}$

5. How many $\frac{2}{3}$'s in 2?

a. $2 \div \frac{2}{3} = \underline{?}$

b. $2 \times \frac{3}{2} = \underline{?}$

6. How many $\frac{3}{4}$'s in 3?

a. $3 \div \frac{3}{4} = \underline{?}$

b. $3 \times \frac{4}{3} = \underline{?}$

7. How many $\frac{1}{4}$'s in $\frac{3}{4}$?

a. $\frac{3}{4} \div \frac{1}{4} = \underline{?}$

b. $\frac{3}{4} \times 4 = \underline{?}$

8. How many $\frac{2}{5}$'s in $\frac{4}{5}$?

a. $\frac{4}{5} \div \frac{2}{5} = \underline{?}$

b. $\frac{4}{5} \times \frac{5}{2} = \underline{?}$

## *Challenge!*

**Guess my number.**

9. Dividing by $\frac{1}{4}$ is the same as multiplying by my number.

10. Dividing by $\frac{3}{4}$ is the same as multiplying by my number.

## KEEPING SKILLS SHARP

**Change to a fraction.**

1. $2\frac{1}{2}$
2. $1\frac{1}{2}$
3. $3\frac{1}{5}$
4. $2\frac{1}{4}$
5. $1\frac{5}{8}$
6. $1\frac{3}{4}$
7. $1\frac{1}{3}$
8. $2\frac{3}{4}$
9. $1\frac{2}{3}$
10. $2\frac{7}{8}$
11. $1\frac{5}{6}$
12. $2\frac{3}{8}$

**Change to a mixed number.**

13. $\frac{3}{2}$
14. $\frac{4}{3}$
15. $\frac{7}{4}$
16. $\frac{5}{2}$
17. $\frac{9}{4}$
18. $\frac{7}{3}$
19. $\frac{11}{4}$
20. $\frac{7}{2}$
21. $\frac{5}{4}$
22. $\frac{5}{3}$
23. $\frac{11}{8}$
24. $\frac{11}{6}$

# Practice

**Give each answer in simplest form.**

| | | |
|---|---|---|
| **1.** $\frac{3}{5} + \frac{1}{5}$ | **2.** $\frac{3}{5} - \frac{1}{5}$ | **3.** $\frac{3}{5} \times \frac{1}{5}$ |
| **4.** $\frac{3}{4} + \frac{1}{4}$ | **5.** $\frac{3}{4} - \frac{1}{4}$ | **6.** $\frac{3}{4} \times \frac{1}{4}$ |
| **7.** $\frac{3}{4} + \frac{1}{2}$ | **8.** $\frac{3}{4} - \frac{1}{2}$ | **9.** $\frac{3}{4} \times \frac{1}{2}$ |
| **10.** $\frac{5}{6} + \frac{1}{3}$ | **11.** $\frac{5}{6} - \frac{1}{3}$ | **12.** $\frac{5}{6} \times \frac{1}{3}$ |
| **13.** $\frac{2}{3} + \frac{1}{2}$ | **14.** $\frac{2}{3} - \frac{1}{2}$ | **15.** $\frac{2}{3} \times \frac{1}{2}$ |
| **16.** $\frac{4}{5} + \frac{1}{2}$ | **17.** $\frac{4}{5} - \frac{1}{2}$ | **18.** $\frac{4}{5} \times \frac{1}{2}$ |
| **19.** $\frac{5}{6} + \frac{4}{5}$ | **20.** $\frac{5}{6} - \frac{4}{5}$ | **21.** $\frac{5}{6} \times \frac{4}{5}$ |
| **22.** $\frac{3}{4} + \frac{1}{6}$ | **23.** $\frac{3}{4} - \frac{1}{6}$ | **24.** $\frac{3}{4} \times \frac{1}{6}$ |
| ★ **25.** $\frac{5}{12} + \frac{3}{10}$ | ★ **26.** $\frac{5}{12} - \frac{3}{10}$ | ★ **27.** $\frac{5}{12} \times \frac{3}{10}$ |

**Give each end number.**

**28.**

**29.**

**30.**

# Problem solving—reading a pictograph

This pictograph shows how many pets some students had.

**Solve.**

1. How many students had 1 pet?

2. How many students had more than 2 pets?

3. How many students had less than 2 pets?

4. How many students were represented in the pictograph?

5. Barbara's Shetland sheepdog cost $\frac{2}{3}$ as much as her poodle, which cost \$180. How much did her Shetland sheepdog cost?

6. Jeffrey has a 120-pound dog. He was told to feed it 1 cup of food a day for each 20 pounds the dog weighs. How many cups of food should he feed the dog in a week?

7. George has three 5-gallon aquariums. He put $\frac{3}{4}$ of a pound of sand in each aquarium. How many pounds of sand did he use?

8. Anne has some plain guppies which cost 50¢ each. She also has some fancy guppies, which cost $2\frac{1}{2}$ times as much. How much did each fancy guppy cost?

## Problem solving

**GINGER SNAPS**
MAKES 10 DOZEN

- $\frac{3}{4}$ cup butter
- 2 cups sugar
- 2 eggs
- $\frac{1}{2}$ cup molasses
- 2 teaspoons vinegar
- 4 cups flour
- 2 teaspoons soda
- 2 teaspoons ginger
- $\frac{1}{2}$ teaspoon cinnamon
- $\frac{1}{4}$ teaspoon cloves

**Solve.**

1. a. How many cookies does a full recipe make?
   b. How many cookies are there in $\frac{1}{2}$ of a full recipe?
2. Find how much of each ingredient would be needed for $\frac{1}{2}$ of the recipe.

| | |
|---|---|
| a. butter | b. sugar |
| c. eggs | d. molasses |
| e. vinegar | f. flour |
| g. soda | h. ginger |
| i. cinnamon | j. cloves |

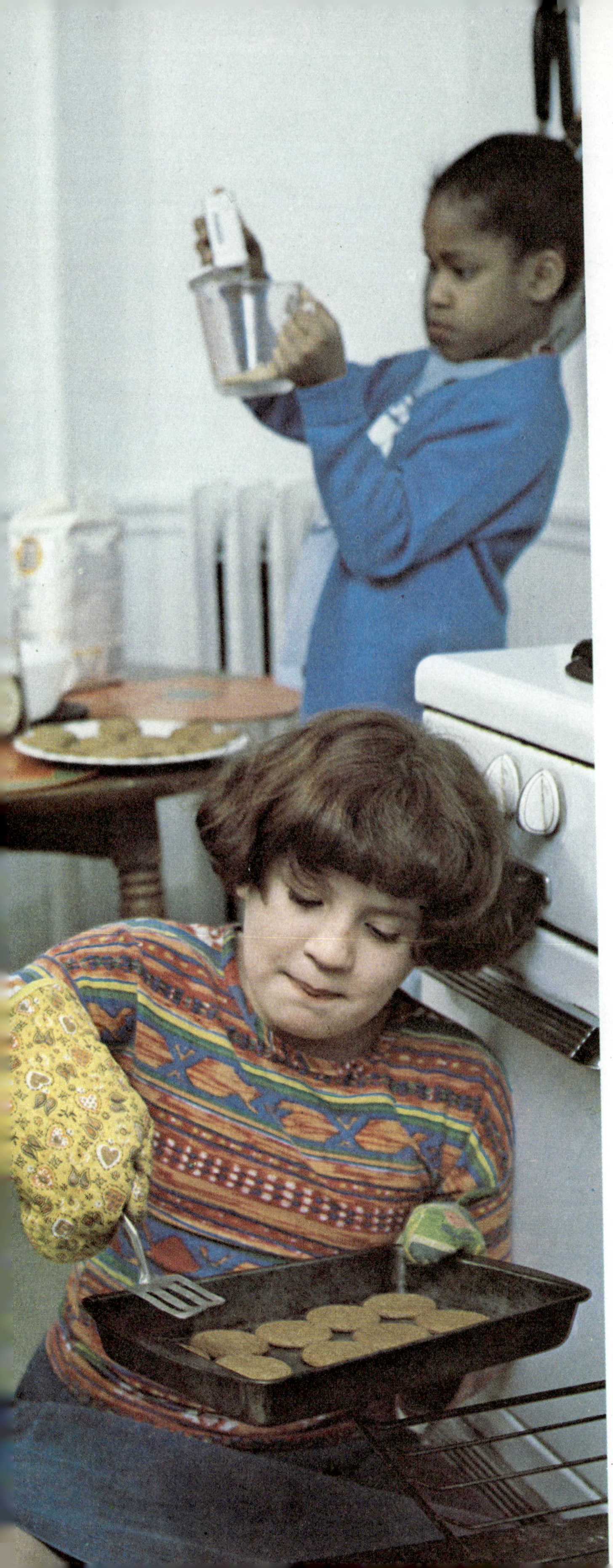

3. a. What fraction of the full recipe is 30 cookies?
   b. How much sugar would be needed for 30 cookies?
   c. How much cinnamon?

4. a. What fraction of a full recipe is 40 cookies?
   b. How much butter would be needed for 40 cookies?
   c. How much flour would be needed for 40 cookies?
   d. How much ginger would be needed for 40 cookies?

5. a. What fraction of a full recipe is 80 cookies?
   b. How much sugar would be needed for 80 cookies?
   c. How much molasses would be needed for 80 cookies?
   d. How much soda would be needed for 80 cookies?

6. Suppose that you want to bake a full recipe of cookies and that you have $\frac{1}{4}$ cup of butter. How much more butter will you need?

7. How many teaspoons of spices (ginger, cinnamon, and cloves) are there in the full recipe?

8. Suppose you used 3 cups of flour to bake some cookies.
   a. What fraction of the full recipe did you bake?
   b. How many cookies did you bake?
   c. How much sugar did you use?
   d. How much ginger did you use?

# CHAPTER CHECKUP

**Complete.** [pages 206–209]

1. $\frac{1}{2}$ of 12 = ?
2. $\frac{1}{4}$ of 20 = ?
3. $\frac{1}{6}$ of 18 = ?
4. $\frac{2}{3}$ of 24 = ?
5. $\frac{3}{4}$ of 36 = ?
6. $\frac{4}{3}$ of 48 = ?

**Give each product in simplest form.** [pages 212–215, 218]

7. $\frac{1}{4} \times \frac{1}{3}$
8. $\frac{2}{3} \times \frac{1}{5}$
9. $\frac{3}{5} \times \frac{1}{2}$
10. $4 \times \frac{1}{2}$
11. $3 \times \frac{3}{4}$
12. $\frac{5}{6} \times 2$
13. $\frac{3}{4} \times \frac{2}{3}$
14. $\frac{2}{3} \times \frac{3}{2}$
15. $\frac{5}{4} \times \frac{8}{5}$

**Solve.** [pages 210–211, 219–221]

16. a. How much will 2 dozen cost?
    b. How much will $\frac{3}{4}$ dozen cost?
    c. How much will $2\frac{3}{4}$ dozen cost?

17. Fred bought $\frac{3}{4}$ of a pound of mints. He ate $\frac{1}{2}$ of them. What fraction of a pound of mints did he eat?

18. Connie made 20 candles. She used $\frac{1}{8}$ of a pound of paraffin for each candle. How much paraffin did she use?

# CHAPTER PROJECT

1. The list shows how Teresa spent her time during a 24-hour school day. Make a list of how you spent your time during a 24-hour school day.

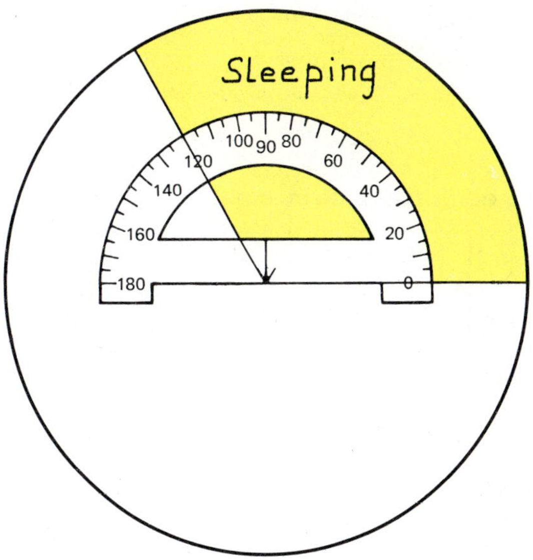


2. Make a circle graph of how you spent your time. *Hint:* Draw a large circle and a radius. If you slept 8 hours, you will need to shade in $\frac{8}{24}$, or $\frac{1}{3}$, of the circle for sleeping. (There are 360° in a circle.)

$$\frac{1}{3} \text{ of } 360° = 120°$$

Measure off a 120° angle and label that part *Sleeping*.

# CHAPTER REVIEW

**Complete.**

1. 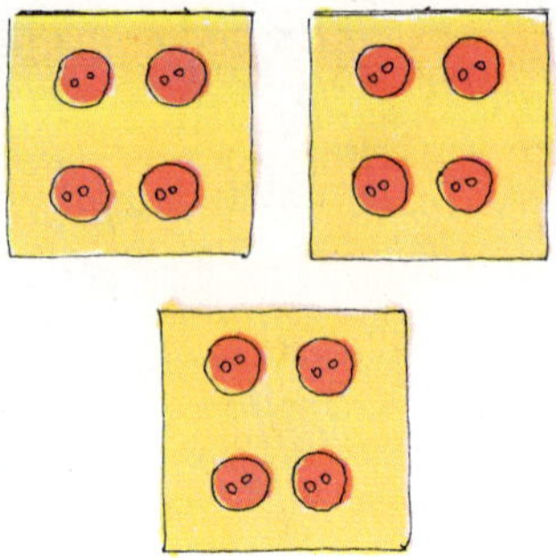

$\frac{1}{3}$ of 12 = ?

2. 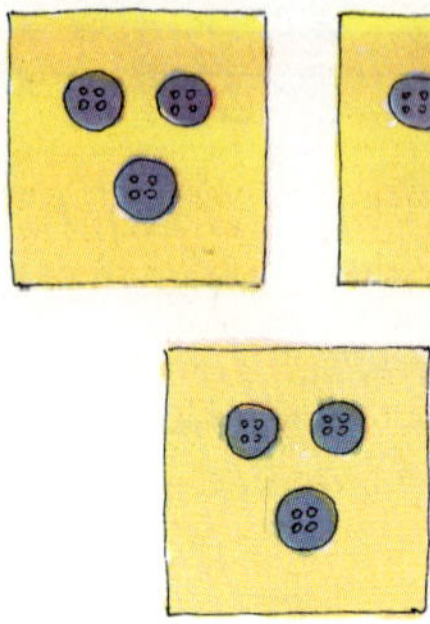

$\frac{2}{3}$ of 9 = ?

3. 

$\frac{3}{4}$ of 20 = ?

4. $\frac{1}{6}$ of 18 = ?

5. $\frac{2}{5}$ of 20 = ?

6. $\frac{4}{3}$ of 24 = ?

**Solve.**

7. 

How many minutes?

a. 2 hours

b. $\frac{1}{4}$ hour

c. $2\frac{1}{4}$ hours

8. 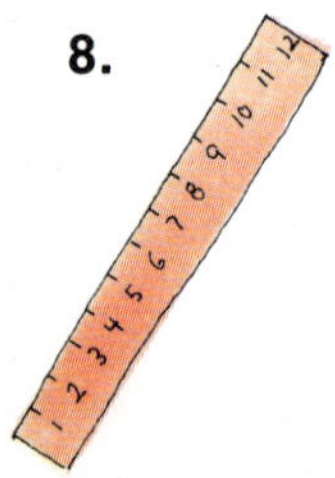

How many inches?

a. 2 feet

b. $\frac{2}{3}$ foot

c. $2\frac{2}{3}$ feet

**Give each product in simplest form.**

9. 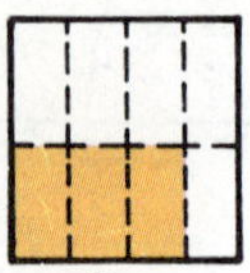

$\frac{1}{2}$ (side), $\frac{3}{4}$ (bottom)

$\frac{1}{2} \times \frac{3}{4}$

10. 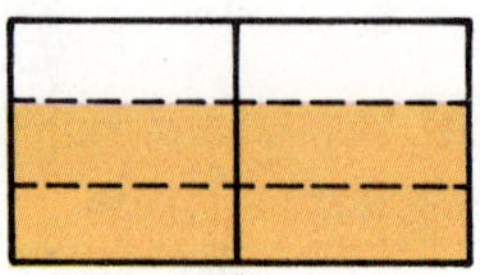

$\frac{2}{3}$ (side), 2 (bottom)

$\frac{2}{3} \times 2$

11. 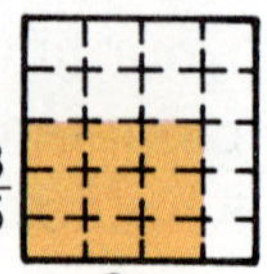

$\frac{3}{5}$ (side), $\frac{3}{4}$ (bottom)

$\frac{3}{5} \times \frac{3}{4}$

12. $\frac{2}{3} \times \frac{3}{4}$

13. $\frac{5}{8} \times \frac{2}{5}$

14. $\frac{1}{3} \times 3$

15. $\frac{7}{8} \times \frac{4}{3}$

16. $\frac{3}{2} \times \frac{2}{3}$

17. $\frac{4}{5} \times \frac{5}{2}$

# CHAPTER CHALLENGE

When we spin this spinner, there are 6 possible outcomes. Since the chance of getting any of the outcomes is the same, we say that the outcomes are **equally likely.**

The **probability** of getting each outcome is $\frac{1}{6}$. For example, when we spin the spinner, the probability that it will stop on 5 is $\frac{1}{6}$.

$$P(5) = \frac{1}{6}$$

Read as "The probability of getting 5 equals $\frac{1}{6}$."

$$P(1 \text{ or } 4) = \frac{1}{6} + \frac{1}{6} = \frac{1}{3}$$

**Give each probability in simplest form.**

1. P(2) = ?
2. P(3) = ?
3. P(2 or 3) = ?
4. P(2, 4, or 6) = ?
5. P(odd number) = ?
6. P(number less than 5) = ?
7. P(number greater than 1) = ?
8. P(factor of 6) = ?
9. P(not a factor of 6) = ?
10. P(multiple of 2) = ?
11. P(not a multiple of 2) = ?
12. P(1, 2, 3, 4, 5, or 6) = ?
13. P(not 1, 2, 3, 4, 5, or 6) = ?

# MAJOR CHECKUP
## STANDARDIZED FORMAT

**Choose the correct letter.**

**1.** 8,376,852 rounded to the nearest thousand is

- **a.** 8,380,000
- **b.** 8,376,000
- **c.** 8,377,000
- **d.** none of these

**2.** Subtract.

$$\begin{array}{r} 58{,}307 \\ -1{,}659 \\ \hline \end{array}$$

- **a.** 57,352
- **b.** 56,748
- **c.** 56,648
- **d.** none of these

**3.** Multiply.

$$\begin{array}{r} 682 \\ \times 153 \\ \hline \end{array}$$

- **a.** 104,346
- **b.** 74,138
- **c.** 100,046
- **d.** none of these

**4.** Divide.

$28\overline{)2893}$

- **a.** 13
- **b.** 13 R9
- **c.** 103 R9
- **d.** none of these

**5.** $\overline{AC}$ is called a

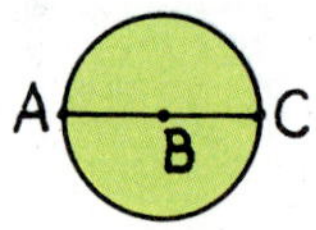


- **a.** diameter
- **b.** center
- **c.** radius
- **d.** none of these

**6.** The number pair for point A is

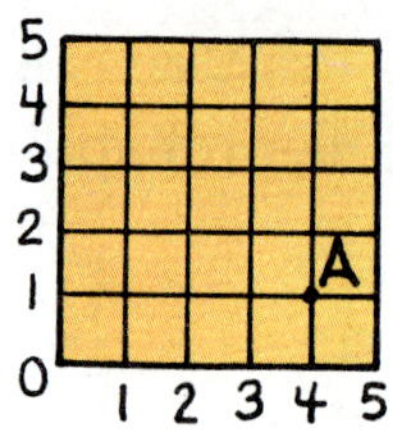


- **a.** (1, 4)
- **b.** (4, 0)
- **c.** (4, 1)
- **d.** none of these

**7.** $\frac{1}{4} + \frac{2}{3} = \underline{?}$

- **a.** $\frac{3}{7}$
- **b.** $\frac{1}{4}$
- **c.** $\frac{5}{6}$
- **d.** none of these

**8.** $\frac{5}{6} - \frac{1}{2} = \underline{?}$

- **a.** $\frac{1}{3}$
- **b.** $\frac{2}{3}$
- **c.** 1
- **d.** none of these

**9.** $\frac{5}{2}$ written in simplest form is

- **a.** $1\frac{1}{2}$
- **b.** $2\frac{1}{2}$
- **c.** $3\frac{1}{2}$
- **d.** none of these

**10.** $\frac{3}{4} \times \frac{2}{5} = \underline{?}$

- **a.** $\frac{3}{10}$
- **b.** $\frac{15}{8}$
- **c.** $\frac{23}{20}$
- **d.** none of these

**11.** Roger baked 24 cookies. He ate $\frac{3}{4}$ of them. How many did he eat?

- **a.** 6
- **b.** 18
- **c.** 32
- **d.** none of these

**12.** Selma spent 71¢ for 2 pencils and a pad. Each pencil cost 19¢. How much was the pad?

- **a.** 31¢
- **b.** 47¢
- **c.** 52¢
- **d.** none of these

# Decimals— Addition and Subtraction

# 8

## READY OR NOT?

**What fraction of the square is shaded?**

1. 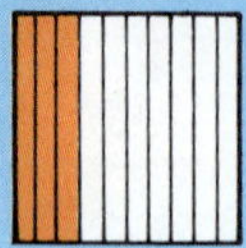

2. 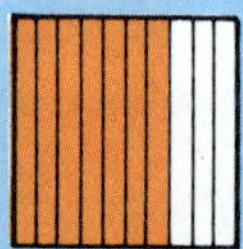

3. 

4. 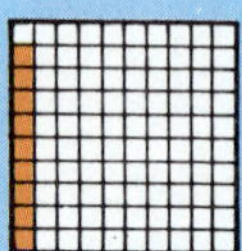

5. 

6. 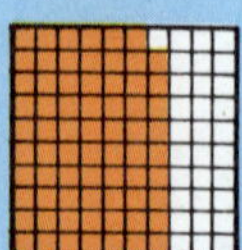

# Tenths and hundredths

Below are three ways of writing about the number of shaded squares.

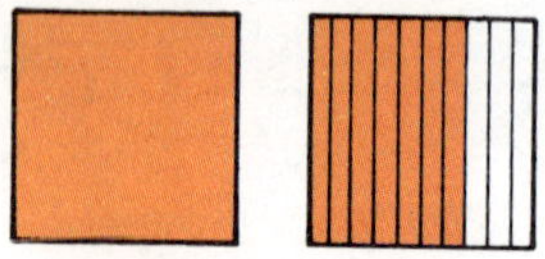

$1\frac{7}{10}$

| Ones | Tenths |
|---|---|
| 1 | 7 |

**1.7**

The last numeral is called a **decimal fraction,** or just a **decimal.** "1.7" is read "one and seven tenths." Study these other decimals.

| Ones | Tenths | Hundredths |
|---|---|---|
| 1 | 0 | 6 |

**1.06**

1 and 6 hundredths

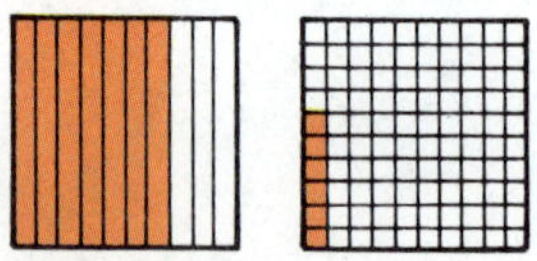

| Ones | Tenths | Hundredths |
|---|---|---|
| 0 | 7 | 6 |

**0.76**

76 hundredths

Notice how the decimal is read. The picture shows that

7 tenths 6 hundredths = 76 hundredths

## EXERCISES

**Give a decimal for the number of shaded squares.**

1. 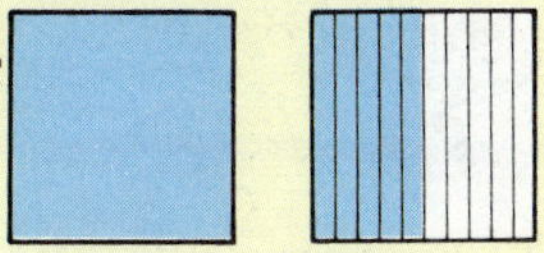

2. 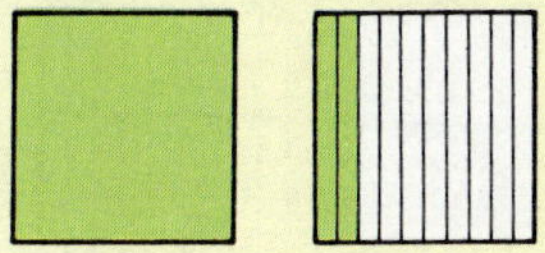

3. 

4. 

5. 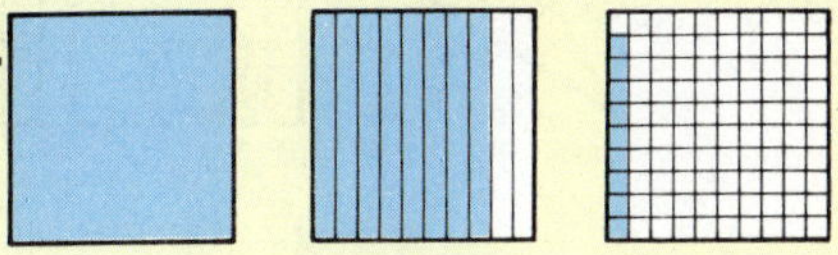

6. 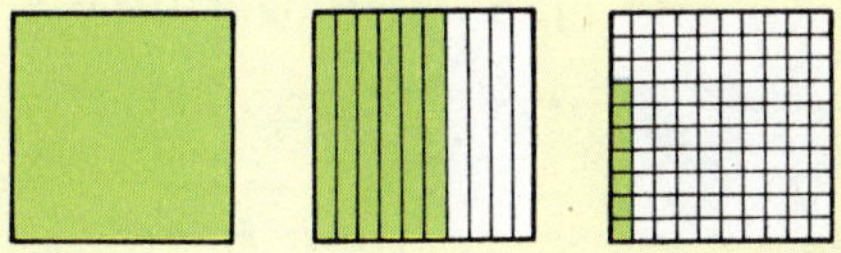

7. 

8. 

**Read each decimal aloud.**

| | | | |
|---|---|---|---|
| **9.** 2.54 | **10.** 25.4 | **11.** 8.7 | **12.** 8.07 |
| **13.** 16.83 | **14.** 9.40 | **15.** 38.5 | **16.** 4.9 |
| **17.** 37.85 | **18.** 62.5 | **19.** 152.63 | **20.** 48.75 |

**Tell which number is greater.**

| | | |
|---|---|---|
| **21.** 0.5, 0.6 | **22.** 0.8, 0.2 | **23.** 0.5, 1.5 |
| **24.** 1.3, 1.8 | **25.** 1.9, 1.6 | **26.** 2.3, 2.7 |
| **27.** 1.05, 1.06 | **28.** 3.24, 3.21 | **29.** 4.37, 4.39 |
| **30.** 24.86, 24.93 | **31.** 46.74, 46.52 | **32.** 75.69, 75.48 |

# Thousandths

| Ones | Tenths | Hundredths | Thousandths |
|---|---|---|---|
| 2 | 8 | 5 | 3 |

**2.853**

two and eight hundred fifty-three thousandths

| Tens | Ones | Tenths | Hundredths | Thousandths |
|---|---|---|---|---|
| 2 | 3 | 0 | 6 | 5 |

**23.065**

twenty-three and sixty-five thousandths

| Tens | Ones | Tenths | Hundredths | Thousandths |
|---|---|---|---|---|
| 5 | 8 | 0 | 0 | 9 |

**58.009**

fifty-eight and nine thousandths

**EXERCISES**

**Read each decimal aloud.**

1. 8.375
2. 691.8
3. 84.214
4. 65.017
5. 48.006
6. 523.74
7. 32.296
8. 7.063
9. 0.659
10. 0.003

**Give the next three numbers.**

**11.** 6.5, 6.6, 6.7, ?, ?, ?

**12.** 15.8, 15.9, 16.0, ?, ?, ?

**13.** 24.7, 24.8, 24.9, ?, ?, ?

**14.** 8.03, 8.04, 8.05, ?, ?, ?

**15.** 3.09, 3.10, 3.11, ?, ?, ?

**16.** 35.07, 35.08, 35.09, ?, ?, ?

**17.** 8.098, 8.099, 8.100, ?, ?, ?

**18.** 14.096, 14.097, 14.098, ?, ?, ?

**Tell which digit is in the**

**19.** tens place.

**20.** ones place.

**21.** hundredths place.

**22.** tenths place.

**23.** thousandths place.

**24.** hundreds place.

## *Challenge!*

**Copy each numeral and place a decimal point so that the statement makes sense.**

These statements were made about the students in Elmwood Elementary School.

**25.** John is 102 years old.

**26.** Alice is 545 inches tall.

**27.** Bill weighs 8275 pounds.

**28.** Ted can high-jump 426 inches.

**29.** Sarah ate 150 sandwiches for lunch.

**30.** Jill drank 25 quarts of milk.

**31.** In one hour, David read 275 pages of a library book.

**32.** In one hour, Mary used 250 pencils.

**33.** Lunch time lasts for 750 hours.

**34.** Recess lasts for 205 minutes.

# Comparing and ordering decimals

At the spring track meet Kim long-jumped 3.03 meters and Diane long-jumped 3.12 meters. We can compare these numbers by using a number line.

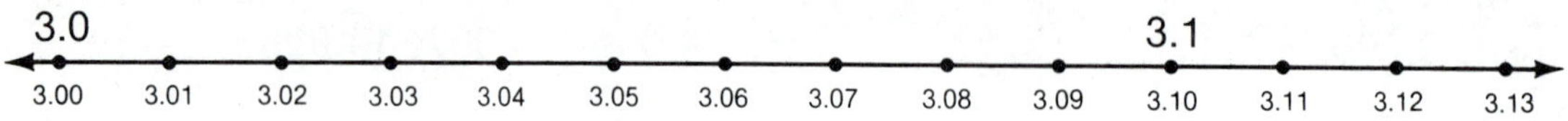


**3.03 < 3.12** **or** **3.12 > 3.03**

is less than — is greater than

## EXERCISES

**Give the number for each lettered point.**

1.
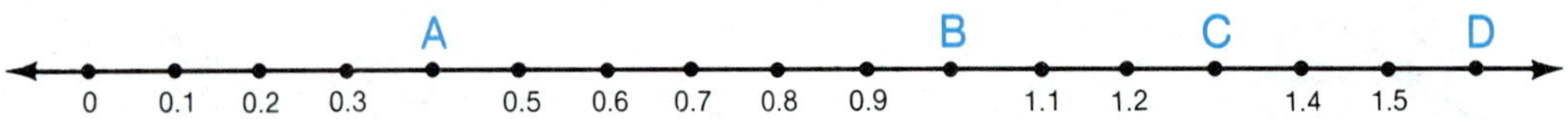


2.
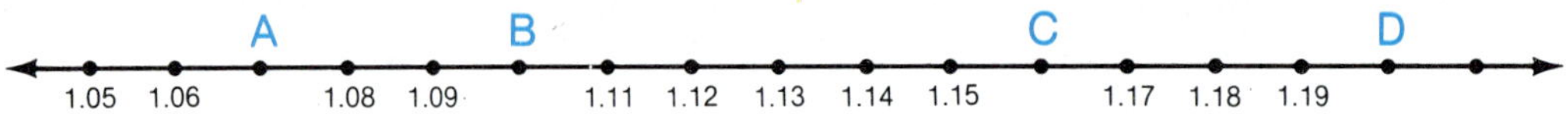


3.

**<, =, or > ?**

**4.** 3.1 ● 3.4 **5.** 5.9 ● 6.0 **6.** 12.0 ● 12.1

**7.** 4.09 ● 4.10 **8.** 4.00 ● 4.0 **9.** 8.90 ● 8.89

**10.** 3.784 ● 3.782 **11.** 6.29 ● 6.287 **12.** 12.080 ● 12.08

**13.** 0.40 ● 0.4 **14.** 4.7 ● 4.70 **15.** 8.200 ● 8.20

**16.** 6 ● 5.99 **17.** 8.9 ● 9 **18.** 11.53 ● 11.54

**19.** 2.999 ● 3.0 **20.** 9.79 ● 9.78 **21.** 7.257 ● 7.263

**22.** 63.4 ● 6.35 **23.** 5.821 ● 58.21 **24.** 5.017 ● 50.17

The table shows the results of the long jump.

| Name | School | Distance in meters |
|---|---|---|
| Kim Davis | Washington | 3.03 |
| Diane Floyd | Valley View | 3.12 |
| Bob Hoover | Jefferson | 3.01 |
| Daniel Jackson | Morningside | 2.97 |
| Manuel Lopez | Wells Corner | 3.15 |
| Richard Smith | Maryville | 2.95 |

**25.** Who won the long-jump event?

**26.** Who came in second? third?

**27.** List the names of the students in the order that they finished.

This table shows the results of the 40-meter dash.

| Name | School | Time in seconds |
|---|---|---|
| John Barker | Valley View | 8.234 |
| Susan Kelley | Morningside | 7.940 |
| Bill Locker | Wells Corner | 8.082 |
| Gerry Manning | Washington | 8.306 |
| Kim Rogers | Maryville | 7.899 |
| Terry Thomas | Jefferson | 8.117 |

**28.** Who had the best time?

**29.** How many ran the dash in less than 8 seconds?

**30.** List the names in the order that they finished.

## KEEPING SKILLS SHARP

**Multiply.**

**1.** $48 \times 6$

**2.** $93 \times 7$

**3.** $614 \times 8$

**4.** $921 \times 6$

**5.** $406 \times 20$

**6.** $384 \times 50$

**7.** $527 \times 53$

**8.** $291 \times 28$

**9.** $921 \times 252$

**10.** $742 \times 326$

**Divide.**

**11.** $4\overline{)840}$

**12.** $3\overline{)174}$

**13.** $5\overline{)2640}$

**14.** $8\overline{)36,527}$

**15.** $24\overline{)38,291}$

**16.** $50\overline{)47,382}$

**17.** $68\overline{)96,715}$

**18.** $86\overline{)44,056}$

# Rounding to the nearest whole number

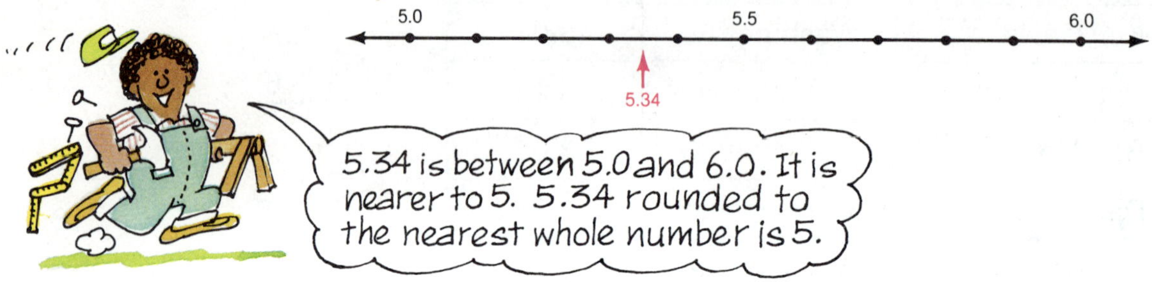


Study these examples.

**EXAMPLE 1.**

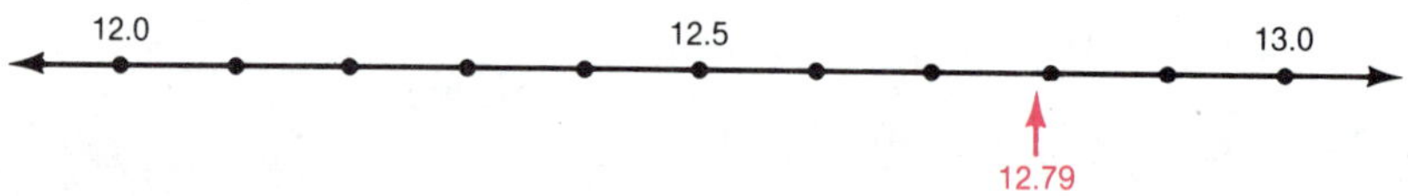


12.79 is nearer to 13.0 than to 12.0.
12.79 rounded to the nearest whole number is 13.

**EXAMPLE 2.**

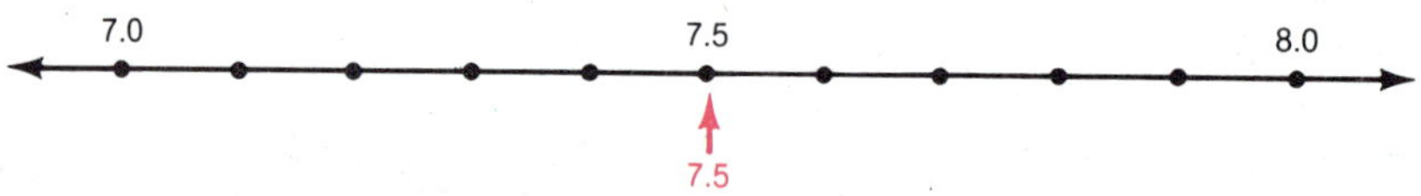


7.5 is exactly halfway between 7.0 and 8.0.
In this case we round *up*.
7.5 rounded to the nearest whole number is 8.

**EXERCISES**

**Round to the nearest whole number. [*Hint:* You may wish to use the number lines shown above.]**

**1. a.** 5.2 **b.** 5.9 **2. a.** 5.38 **b.** 5.50

**3. a.** 12.8 **b.** 12.5 **4. a.** 12.49 **b.** 12.63

**5. a.** 7.4 **b.** 7.7 **6. a.** 7.50 **b.** 7.09

7. 9.3 8. 6.9 9. 1.5 10. 9.0 11. 4.5

12. 8.2 13. 8.3 14. 8.4 15. 8.5 16. 8.6

17. 8.7 18. 13.4 19. 13.43 20. 13.45 21. 13.48

22. 13.49 23. 13.50 24. 13.51 25. 13.58 26. 13.59

263.52

263.52 rounded to the nearest whole number is 264.

**Round to the nearest whole number.**

27. 5.73 28. 46.39 29. 85.46 30. 243.58

31. 49.674 32. 86.583 33. 42.486 34. 128.411

35. 96.514 36. 426.883 37. 2.488 38. 62.703

**Round each price to the nearest dollar.**

# Adding decimals

You can find the sum of two decimals by lining up the decimal points and adding in columns just as you do with whole numbers.

**EXAMPLE 1.** **2.52 + 1.24 = ?**

$$\begin{array}{r} 2.52 \\ +1.24 \\ \hline 3.76 \end{array}$$

If a sum is 10 or greater, regroup!

**EXAMPLE 2.** **2.486 + 3.592 = ?**

**Step 1.** Add thousandths.

$$\begin{array}{r} 2.486 \\ +3.592 \\ \hline 8 \end{array}$$

**Step 2.** Add hundredths and regroup.

$$\begin{array}{r} 1\phantom{6} \\ 2.486 \\ +3.592 \\ \hline 78 \end{array}$$

**Step 3.** Add tenths and regroup.

$$\begin{array}{r} 1\phantom{.}1\phantom{86} \\ 2.486 \\ +3.592 \\ \hline .078 \end{array}$$

**Step 4.** Add ones.

$$\begin{array}{r} 1\phantom{.}1\phantom{86} \\ 2.486 \\ +3.592 \\ \hline 6.078 \end{array}$$

**EXERCISES**

**Give each sum.**

1. 43.6 + 21.0
2. 3.64 + 5.25
3. 7.08 + 2.10
4. 53.6 + 44.3
5. 0.638 + 0.211
6. 38.2 + 11.5
7. 4.78 + 3.19
8. 7.42 + 8.06
9. 5.431 + 2.081
10. 7.385 + 2.810

11. 633.4 + 683.3

12. 28.19 + 59.74

13. 7.416 + 3.829

14. 52.17 + 28.47

15. 8.382 + 3.695

16. 372.8 + 636.9

17. 33.96 + 97.49

18. 49.72 + 16.05

19. 65.03 + 8.29

20. 16.429 + 34.747

21. 3.65 + 2.53

22. 98.7 + 35.6

23. 7.28 + 8.59

24. 12.6 + 8.9

25. 25.38 + 9.67

26. 3.682 + 4.917

27. 9.038 + 6.952

28. 48.72 + 5.99

**Solve.**

29. Joanne wrote a computer program that would balance her checkbook in 14.59 seconds. Marian's program can do the same thing 4.13 seconds faster. How long does Marian's program take?

30. The Computer Outlet sells the computer that Mark wants for $1207.45. Mark knows the same computer sells for $192.56 less at Johnson's Computers. Can he buy it for less than $1000 at Johnson's?

## *Challenge!*

**Copy. Then put decimal points in the addends to get the answers shown.**

31. 384 + 297 = 68.1

32. 729 + 531 = 78.21

33. 18 + 395 = 39.68

34. 3785 + 293 = 407.8

35. 8076 + 5997 = 867.57

36. 136 + 4987 = 18.587

# Subtracting decimals

You can find the difference of two decimals by lining up the decimal points and subtracting in columns, just as you do with whole numbers.

**EXAMPLE 1.** **2.84 − 1.13 = ?**

$$\begin{array}{r} 2.84 \\ -1.13 \\ \hline 1.71 \end{array}$$

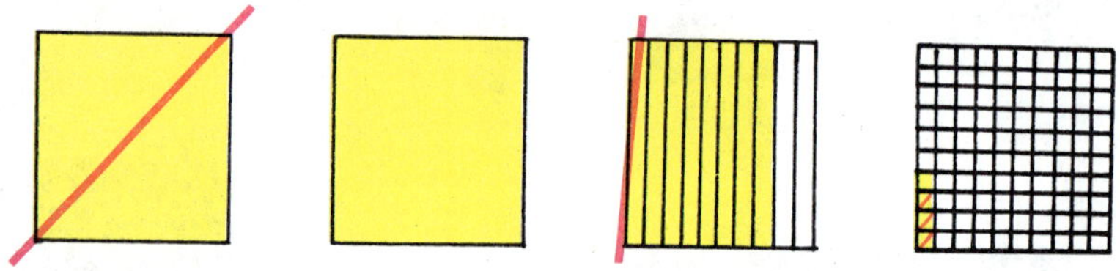

Sometimes you have to regroup before you can subtract.

**EXAMPLE 2.**

$$\begin{array}{r} 7.204 \\ -1.265 \\ \hline \end{array}$$

**Step 1.** Regroup 1 tenth for 10 hundredths.

$$\begin{array}{r} 7.204 \\ -1.265 \\ \hline \end{array}$$

**Step 2.** Regroup 1 hundredth for 10 thousandths.

$$\begin{array}{r} 7.204 \\ -1.265 \\ \hline \end{array}$$

**Step 3.** Subtract thousandths.

$$\begin{array}{r} 7.204 \\ -1.265 \\ \hline 9 \end{array}$$

**Step 4.** Subtract hundredths.

$$\begin{array}{r} 7.204 \\ -1.265 \\ \hline 39 \end{array}$$

**Step 5.** Regroup 1 one for 10 tenths.

$$\begin{array}{r} 7.204 \\ -1.265 \\ \hline 39 \end{array}$$

**Step 6.** Subtract tenths.

$$\begin{array}{r} 7.204 \\ -1.265 \\ \hline .939 \end{array}$$

**Step 7.** Subtract ones.

$$\begin{array}{r} 7.204 \\ -1.265 \\ \hline 5.939 \end{array}$$

## EXERCISES

**Give each difference.**

| | | | | |
|---|---|---|---|---|
| **1.** $\begin{array}{r} 39.5 \\ -16.2 \\ \hline \end{array}$ | **2.** $\begin{array}{r} 5.98 \\ -2.46 \\ \hline \end{array}$ | **3.** $\begin{array}{r} 9.78 \\ -3.58 \\ \hline \end{array}$ | **4.** $\begin{array}{r} 68.67 \\ -24.35 \\ \hline \end{array}$ | **5.** $\begin{array}{r} 8.967 \\ -2.531 \\ \hline \end{array}$ |
| **6.** $\begin{array}{r} 75.2 \\ -38.9 \\ \hline \end{array}$ | **7.** $\begin{array}{r} 8.03 \\ -5.26 \\ \hline \end{array}$ | **8.** $\begin{array}{r} 4.16 \\ -2.57 \\ \hline \end{array}$ | **9.** $\begin{array}{r} 50.3 \\ -21.4 \\ \hline \end{array}$ | **10.** $\begin{array}{r} 3.00 \\ -1.86 \\ \hline \end{array}$ |
| **11.** $\begin{array}{r} 435.1 \\ -167.6 \\ \hline \end{array}$ | **12.** $\begin{array}{r} 2.060 \\ -1.784 \\ \hline \end{array}$ | **13.** $\begin{array}{r} 63.40 \\ -15.64 \\ \hline \end{array}$ | **14.** $\begin{array}{r} 40.35 \\ -21.84 \\ \hline \end{array}$ | **15.** $\begin{array}{r} 786.2 \\ -195.8 \\ \hline \end{array}$ |
| **16.** $\begin{array}{r} 6.756 \\ -5.851 \\ \hline \end{array}$ | **17.** $\begin{array}{r} 32.82 \\ -16.95 \\ \hline \end{array}$ | **18.** $\begin{array}{r} 892.4 \\ -365.9 \\ \hline \end{array}$ | **19.** $\begin{array}{r} 2.767 \\ -1.898 \\ \hline \end{array}$ | **20.** $\begin{array}{r} 7.631 \\ -4.594 \\ \hline \end{array}$ |

**21.** 8.96 − 4.09 **22.** 12.35 − 10.70 **23.** 42.07 − 29.38

**24.** 50.03 − 25.64 **25.** 5.242 − 3.557 **26.** 37.00 − 16.75

**Solve.**

**27.** **a.** What is the total weight?

**b.** What is the total cost?

**c.** How much more does the larger melon weigh?

**d.** How much less does the smaller melon cost?

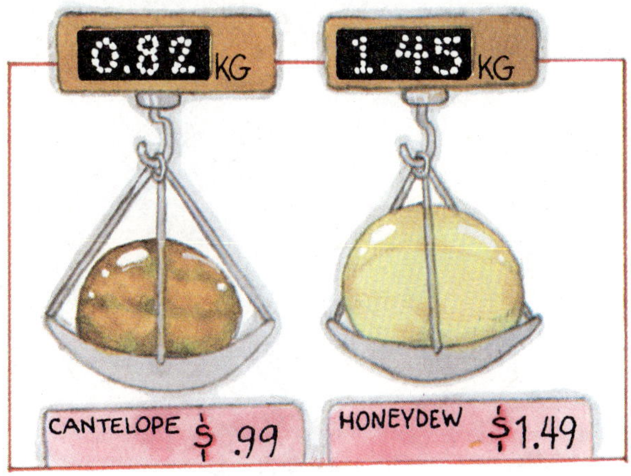


## *Challenge!*

**Copy and complete.**

| | | | |
|---|---|---|---|
| **28.** $\begin{array}{r} 5.63 \\ -\blacksquare.8\blacksquare \\ \hline 2.79 \end{array}$ | **29.** $\begin{array}{r} 4.\blacksquare69 \\ -1.7\blacksquare\blacksquare \\ \hline 2.485 \end{array}$ | **30.** $\begin{array}{r} 7\blacksquare.63\blacksquare \\ -\blacksquare5.0\blacksquare4 \\ \hline 54.564 \end{array}$ | **31.** $\begin{array}{r} 4\blacksquare.0\blacksquare5 \\ -\blacksquare2.84\blacksquare \\ \hline 35.193 \end{array}$ |

# More about adding and subtracting decimals

Remember to line up the decimal points when adding or subtracting decimals. In these problems you can write zeros before you add.

5.38 + 0.496 = ?

$$\begin{array}{r} 5.380 \\ +0.496 \\ \hline 5.876 \end{array}$$

28 + 15.84 + 0.73 = ?

$$\begin{array}{r} 28.00 \\ 15.84 \\ +\ \ 0.73 \\ \hline 44.57 \end{array}$$

9.28 − 3.775 = ?

$$\begin{array}{r} 9.280 \\ -3.775 \\ \hline 5.505 \end{array}$$

24 − 15.45 = ?

$$\begin{array}{r} 24.00 \\ -15.45 \\ \hline 8.55 \end{array}$$

**EXERCISES**

**Give each sum.**

1. $\begin{array}{r} 3.684 \\ +2.85 \\ \hline \end{array}$

2. $\begin{array}{r} 0.430 \\ +6.597 \\ \hline \end{array}$

3. $\begin{array}{r} 4.963 \\ +2.84 \\ \hline \end{array}$

4. $\begin{array}{r} 6.00 \\ +3.25 \\ \hline \end{array}$

5. $\begin{array}{r} 112.625 \\ +14. \\ \hline \end{array}$

6. $\begin{array}{r} 3.85 \\ 2.1 \\ +4.61 \\ \hline \end{array}$

7. $\begin{array}{r} 13.82 \\ 4.5 \\ +6.3 \\ \hline \end{array}$

8. $\begin{array}{r} 8.375 \\ 14.5 \\ +2.950 \\ \hline \end{array}$

9. $\begin{array}{r} 5 \\ 7.5 \\ +2.6 \\ \hline \end{array}$

10. $\begin{array}{r} 3.625 \\ 2 \\ +5.75 \\ \hline \end{array}$

11. 2.837 + 5.42
12. 39.7 + 16.985
13. 6.584 + 25.3

14. 4.06 + 2.385 + 0.4
15. 17.2 + 3.5 + 4.73
16. 5.55 + 5.5 + 0.5

17. $16 + $3.80
18. $2.75 + $48
19. $63 + $58.75

**Give each difference.**

20. $\begin{array}{r} 52.68 \\ -5.4 \\ \hline \end{array}$
21. $\begin{array}{r} 6.9 \\ -3.42 \\ \hline \end{array}$
22. $\begin{array}{r} 9.36 \\ -4.8 \\ \hline \end{array}$
23. $\begin{array}{r} 92.8 \\ -15.34 \\ \hline \end{array}$
24. $\begin{array}{r} 32. \\ -25.85 \\ \hline \end{array}$

25. $\begin{array}{r} 4.385 \\ -2.63 \\ \hline \end{array}$
26. $\begin{array}{r} 15 \\ -8.64 \\ \hline \end{array}$
27. $\begin{array}{r} 9 \\ -3.625 \\ \hline \end{array}$
28. $\begin{array}{r} 7.425 \\ -3.8 \\ \hline \end{array}$
29. $\begin{array}{r} 734 \\ -16.85 \\ \hline \end{array}$

30. 5.8 − 4.25
31. 15.74 − 3.618
32. 49.5 − 37.837

33. $38 − $3.65
34. $74 − $52.80
35. $74.36 − $36

**Solve.**

36. A $20 pair of blue jeans was on sale for $13.95. By how much was the price reduced?

37. A $15 shirt was on sale for $9.95. By how much was the price reduced?

38. Jim had $4.82 and he earned $6 more raking leaves. How much more money does he need to buy a $20 pair of jeans?

39. Alicia had $30. She bought a pair of jeans for $13.95 and a shirt for $9.95. Could she buy a $10 belt that was on sale for $7.95?

## KEEPING SKILLS SHARP

**Add or subtract. Write answers in simplest form.**

1. $\frac{2}{3} + \frac{2}{3}$
2. $\frac{5}{8} + \frac{3}{4}$
3. $\frac{3}{4} + \frac{1}{2}$
4. $\frac{7}{8} + \frac{3}{4}$

5. $\frac{5}{6} - \frac{1}{3}$
6. $\frac{3}{4} - \frac{1}{2}$
7. $\frac{1}{2} - \frac{1}{3}$
8. $\frac{3}{4} - \frac{3}{8}$

# Computer timing

Races are often timed by computers. The timing usually is started by breaking a light beam that shines across the track. The timing is ended when the light beam is broken a second time. The time is reported in minutes, seconds, and hundredths of a second.

**25:12:78**

minutes — seconds — hundredths of a second

## EXERCISES

1. In a stock-car race, the times were recorded at the beginning of a lap and at the end of the lap. Here are the times:

| Car | Start of lap | End of lap |
|---|---|---|
| X | 12:19:39 | 13:05:73 |
| Y | 12:33:19 | 13:05:22 |
| Z | 13:01:06 | 13:56:33 |

   a. Which car was leading at the start of the lap?
   b. Which car was leading at the end of the lap?
   c. Which car had the fastest lap?
   d. How long did it take the fastest car to complete the lap?

2. At the start of the last lap of a mile race, a runner passed a timer at 3:18:36. She ran the last lap in 1 minute 42.78 seconds. What time did the timer record as the runner finished the race?

SKILL GAME

# TARGET NUMBER

Kathy came within 0.016 of the target number 10.

**How close did these players come to the target number?**

1. $(3.486 + 9.059) - 3.742 \rightarrow 10$
2. $(7.539 + 6.740) - 2.103 \rightarrow 10$
3. $(7.149 + 2.103) - 2.859 \rightarrow 10$
4. $(6.328 + 9.417) - 5.918 \rightarrow 10$

**Play the game.**

1. Make a card for each of the digits 0 through 9.
2. Choose a leader.
3. Draw a table like this: $(\square.\square\square\square + \square.\square\square\square) - \square.\square\square\square \rightarrow 10$
4. Without looking, the leader picks a card. Each player writes the digit in any square of the table.
5. The card is replaced in the deck, and step 4 is repeated until all squares have been filled in.
6. The player who gets closest to the target number (10) wins the game.

# Problem solving

Jane and Robert traveled in Canada with their parents. Here are some problems they solved during the trip.

**1.** Got Up

Left 1 hour 35 minutes later.
What time did they leave?

**2.** Stopped for Gasoline

Tank holds 78 liters

Took

to fill.

How much gasoline was in the tank before it was filled?

**3.** Looked at Map

How far did they have to go?

**4.** Stopped to Eat

Robert had

Gave Clerk

How much change?

**5.** Checked Fuel Gauge
Tank holds 78 liters.
How much gasoline is in the tank?

**6.** Weighed Largest Fish
How much more did Jane's fish weigh?

**7.** Saw Road Sign
How far is it from Adams to Springville?

**8.** Checked Odometer

Beginning of trip

End of trip

How many kilometers did they drive?

**Make up a problem.**

# CHAPTER CHECKUP

[pages 228–231]

**In 159.378, which digit is in the**

1. hundreds place?
2. hundredths place?
3. thousandths place?
4. tens place?
5. tenths place?
6. ones place?

**<, =, or >?** [pages 232–333]

7. 5.6 ● 6.5
8. 4.2 ● 4.20
9. 3.96 ● 3.9
10. 12.783 ● 12.738
11. 5.89 ● 5.9
12. 4.359 ● 4.36

**Add.** [pages 236–237, 240–241]

13. $\begin{array}{r} 5.4 \\ +2.3 \\ \hline \end{array}$
14. $\begin{array}{r} 6.8 \\ +1.4 \\ \hline \end{array}$
15. $\begin{array}{r} 7.03 \\ +6.59 \\ \hline \end{array}$
16. $\begin{array}{r} 2.834 \\ +5.719 \\ \hline \end{array}$
17. $\begin{array}{r} 35.26 \\ 42.69 \\ +19.25 \\ \hline \end{array}$
18. 36.8 + 29.25
19. 5.36 + 4
20. 74.296 + 3.1

**Subtract.** [pages 238–241]

21. $\begin{array}{r} 6.35 \\ -3.12 \\ \hline \end{array}$
22. $\begin{array}{r} 25.8 \\ -16.3 \\ \hline \end{array}$
23. $\begin{array}{r} 4.27 \\ -1.89 \\ \hline \end{array}$
24. $\begin{array}{r} 6.302 \\ -1.748 \\ \hline \end{array}$
25. $\begin{array}{r} 5.00 \\ -3.72 \\ \hline \end{array}$
26. 93.8 − 64.71
27. 73.5 − 29.678
28. 7 − 3.284

**Solve.** [pages 244–245]

29. What is the total cost?
30. What is the total weight?
31. How much more do the cashews weigh?
32. How much less do the peanuts cost?

# CHAPTER PROJECT

Sarah and a friend decided to see how well they could estimate a minute. They got a stopwatch and Sarah told her friend when to start. When she thought that a minute had passed, Sarah said "Stop!" The actual time was 51.8 seconds. By how much did Sarah miss one minute?

1. Do this experiment in your class. Each student should have a turn estimating. Keep a record of your class results like this:

| ONE-MINUTE GUESSES | | | |
|---|---|---|---|
| NAME | TIME | ERROR IN SECONDS | ERROR ROUNDED TO NEAREST SECOND |
| SARAH | 51.8 | 8.2 | 8 |
| | | | |

2. Make a bar graph of the errors rounded to the nearest second.

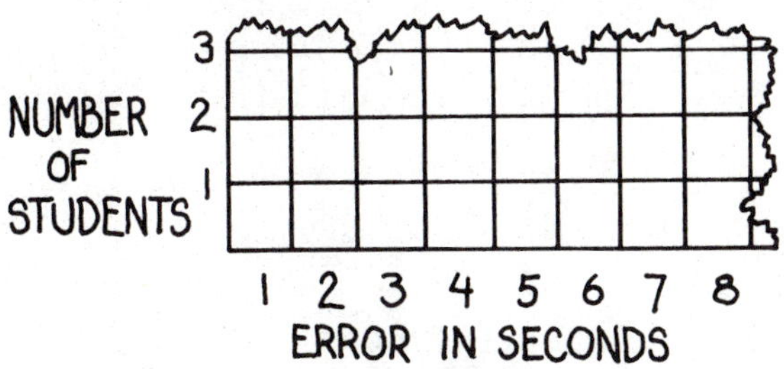


3. List some things that are shown on your graph.

4. Repeat the experiment and compare your results.

# CHAPTER REVIEW

**Tell which digit is in the**

1. tenths place.
2. hundredths place.
3. ones place.
4. thousandths place.

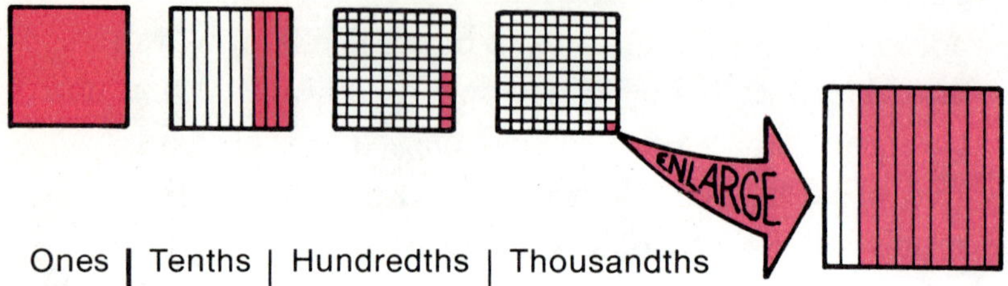


| Ones | Tenths | Hundredths | Thousandths |
|---|---|---|---|
| **1** | **3** | **5** | **8** |

**<, =, or >?**

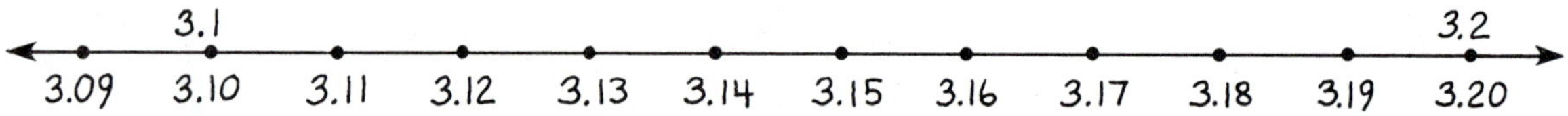


**5.** 3.2 ● 3.1 **6.** 3.11 ● 3.18 **7.** 3.2 ● 3.15

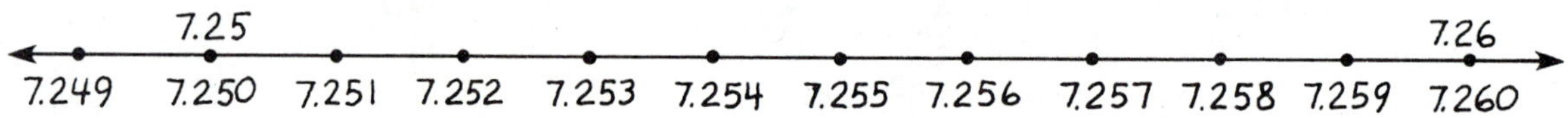


**8.** 7.25 ● 7.250 **9.** 7.256 ● 7.253 **10.** 7.257 ● 7.26

**Add.**

**11.** 5.68 + 3.59 = 9.27 **12.** 3.058 + 6.749 **13.** 89.67 + 37.84 **14.** 42.9 + 5.78 **15.** 67.421 + 8.7

**16.** 5.893 + 2.70 **17.** 31.048 + 9.6 **18.** 59.831 + 74.665

**Subtract.**

**19.** 59.14 − 17.86 = 41.28 **20.** 7.802 − 3.194 **21.** 68.315 − 9.527 **22.** 5 − 2.84 **23.** 59.1 − 28.426

**24.** 432.56 − 18.3 **25.** 69.028 − 18.655 **26.** 44.3 − 18.57

**27.** 98.31 − 17.485 **28.** 8.352 − 1.65 **29.** 62.37 − 58.21

# CHAPTER CHALLENGE

Use this decoder to find the answer.

| A | C | D | E | I | K |
|---|---|---|---|---|---|
| 78.24 | 27.6 | 6.095 | 18.375 | 6.95 | 6.102 |
| **L** | **N** | **Q** | **R** | **T** | **U** |
| 69.5 | 78.42 | 79.79 | 27.675 | 76.144 | 78.8 |

1. the sum of 36.83 and 42.96
2. the difference of 103.6 and 24.8
3. the difference of 128.1 and 49.86
4. the sum of 18.095 and 9.58
5. the sum of 15.8, 23.75, and 36.594
6. the difference of 35 and 16.625
7. the difference of 9.58 and 37.255

8. 26.5 greater than 51.74
9. equal to 78.420
10. 3.85 less than 9.945

11. 2 hundredths greater than 78.4
12. 4 hundredths less than 6.99
13. 5 tenths less than 28.1
14. 7 thousandths greater than 6.095
15. 75 thousandths greater than 18.3
16. ten times as great as 6.95

b c d   2. a b c d   a b c d   a b c d   5. a b c d   6. a b c

b c d   8. a b c d   9. a b c d   10. a b c d   11. a b c d   12. a b c

# MAJOR CHECKUP

## STANDARDIZED FORMAT

**Choose the correct letter.**

**1.** 195,832 rounded to the nearest ten thousand is

**a.** 200,000
**b.** 196,000
**c.** 190,000
**d.** none of these

**2.** Multiply.

$$\begin{array}{r} 698 \\ \times 53 \\ \hline \end{array}$$

**a.** 36,574
**b.** 5584
**c.** 31,874
**d.** none of these

**3.** Divide.

$49\overline{)10,102}$

**a.** 26 R8
**b.** 206 R8
**c.** 260 R8
**d.** none of these

**4.** The measure of $\angle AOC$ is

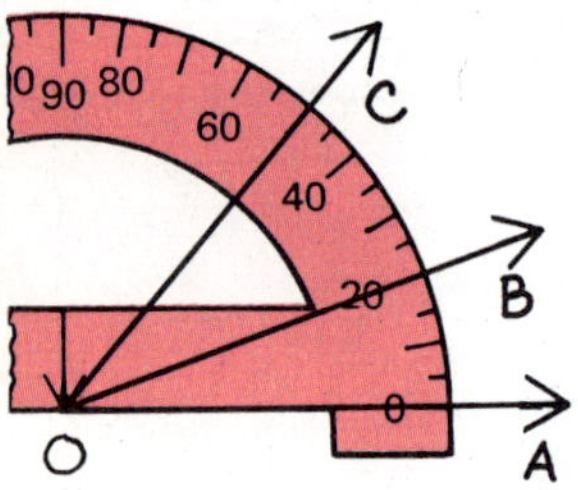


**a.** 50°
**b.** 120°
**c.** 20°
**d.** none of these

**5.** $\frac{2}{5} + \frac{1}{2} = \underline{?}$

**a.** $\frac{4}{7}$
**b.** $\frac{9}{10}$
**c.** $\frac{2}{5}$
**d.** none of these

**6.** $\frac{7}{8} - \frac{1}{3} = \underline{?}$

**a.** $\frac{6}{5}$
**b.** $\frac{6}{24}$
**c.** $\frac{13}{24}$
**d.** none of these

**7.** $2\frac{3}{4}$ written as a fraction is

**a.** $\frac{9}{4}$
**b.** $\frac{5}{2}$
**c.** $\frac{11}{4}$
**d.** none of these

**8.** $\frac{5}{8} \times \frac{2}{3} = \underline{?}$

**a.** $\frac{5}{12}$
**b.** $\frac{7}{24}$
**c.** $\frac{7}{11}$
**d.** none of these

**9.** Change to a standard decimal: two hundred and two thousandths

**a.** 200.002
**b.** 0.202
**c.** 202,000
**d.** none of these

**10.** $14.87 - 3.88 = \underline{?}$

**a.** 11.09
**b.** 10.99
**c.** 18.75
**d.** none of these

**11.** Apples are 25¢ each. Pears are 20¢ each. How much do 3 pears cost?

**a.** 45¢   **b.** 60¢
**c.** 75¢   **d.** none of these

**12.** Each bus can take 32 passengers. How many buses are needed for 100 passengers?

**a.** 2   **b.** 3
**c.** 4   **d.** none of these

# Measurement

# 9

# Time

We write 7:15

We can say: seven fifteen *or* fifteen minutes after seven *or* quarter past seven

A.M. is used for times after 12:00 midnight and before 12:00 noon.

If the time shown on the clock is in the morning, we write 7:15 A.M.

P.M. is used for times after 12:00 noon and before 12:00 midnight.

If the time shown on the clock is in the evening, we write 7:15 P.M.

## EXERCISES

**Give the time. Be sure to write A.M. or P.M.**

**1.**

Lunch time.

**2.**

School's out.

**3.**

Time to get up.

**4.**

Time for bed.

**5.**

Let's watch the football game.

**6.**

I'm getting up early to go fishing.

**Daylight or dark?**

7. 2:06 A.M. 8. 12:00 noon 9. 3:15 P.M. 10. 12:00 midnight

**EXAMPLE 1.**

4 days 5 hours = ? hours

To change days to hours, multiply by 24.

*Solution:*

$$\begin{array}{r} 24 \\ \times 4 \\ \hline 96 \\ +5 \\ \hline 101 \end{array}$$

So, 4 days 5 hours = 101 hours

**EXAMPLE 2.**

582 minutes = ? hours ? minutes

To change minutes to hours, divide by 60.

*Solution:*

$$\begin{array}{r} 9 \\ 60\overline{)582} \\ -540 \\ \hline 42 \end{array}$$

So, 582 minutes = 9 hours 42 minutes

**Multiply or divide?**

11. To change minutes to seconds, ? by 60.
12. To change seconds to minutes, ? by 60.
13. To change hours to days, ? by 24.
14. To change days to hours, ? by 24.

## Challenge!

**Solve.**

15. The school computer lab is open from 8 A.M. to 4 P.M. each weekday. There are 15 computers in the lab. How many students can work on a computer 1 hour each week?

16. Ms. Jones has one computer and 35 students in her classroom. She wants each student to do 10 minutes of math practice on the computer. The school day is 9 A.M. to 3:30 P.M. Can each student get a turn each day? Would students need to be doing computer math practice during the half-hour lunch period?

# More about time

A nature club started hiking at 8:40 A.M. They planned to hike for 1 hour and 30 minutes before taking their first break. Here is a way to find the time when they took their first break.

**8:40** **9:40** **10:10**

The club started hiking again at 10:25. At 1:15 they stopped for lunch. Here is a way to find how long they hiked from the end of the break to lunch.

Count ahead by hours and then by minutes.

## EXERCISES

**Give the time that is**

1. 40 minutes later than 6:15.
2. 45 minutes later than 2:15.
3. 19 minutes later than 9:45.
4. 1 hour and 20 minutes later than 12:35.
5. 2 hours and 15 minutes later than 4:08.
6. 50 minutes earlier than 1:05.
7. 1 hour and 40 minutes earlier than 7:30.
8. 3 hours and 20 minutes earlier than 5:10.

**How much time?**

9. 5:30 to 6:15
10. 2:20 to 3:15
11. 4:45 to 6:30
12. 7:08 to 9:20
13. 10:45 to 1:30
14. 1:30 to 10:45

**Solve.**

15. It is 6:50. The movie starts in 1 hour and 10 minutes. What time does the movie start?

16. Maria started playing her trumpet at 4:30. She played it for $1\frac{1}{4}$ hours. What time did she finish?

17. Jan just missed the 9:07 train. The next train is at 10:48. How long will she have to wait?

18. John went to bed at 9:40 P.M. and got up at 7:20 A.M. How long did he sleep?

19. On Saturday Dianne and her father spent $4\frac{1}{2}$ hours painting a fence. If they started at 9:20 A.M., what time did they finish?

★ 20. The time that Ms. Kelly works beyond 8 hours is considered overtime. One day she worked from 7:30 A.M. to 6:45 P.M. If she took $\frac{1}{2}$ hour for lunch, how much overtime did she work that day?

# Millimeters and centimeters

The **centimeters** are numbered on the ruler. Each centimeter on the ruler is divided into 10 equal parts called **millimeters.**

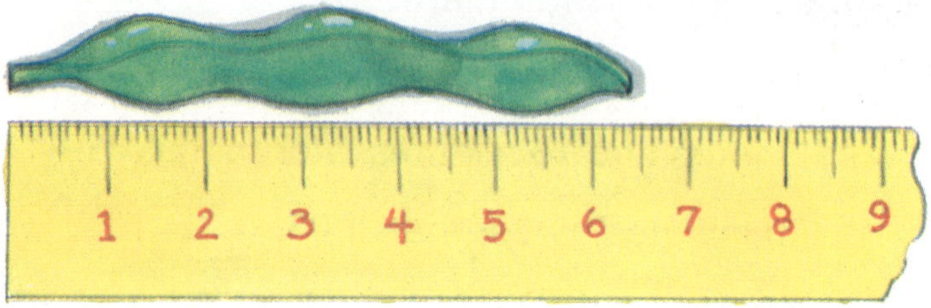

Measured to the nearest centimeter, the bean is 6 cm long.
Measured to the nearest millimeter, the bean is 64 mm long.

**64 mm = 6.4 cm**

**EXERCISES**

**Measure to the nearest centimeter.**

**1.** 

**2.** 

**3.** 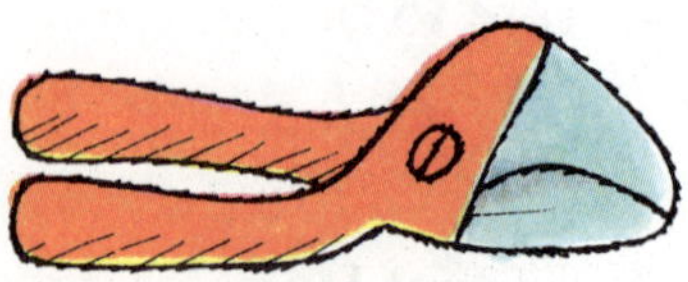

**4.** 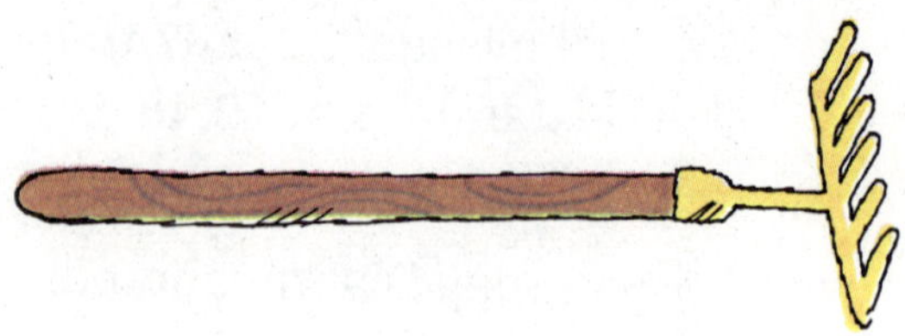

**5.** 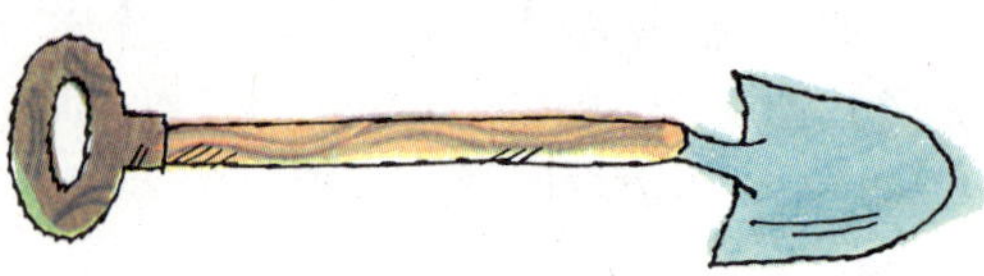

**6.** 

**Measure to the nearest millimeter.**
**Give your answer in both millimeters and centimeters.**

**7.** 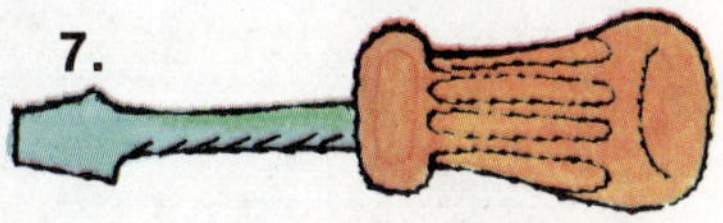

**8.** 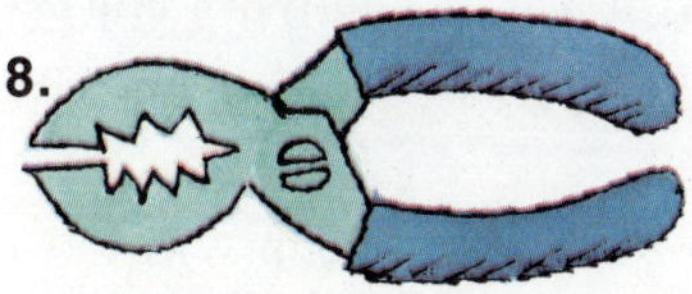

**9.** 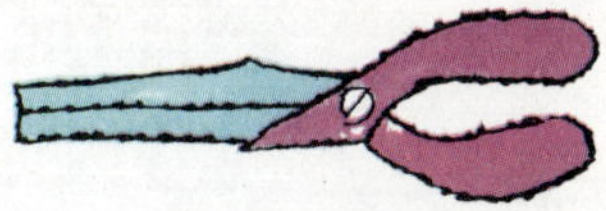

**10.** 

**11.** 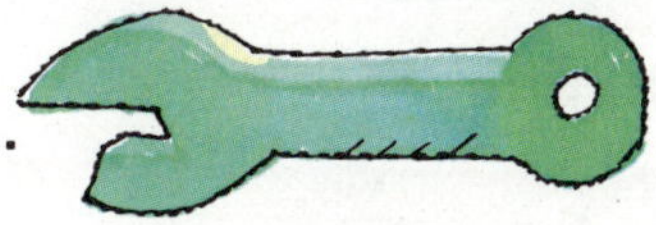

**12.** 

**13.** 

**14.** 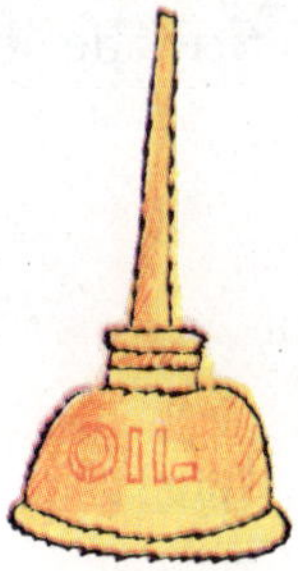


**15.** 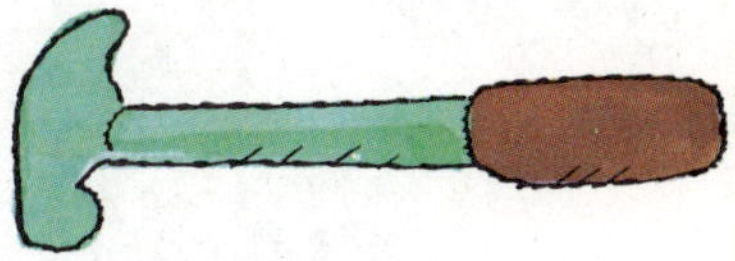

**Measure to the nearest centimeter.**

**16.** the length of your index finger

**17.** the width of your hand span

**18.** your height

**19.** the distance around your waist

**Draw segments of**

**20.** 8 cm **21.** 8.2 cm **22.** 7.4 cm **23.** 6.3 cm **24.** 63 mm

**Complete.**

**25.** 23 mm = _?_ cm **26.** 40 mm = _?_ cm **27.** 5 mm = _?_ cm

**28.** 3 cm = _?_ mm **29.** 2.5 cm = _?_ mm **30.** 12.6 cm = _?_ mm

# Meters and kilometers

The basic unit for measuring length in the metric system is the **meter.**

The door is about 1 meter wide.

1 meter (m) = 100 cm

1 m = 1000 mm

The door is 96 cm wide.
We could also write the length as 0.96 m.

96 cm = 0.96 m

**EXERCISES**

**Measure to the nearest hundredth of a meter.**

**1.** the height of the door

**2.** the length of the chalkboard

**3.** the width of the hall

**4.** the bulletin board

**Complete. If you need to, look at a centimeter tape.**

5. 125 cm = ? m
6. 150 cm = ? m
7. 200 cm = ? m
8. 1 m = ? cm
9. 1.4 m = ? cm
10. 2.34 m = ? cm

1 kilometer (km) = 1000 m

**Solve.**

11. How far is it to Niagara Falls?
12. How far is it from Utica to Rochester?
13. How far is it from Rochester to Niagara Falls?

**Centimeter, meter, or kilometer?**
**Tell which unit you would use to measure**

14. a pencil.
15. the distance to the moon.
16. a tall building.
17. the length of an airplane flight.
18. a city block.
19. the thickness of a dictionary.

**Complete.**

20. 2 km = ? m
21. 2.5 km = ? m
22. 3.48 km = ? m

## *Challenge!*

**The decimal point is missing from each 3-digit numeral. Copy the numeral and place the decimal point so that each statement makes sense.**

23. Jacqueline's math book is 205 centimeters wide.
24. Kazumi can high-jump 106 meters.
25. Brenda hiked 112 kilometers in 4 hours.

26. a. Estimate in meters the length of your classroom.
    b. Keep a record of your classmates' estimates.
    c. Measure the distance to see who had the closest estimate.

# Perimeter

The distance around a figure is called the **perimeter** of the figure.

The perimeter of the picture is 70 centimeters.

The perimeter of a figure can be found by adding the lengths of its sides.

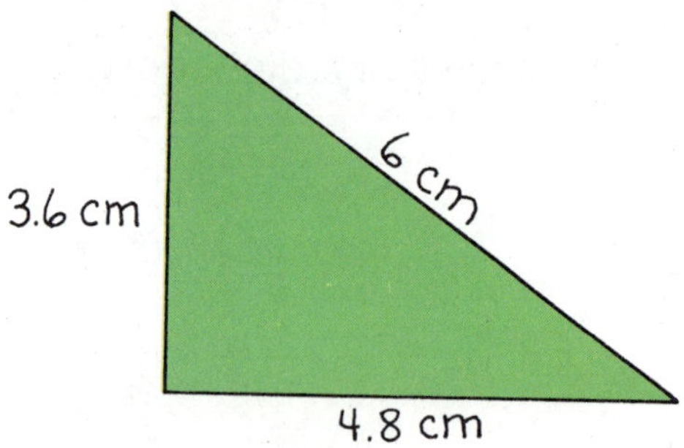


$$\begin{array}{r} \scriptstyle 1 \\ 3.6 \text{ cm} \\ 4.8 \text{ cm} \\ +6.0 \text{ cm} \\ \hline 14.4 \text{ cm} \end{array}$$

The perimeter of the triangle is 14.4 centimeters.

**EXERCISES**

**Give the perimeter of each figure.**

**1.**

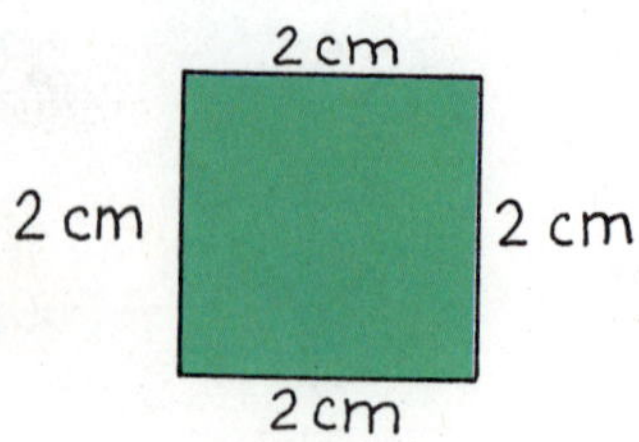


**2.**

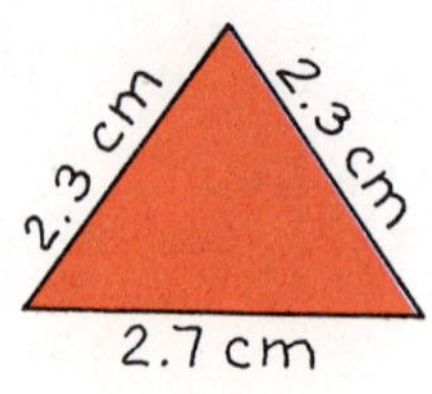


**3.**

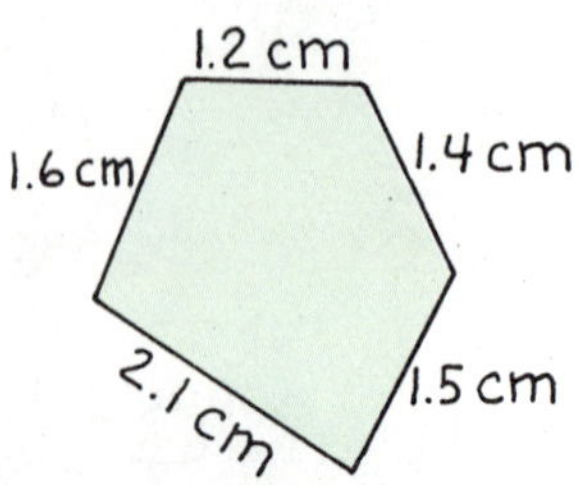


**4.**

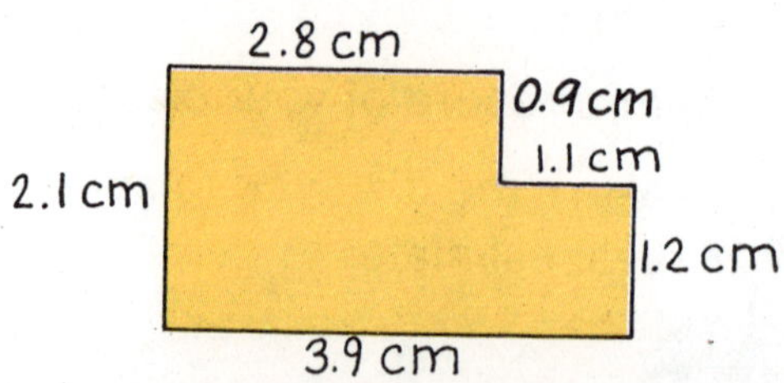

*Remember:* The units must be the same before you add the numbers.

**5.**

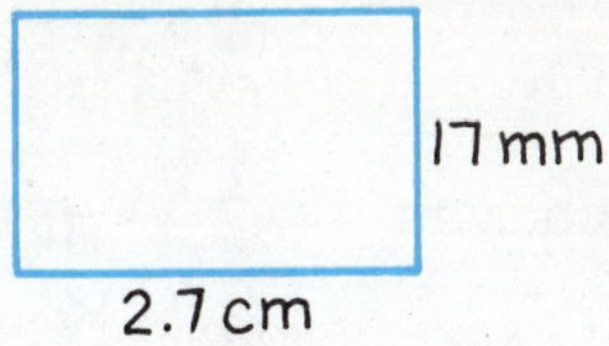


**6.**

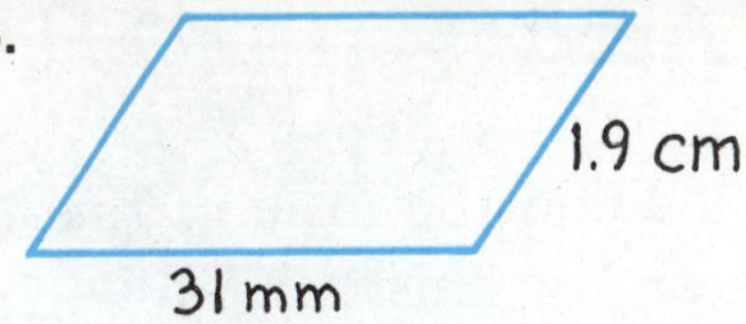


**Copy and complete these tables.**

**7.**

| *Squares* | |
|---|---|
| Length of each side | Perimeter |
| 32 mm | ? |
| 15 m | ? |
| 2.5 cm | ? |
| 5.25 m | ? |

**8.**

| *Rectangles* | | |
|---|---|---|
| Length | Width | Perimeter |
| 4 cm | 2 cm | ? |
| 18 mm | 12 mm | ? |
| 12.4 m | 7.6 m | ? |
| 24.3 m | 18.7 m | ? |

**9.** Measure the length and width of a chalkboard to the nearest cm. Compute the perimeter.

**10.** Estimate the perimeter of your classroom to the nearest meter. Check your estimate by measuring.

**11.** The perimeter of a circle is called the **circumference** of the circle. Find a round object and measure its circumference.

**Solve.**

**12.** How much fencing is needed for a rectangular field that is 185 meters long and 76 meters wide?

**13.** A square park has a perimeter of 6812 meters. How long is each side?

## KEEPING SKILLS SHARP

**Give each product in simplest form.**

**1.** $\frac{3}{8} \times \frac{1}{2}$ **2.** $\frac{1}{2} \times \frac{1}{3}$ **3.** $\frac{3}{4} \times \frac{1}{5}$ **4.** $\frac{5}{6} \times \frac{3}{4}$

**5.** $\frac{7}{8} \times 4$ **6.** $\frac{4}{3} \times \frac{3}{4}$ **7.** $\frac{3}{4} \times 4$ **8.** $\frac{9}{2} \times \frac{4}{3}$

# Area

To find the **area** of this region, you can use this unit

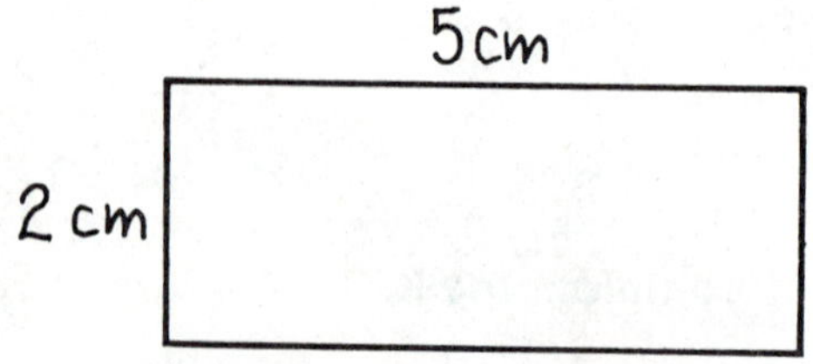

1 square centimeter

and count the units (square-centimeter tiles) that it takes to cover the region.

Here is another way to find the **area of a rectangle.**

$$\text{Area} = \text{length} \times \text{width}$$
$$= 5 \text{ cm} \times 2 \text{ cm}$$
$$= 10 \text{ square centimeters}$$

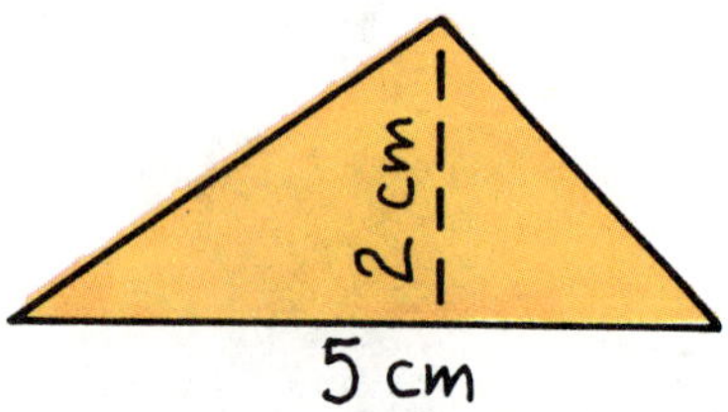


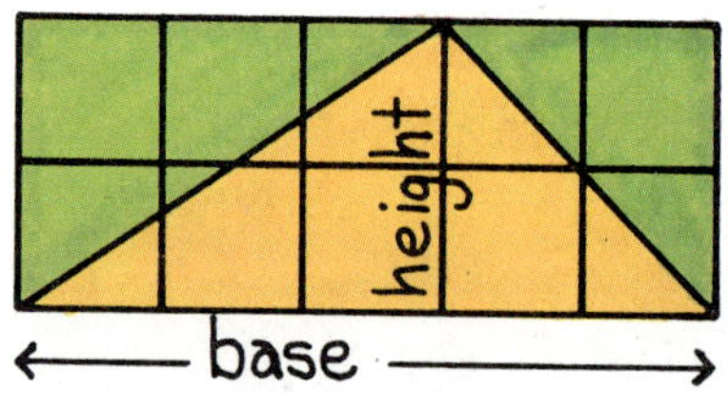


Notice that the area of the triangle is $\frac{1}{2}$ the area of the rectangle.

Here is how to find the **area of a triangle.**

$$\text{Area} = \frac{1}{2} \times \text{base} \times \text{height}$$
$$= \frac{1}{2} \times 5 \text{ cm} \times 2 \text{ cm}$$
$$= 5 \text{ square centimeters}$$

## EXERCISES

**Give each area.**

**1.**

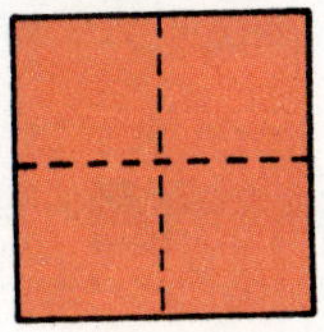

**2.**

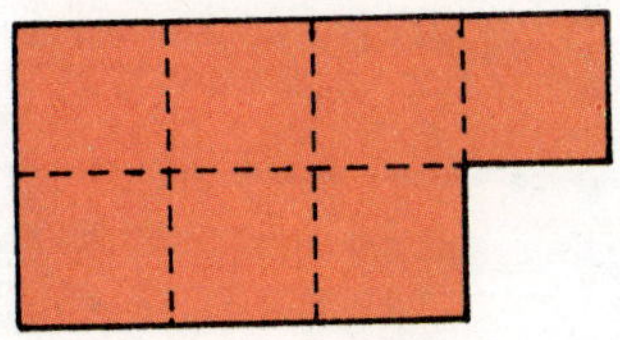

**3.**

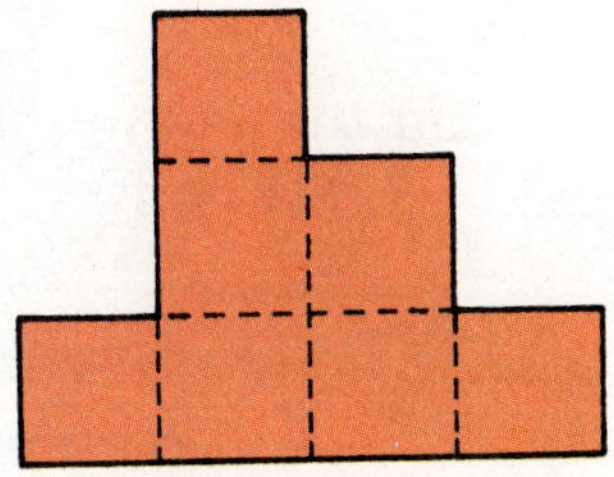

**4.**

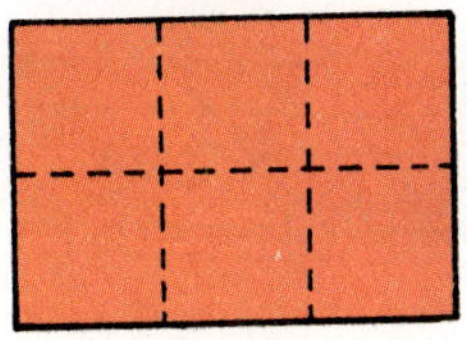

**5.**

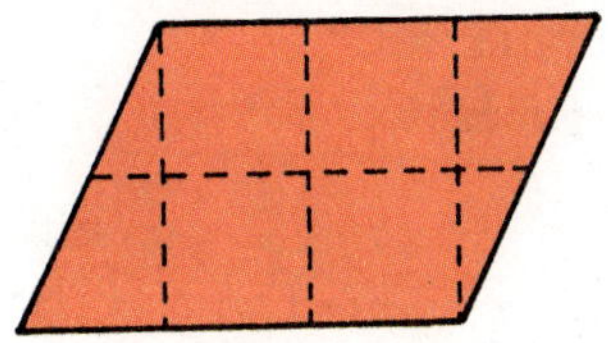

**6.**

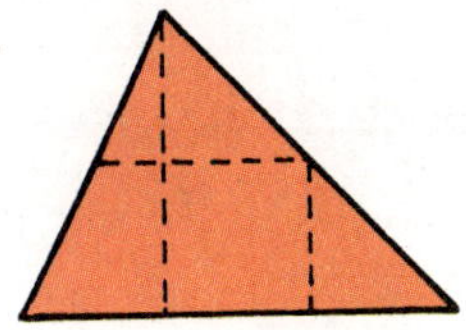

**Give each area.**

**7.**

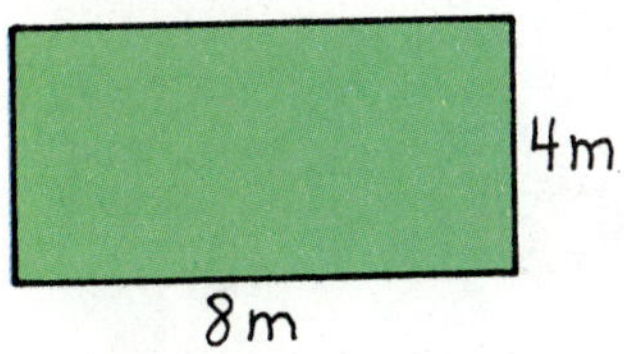


**8.**

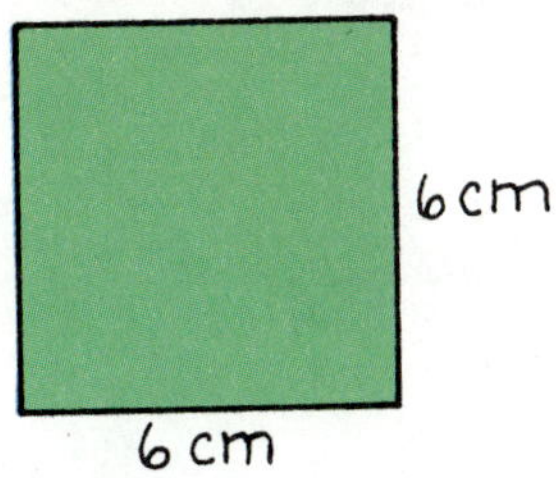


**9.**

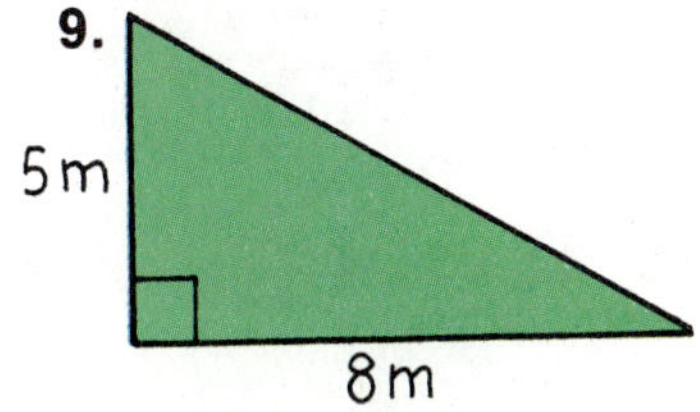


**10.**

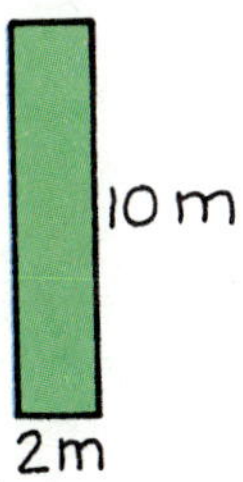


**11.**

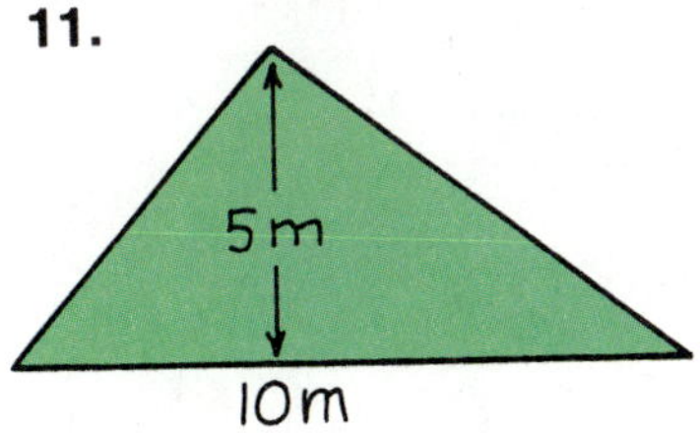


**12.**

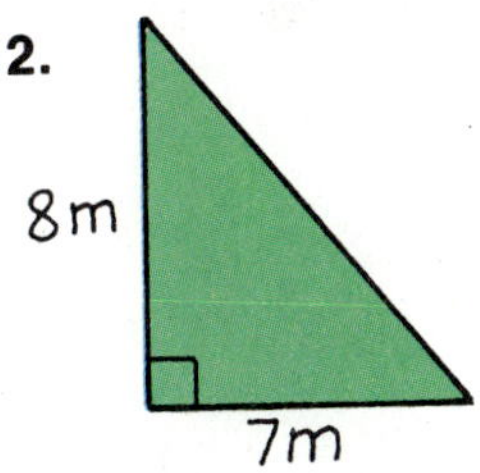


**13.** **a.** Measure to the nearest cm the length and width of your math book cover.
**b.** Compute the area.

**14.** **a.** Measure to the nearest meter the length and width of your classroom.
**b.** Compute the area.

**Solve.**

**15.** The base of a triangular piece of tile is 64 millimeters. Its height is 86 millimeters. What is its area?

★ **16.** The area of a rectangular piece of paper is 180 square centimeters. It is 15 centimeters long. What is its width?

# Volume

To find the **volume** of this box, you can use this unit

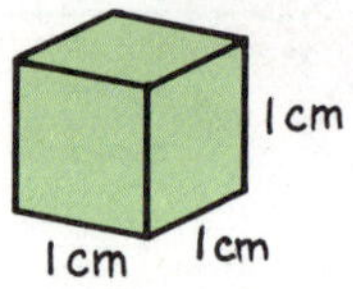


1 cubic centimeter

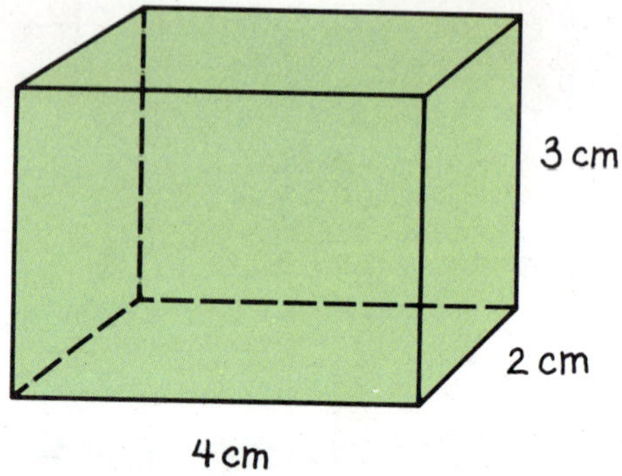


and count the units (cubic-centimeter blocks) that it takes to fill the box.

There are 4 × 2 blocks in a layer.

There are 3 layers.

So, it takes 4 × 2 × 3, or 24, blocks to fill the box.

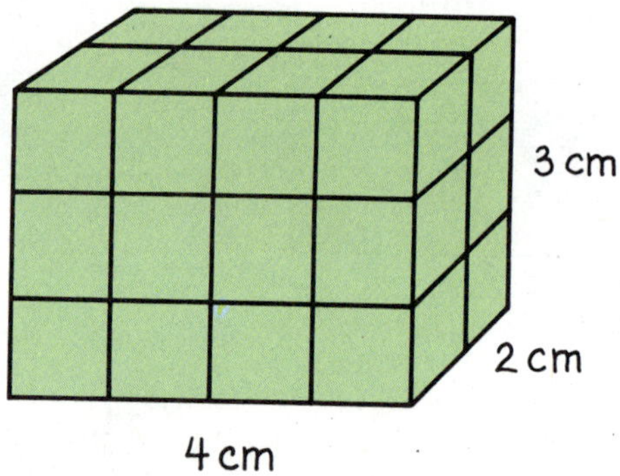


Here is how to find the **volume of a box.**

Volume = length × width × height

= 4 cm × 2 cm × 3 cm

= 24 cubic centimeters

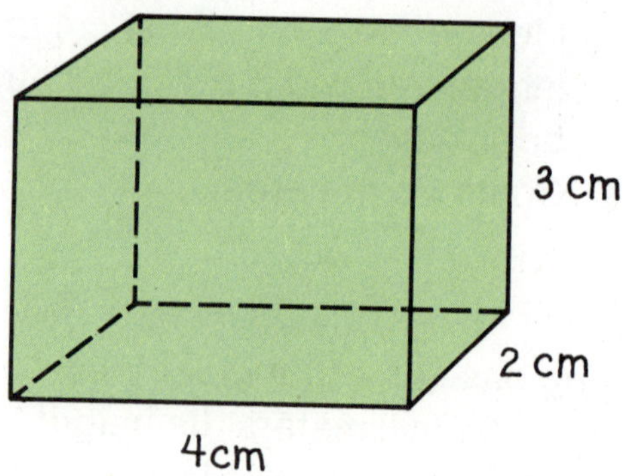

## EXERCISES

**Give each volume.**

**1.**

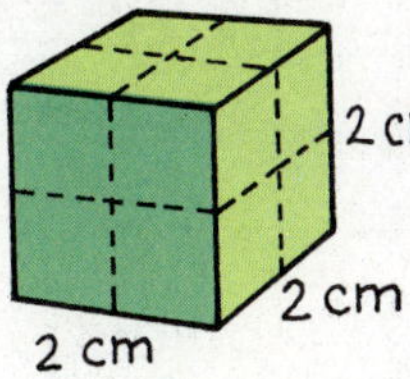


**2.**

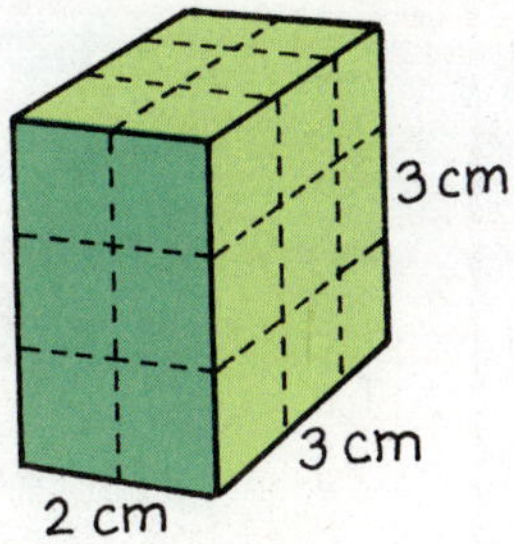


**3.**

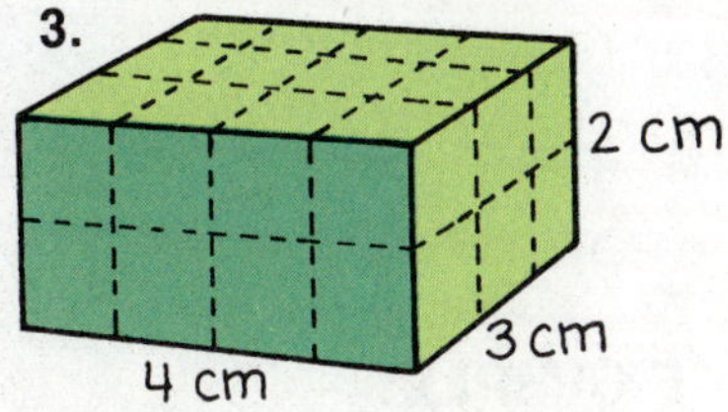


**4.**

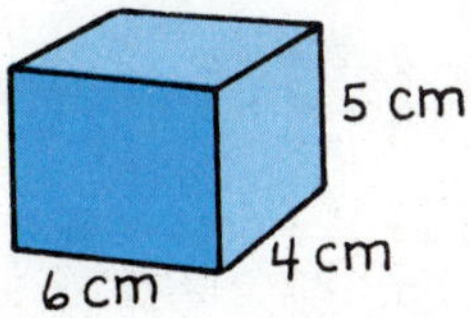


**5.**

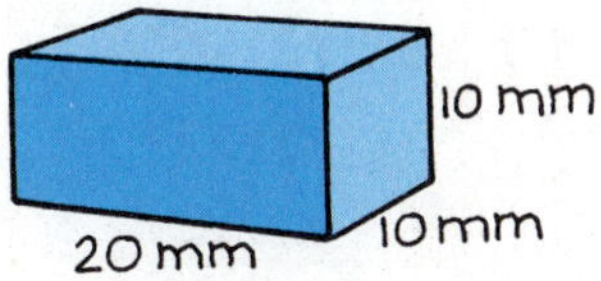


**6.**

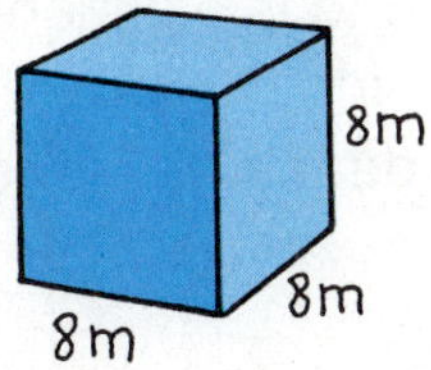


**7.**

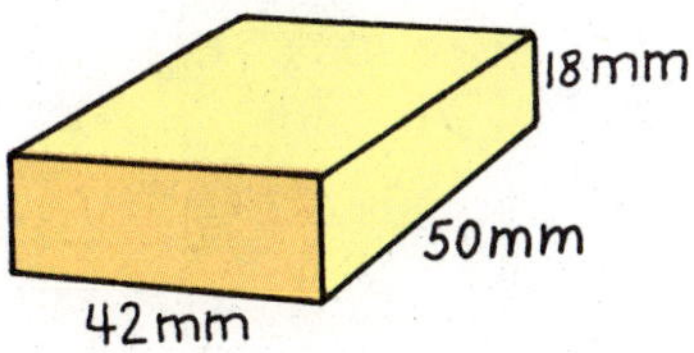


**8.**

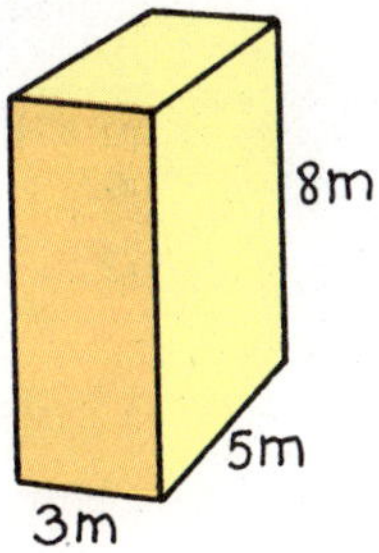


**9.**

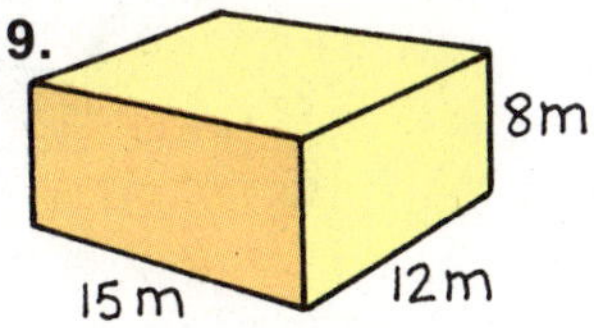


**10.** **a.** Measure to the nearest centimeter the length, width and height of a desk drawer.
**b.** Compute its volume.

**11.** **a.** Measure to the nearest centimeter the length, width, and height of a bookcase.
**b.** Compute its volume.

**Solve.**

**12.** An aquarium is 42 centimeters long, 18 centimeters wide, and 20 centimeters high. How many cubic centimeters of water wil it hold?

★ **13.** The volume of a sandbox is 3 cubic meters. The sandbox is 3 meters long and 0.5 meter deep. How wide is the sandbox?

# Problem solving—drawing pictures

**EXAMPLE.**

The club members play baseball in a field that is 50 meters wide and 60 meters long. The baseball diamond is a square with 27 meters on a side. How much of the field is outside the diamond?

1. Study and understand.

   Draw a simple picture.

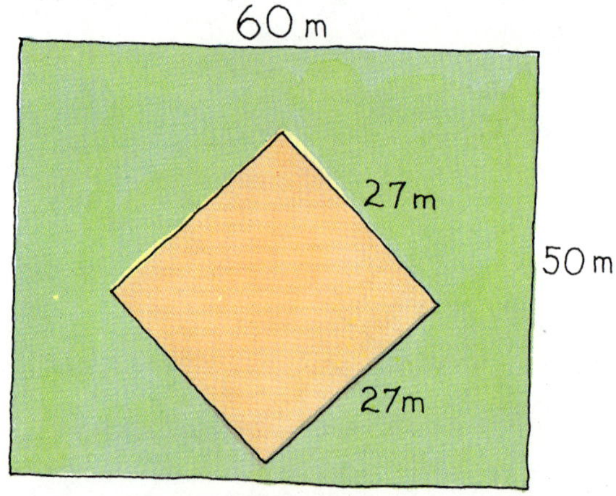


2. Plan and do.

   Use your picture to plan what to do.

You need to take away one area from another area. So, you need to multiply and subtract.

$$\begin{array}{r} 60 \\ \times 50 \\ \hline 3000 \end{array} \qquad \begin{array}{r} 27 \\ \times 27 \\ \hline 189 \\ 54\phantom{0} \\ \hline 729 \end{array} \qquad \begin{array}{r} 3000 \\ -\ 729 \\ \hline 2271 \end{array}$$

3. Answer and check.

2271 square meters

**Solve.**

1. The club members are going to put a line of stones around their clubhouse. The stones will form a 5 m by 6 m rectangle, except for an opening 1 m wide. How long will the line of stones be?

2. One clubhouse wall is to have a door opening cut out of it. The wall is 2 m high and 4 m long. The door opening is to be 1 m high and 0.5 m wide. How much of the wall will be left after the opening is cut out?

3. Another clubhouse wall is to have a window opening cut out of it. The wall is 200 cm high and 400 cm long. The window opening is to be 50 cm long and 50 cm high. The window is to be centered in the wall.
   a. How much of the wall will be left after the opening is cut out?
   b. How far from the floor should the window be? How far from each side?

4. On a bike trip the club members came to an intersection. A sign showed them that Rochester was 45 km to the left and Eden was 38 km to the right.
   a. How much farther was it to Rochester than to Eden?
   b. How far apart were Eden and Rochester?

5. On a bike trip the club members rode 5 km east, then 6 km north, then 10 km east, then 6 km south, and then 17 km east. At the end of their trip, how far from their starting point were they?

6. The club members went on a loop hike that was 12 km long. They hiked 5 km to the pond and then 3 km to the waterfall. How many more kilometers did they have to go to finish the loop hike?

7. The club members are building a fence for a vegetable garden. The garden is a rectangle that is 3 m long and 2 m wide. Sticks will be put in the ground every half meter to hold the fence wire. How many sticks will be needed?

★ 8. Larry made a frame for a picture of the club members. The frame was 40 cm long and 30 cm wide. The wood in the frame was 2 cm wide. What area was left for the picture?

# Liquid volume and weight—metric

1 **milliliter** of water will fill this cube:

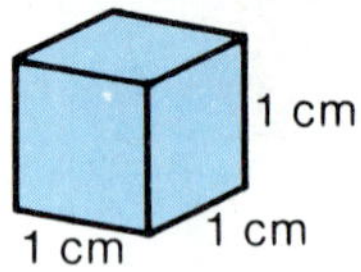


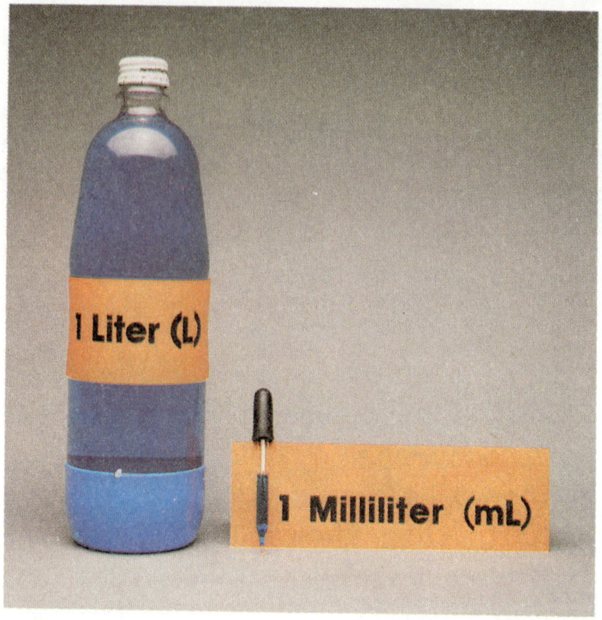


The **liter** is used to measure larger liquid volumes.

1000 milliliters (mL) = 1 liter (L)

1 milliliter = 0.001 liter

**EXERCISES**

**Choose the answer that seems right.**

**1.** 

The amount of ink in a pen is

**a.** 2 mL **b.** 2 L

**2.** 

The amount of gasoline in a car tank is

**a.** 60 mL **b.** 60 L

**3.** 

The amount of orange juice in a glass is

**a.** 200 mL **b.** 200 L

**Complete.**

**4.** 

500 mL or ? L

**5.** 

750 mL or ? L

**6.** 

1.5 L or ? mL

**7.** 2000 mL = ? L

**8.** 375 mL = ? L

**9.** 2.4 L = ? mL

**10.** 4000 mL = ? L

**11.** 105 mL = ? L

**12.** 3.75 L = ? mL

**PROJECT**

Collect some containers and estimate the volume of each.

Check your estimates by filling each container.

The water in this cube weighs 1 **gram (g)**.

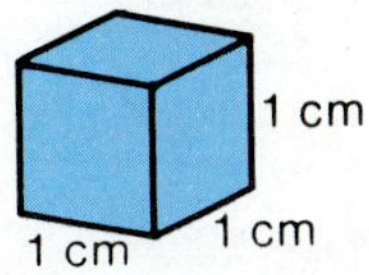


**The kilogram** is used to measure heavier objects.

1000 grams (g) = 1 kilogram (kg)
1 gram = 0.001 kilogram

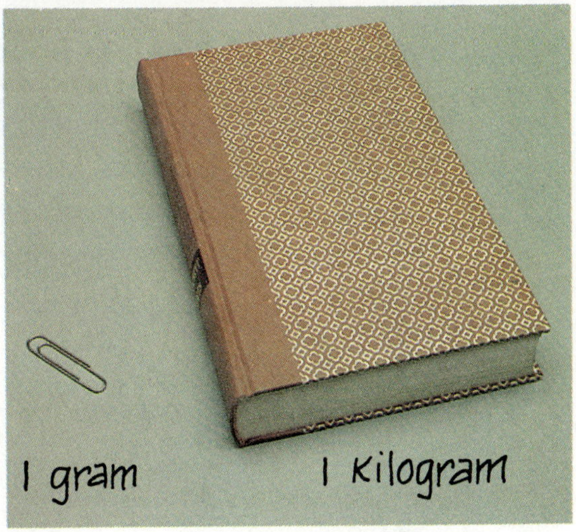


## EXERCISES

**Choose the answer that seems right.**

**13.** 

The sandwich weighs

**a.** 175 g **b.** 175 kg

**14.** 

The penny weighs

**a.** 3 g **b.** 3 kg

**15.** 

The bicycle weighs

**a.** 12 g **b.** 12 kg

**Complete.**

**16.** 

550 g or ? kg

**17.** 

1250 g or ? kg

**18.** 

3.4 kg or ? g

**19.** 1500 g = ? kg

**20.** 248 g = ? kg

**21.** 0.043 kg = ? g

**22.** 3200 g = ? kg

**23.** 500 g = ? kg

**24.** 0.03 kg = ? g

**PROJECT**

Estimate the weights of some objects in your classroom.

Check your estimates with a kilogram scale.

# Inch

Now we will measure length using the customary system. The **inch** (in.) is a unit in the customary system for measuring length.

The ruler is marked in $\frac{1}{16}$ inches. It can be used to measure the length of the needle to the nearest inch, $\frac{1}{2}$ inch, $\frac{1}{4}$ inch, $\frac{1}{8}$ inch, or $\frac{1}{16}$ inch.

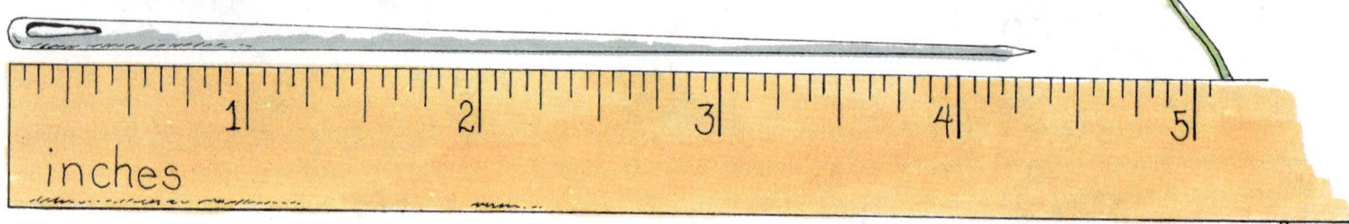

The length of the needle measured

to the nearest inch is 4 in. (4″).

to the nearest $\frac{1}{2}''$ is $4\frac{1}{2}''$.

to the nearest $\frac{1}{4}''$ is $4\frac{1}{4}''$.

to the nearest $\frac{1}{8}''$ is $4\frac{3}{8}''$.

to the nearest $\frac{1}{16}''$ is $4\frac{5}{16}''$.

**EXERCISES**

**Draw segments of these lengths.**

**1.** $2\frac{1}{2}$ in.

**2.** 3 in.

**3.** $1\frac{3}{4}$ in.

**4.** $2\frac{5}{8}$ in.

**5.** $3\frac{1}{4}$ in.

**6.** $1\frac{1}{16}$ in.

**7.** $2\frac{3}{16}$ in.

**8.** $\frac{7}{8}$ in.

**9.** $1\frac{3}{8}$ in.

**10.** $3\frac{5}{16}$ in.

**11.** $2\frac{15}{16}$ in.

**12.** $\frac{11}{16}$ in.

**Measure the length of each object to the nearest $\frac{1}{4}$ in., $\frac{1}{8}$ in., and $\frac{1}{16}$ in.**

**13.** 

**14.** 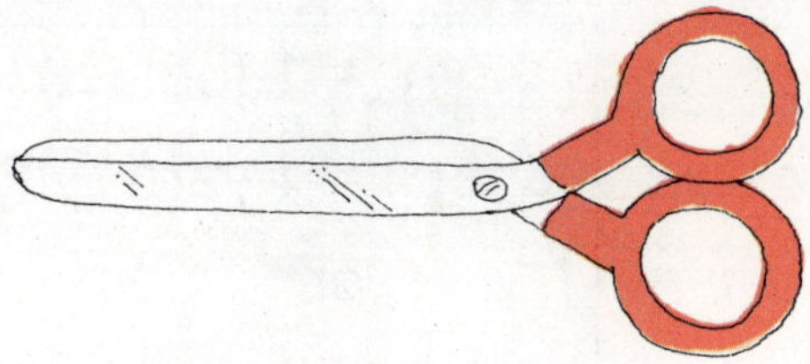

**15.** 

**Measure**

**16.** the length of your pencil to the nearest $\frac{1}{8}''$.

**17.** the width of your math book to the nearest $\frac{1}{4}''$.

**18.** the diameter of (distance across) a penny to the nearest $\frac{1}{16}''$.

**Which is longer?**

**19.** $1\frac{1}{2}''$ or $1\frac{1}{4}''$

**20.** $2\frac{1}{4}''$ or $2\frac{3}{8}''$

**21.** $4\frac{1}{2}''$ or $4\frac{7}{16}''$

**22.** $3\frac{7}{8}''$ or $3\frac{9}{16}''$

**23.** $5\frac{13}{16}''$ or $5\frac{3}{4}''$

**24.** $6\frac{5}{8}''$ or $5\frac{11}{16}''$

**Solve.**

**25.** The thickness of a certain book is $\frac{7}{8}$ inch. How tall would a stack of 12 of these books be?

**26.** A rectangular picture is $7\frac{1}{2}$ inches long and 5 inches wide. What is its perimeter?

## KEEPING SKILLS SHARP

**Give each sum or difference in simplest form.**

**1.** $\frac{3}{5} + \frac{1}{2}$

**2.** $\frac{3}{8} + \frac{1}{2}$

**3.** $\frac{3}{4} + \frac{1}{6}$

**4.** $\frac{3}{8} - \frac{1}{8}$

**5.** $\frac{2}{3} - \frac{1}{4}$

**6.** $\frac{5}{6} - \frac{5}{9}$

**7.** $\frac{1}{3} + \frac{2}{5}$

**8.** $\frac{7}{10} + \frac{1}{2}$

**9.** $\frac{1}{3} + \frac{1}{2}$

**10.** $\frac{9}{10} - \frac{1}{2}$

**11.** $\frac{5}{6} - \frac{1}{2}$

**12.** $\frac{5}{6} - \frac{3}{4}$

# Inch, foot, yard, and mile

| 12 in. = 1 foot (ft) |
|---|
| 3 ft (′) = 1 yard (yd) |
| or |
| 36 in. = 1 yd |

| 5280 ft = 1 mile (mi) |
|---|
| or |
| 1760 yd = 1 mi |

Sometimes you will need to change from one unit to another. Study these examples.

**EXAMPLE 1.**

5 ft 4 in. = ? in.

To change feet to inches, multiply by 12.

Solution:

$$\begin{array}{r} 12 \\ \times 5 \\ \hline 60 \\ +4 \\ \hline 64 \end{array}$$

So, 5 ft 4 in. = 64 in.

**EXAMPLE 2.**

196 ft = ? yd ? ft

To change feet to yards, divide by 3.

Solution:

$$\begin{array}{r} 65 \\ 3\overline{)196} \\ -18 \\ \hline 16 \\ -15 \\ \hline 1 \end{array}$$

So, 196 ft = 65 yd 1 ft.

**EXERCISES**

**Multiply or divide?**

1. To change from feet to inches, ? by 12.
2. To change from inches to feet, ? by 12.
3. To change from feet to yards, ? by 3.
4. To change from yards to feet, ? by 3.
5. To change from inches to yards, ? by 36.
6. To change from yards to inches, ? by 36.

**Complete.**

7. 8 ft = ? in.
8. 48 ft = ? yd
9. 132 in. = ? ft
10. 13 in. = ? ft ? in.
11. 15 yd = ? ft
12. 79 ft = ? yd ? ft
13. 2 mi = ? ft
14. 5 mi = ? yd

15.

Blue Whale
Largest ever measured
Length: 110′

? yd ? ft

16.

African Bush Elephant
Largest ever measured
Length: 33′2″

? in.

**Make the measurement. Then copy and complete the sentence.**

17. Your height
? in. = ? ft ? in.

18. Your waist
? in. = ? ft ? in.

19. The height of your classroom
? ft = ? yd ? ft

20. The length of your classroom
? ft = ? yd ? ft

★ **Add.** (Regroup 12 in. for 1 ft) (Regroup 3 ft for 1 yd)

21. 5 ft 9 in. + 3 ft 6 in. (1 carried; 3 in.)

22. 7 ft 8 in. + 6 ft 6 in.

23. 9 yd 2 ft + 3 yd 2 ft

24. 8 yd + 15 yd 2 ft

★ **Subtract.** (Regroup 1 ft for 12 in.)

25. 8 ft 2 in. − 2 ft 5 in. (7 ft 14 in.; 9 in.)

26. 6 ft − 2 ft 7 in.

27. 5 yd 1 ft − 1 yd 2 ft

28. 6 yd − 4 yd 1 ft

# Perimeter, area, and volume

You can add to find the perimeter.

```
   2
 5 ft  6 in.
 3 ft 10 in.
 5 ft  6 in.
+3 ft 10 in.
18 ft  8 in.
```

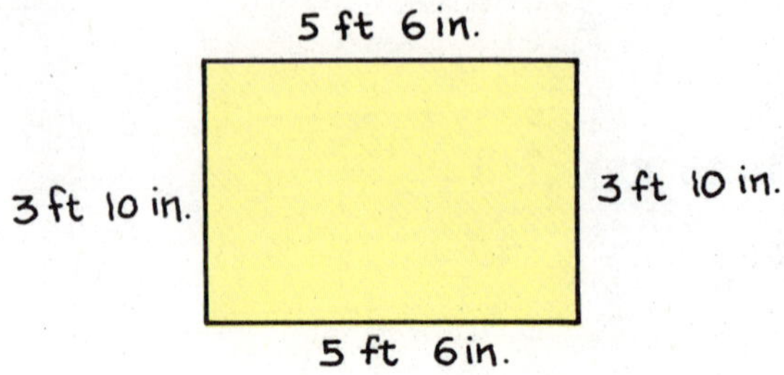


You can multiply to find the area.

Area = length × width
Area = 12 yd × 3 yd
Area = 36 square yards

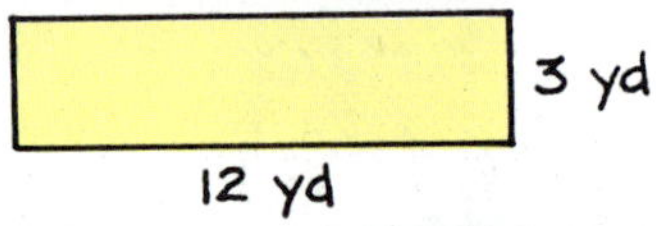


You can multiply to find the volume.

Volume = length × width × height
Volume = 5 yd × 6 yd × 4 yd
Volume = 120 cubic yards

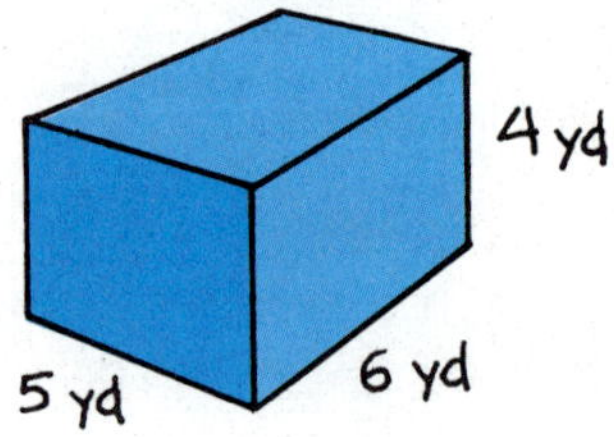


**EXERCISES**

**Give the perimeter of each rectangle.**

**1.** 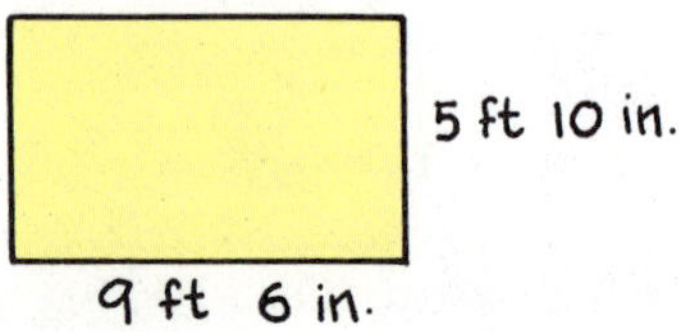


**2.** 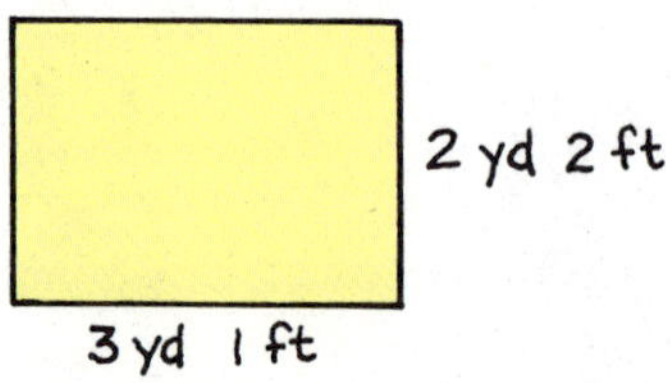


**3.** 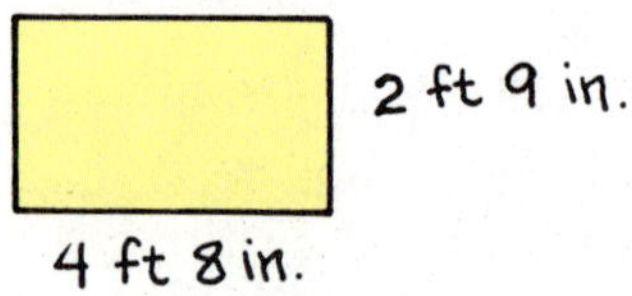


**Give the area of each rectangle.**

**4.** 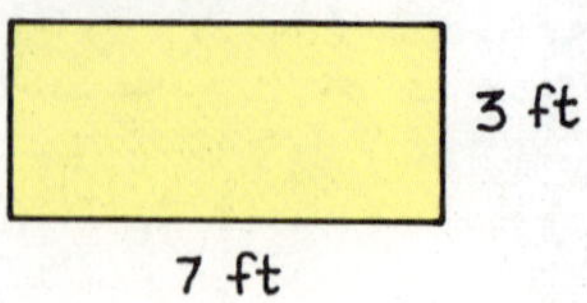


**5.** 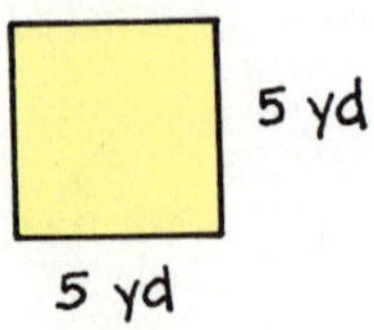


**6.** 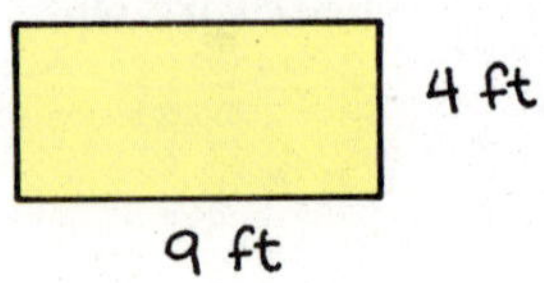


**Give the volume of each rectangular solid.**

**7.** 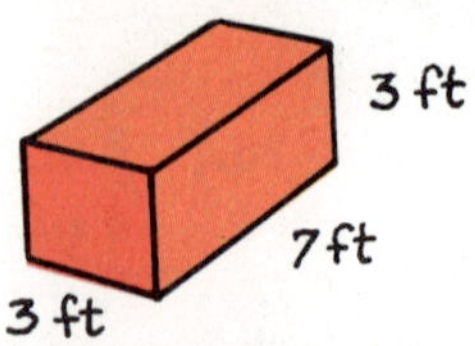


**8.** 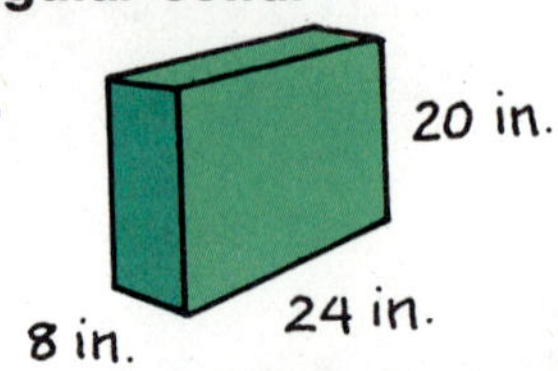


**9.** 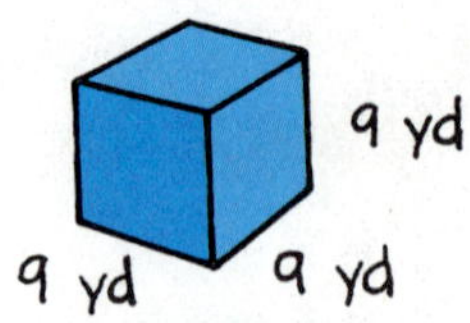

**Is the question about length, area, or volume?**

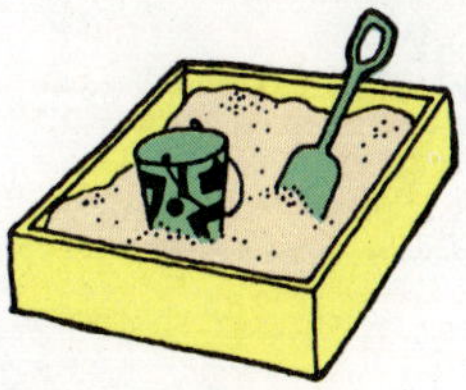

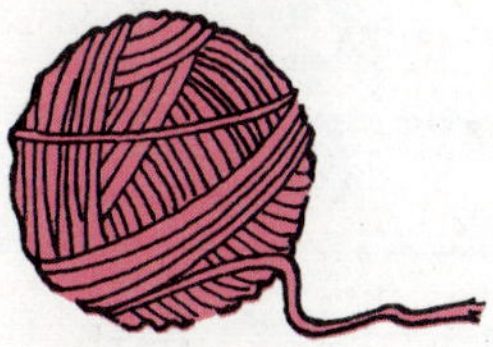

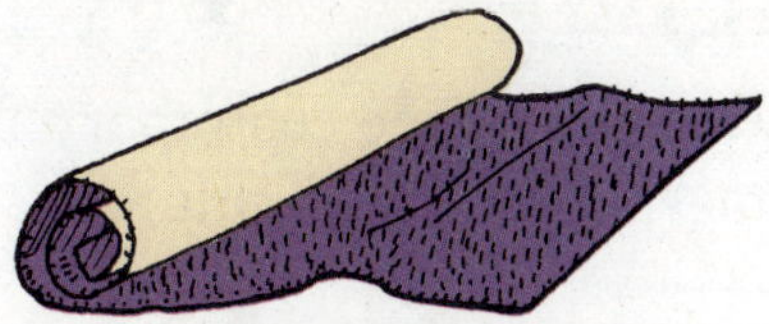

10. How much sand will the box hold?

11. How much string is on the spool?

12. How much carpet is in the roll?

**Solve.**

13. A rectangular floor is 20 feet long and 13 feet wide. What is its perimeter? Its area?

14. The area of a rectangular patio is 72 square feet. Its length is 9 feet. What is its width?

15. How many square inches in a square foot? In a square yard?

16. How many cubic inches in a cubic foot? In a cubic yard?

17. The perimeter of a square garden is 40 feet. What is its area?

18. The area of a square is 49 square feet. What is its perimeter?

19. The area of a rectangular carpet is 54 square yards. Its width is 6 yards. What is its perimeter?

20. The width of a rectangle is 5 yards. Its length is twice its width. What is its area?

## *Challenge!*

**Each figure is made from rectangles. Find each area.**

21.

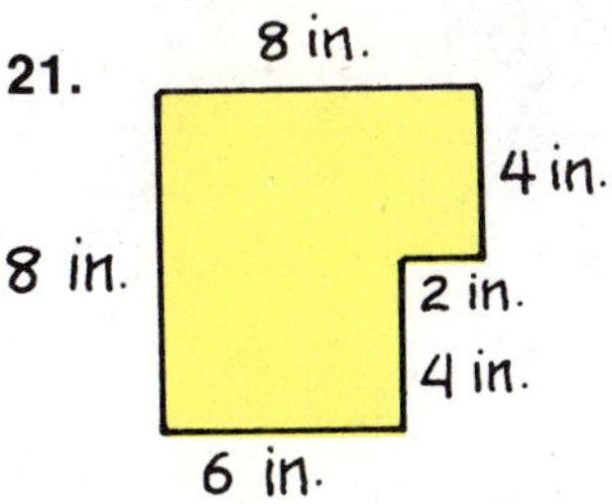


22.

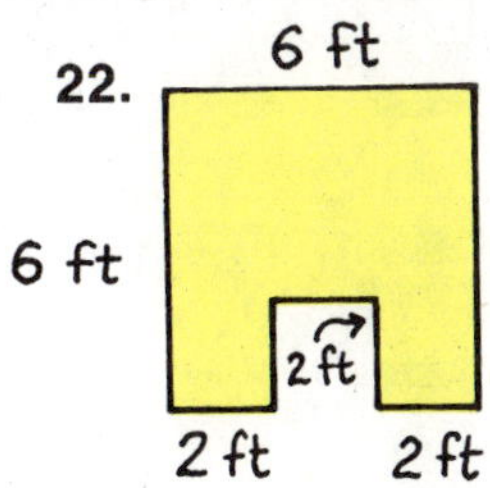


23.

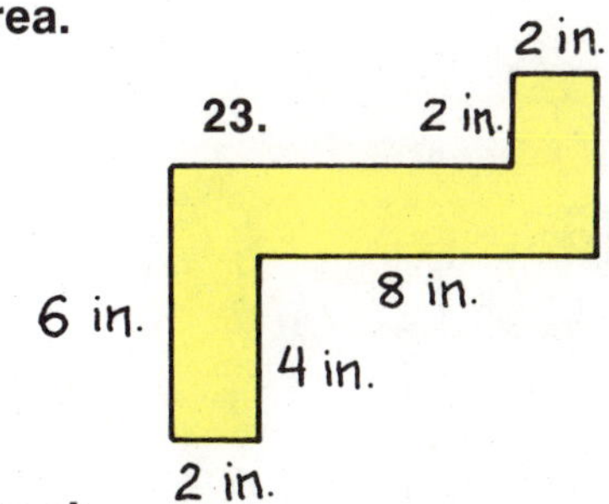


**Each solid is made from rectangular solids. Find each volume.**

24.

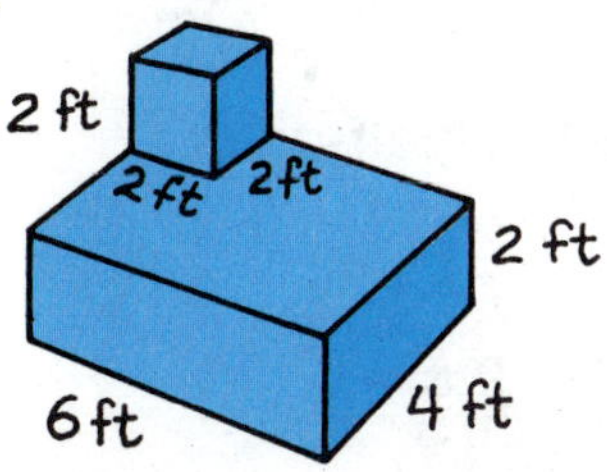


25.

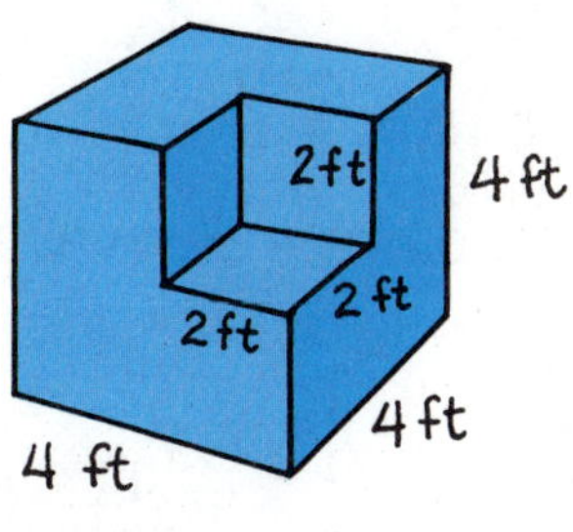


26.

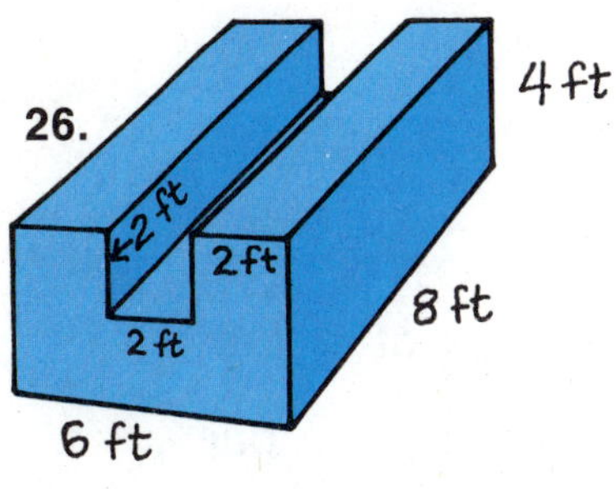

# Liquid volume and weight

Here are some customary units for measuring liquid volumes.

1 cup (c)

2 c = 1 pint (pt)

2 pt = 1 quart (qt)

2 qt = 1 half-gallon (half-gal)

4 qt = 1 gallon (gal)

**EXERCISES**

**Complete.**

**1.**

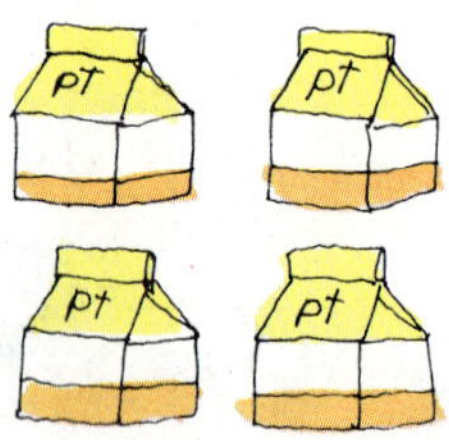


? qt

**2.**

? pt

**3.**

? gal

**4.** 2 qt = ? pt

**5.** 6 c = ? pt

**6.** 3 half-gal = ? qt

**7.** 6 qt = ? half-gal

**8.** 8 pt = ? qt

**9.** 2 gal = ? qt

**10.** 10 pt = ? c

**11.** 12 qt = ? gal

**12.** 8 half-gal = ? gal

**13.** $2\frac{1}{2}$ qt = ? pt

**14.** $3\frac{1}{2}$ qt = ? pt

**15.** $2\frac{3}{4}$ gal = ? qt

**16.** 9 qt = ? gal

**17.** 11 pt = ? qt

**18.** 9 c = ? pt

Here are some units for measuring weight.

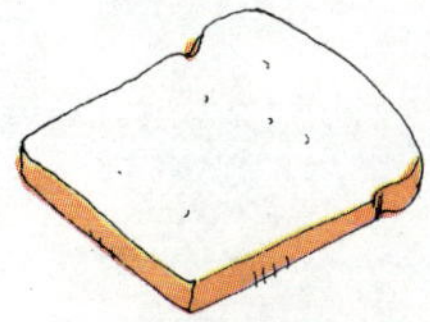

The **ounce** (oz) is used to measure small units.

about 1 oz

16 oz = 1 pound (lb)

The **pound** is used to measure heavier weights.

about 1 lb

2000 lb = 1 ton

The **ton** is used to measure quite heavy objects.

about 1 ton

**Choose the correct weight.**

**19.**

72 oz
72 lb
72 tons

**20.**

40 oz
40 lb
40 tons

## *Challenge!*

**Solve.**

**21.** **a.** Weigh yourself in pounds.

**b.** Is the combined weight of all the students in your class less than or greater than a ton?

**22.** If bricks weigh 40 ounces each, how many bricks would be needed to balance a $1\frac{1}{2}$-ton elephant?

## Problem solving

Architects design buildings where people work and live. One of the things that an architect considers when designing a building is the floor plan. Here is the floor plan of a three-bedroom house.

1. What is the length of the family room?

2. What is the width of the smallest bedroom?

3. a. A carpenter wants to put a molding around the ceiling of the smallest bedroom. How much molding is needed?
   b. If the molding comes in 5-foot lengths, how many lengths should the carpenter order?

4. a. The living room floor is to be carpeted. How many square feet of carpeting are needed?
   b. How many square yards of carpeting are needed? Round your answer up to the next yard.
   c. If the carpet costs \$9.50 a square yard, what will the total cost be?

5. The floor in Bedroom A is to be covered with carpet tiles that are 1 foot square. How many tiles are needed?

6. The family room floor is to be covered with square tiles that measure 8 inches on a side. How many tiles are needed?

7. A heating and cooling contractor needs to find the volume of the house. If the ceilings are 8 feet high, what is the volume of the living room?

★ 8. The walls and ceiling of the largest bedroom are going to be painted. How many square feet are to be painted? (Don't consider doors and windows.)

★ 9. A sheet of dry wall is 4 feet by 8 feet. How many sheets of dry wall will be needed for the walls of the living room? (Don't consider doors and windows.)

# CHAPTER CHECKUP

**Write A.M. or P.M.** [pages 252–255]

**1.** Time to go shopping: 2:30

**2.** Time to watch the sunrise: 6:20

**Complete.** [pages 252–253]

**3.** 4 days 7 hours = ? hours

**4.** 8 hours 5 minutes = ? minutes

**5.** 3 minutes = ? seconds

**6.** 253 minutes = ? hours ? minutes

**Complete.** [pages 256–259, 268–269]

**7.** 1 m = ? cm

**8.** 1 cm = ? mm

**9.** 3 km = ? m

**10.** 3.8 km = ? m

**11.** 1 liter = ? mL

**12.** 425 mL = ? liter

**13.** 1.2 kg = ? g

**14.** 0.45 kg = ? g

**15.** 500 g = ? kg

**Complete.** [pages 256–257, 260–265]

**16.** Measure the rectangle with a centimeter ruler.

**a.** The length of the rectangle is ? cm.

**b.** The width is ? cm.

**c.** The perimeter ? cm.

**d.** The area is ? square cm.

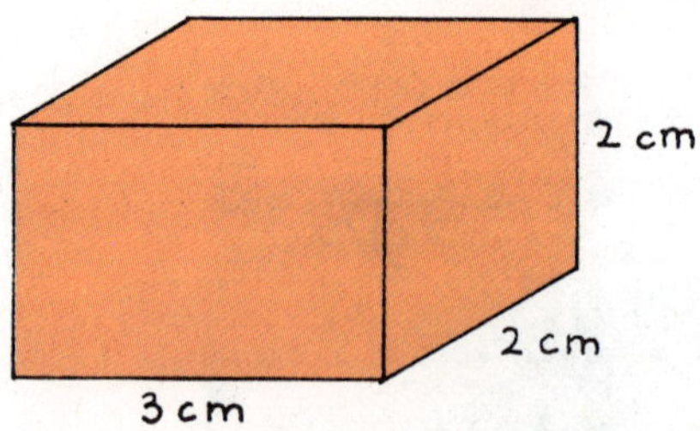


**17.** The volume is ? cubic cm.

[pages 270–271]

**Measure the length of the ribbon to the nearest**

**18.** $\frac{1}{4}$ inch.

**19.** $\frac{1}{8}$ inch.

**20.** $\frac{1}{16}$ inch.

**Complete.** [pages 272–273, 276–277]

**21.** 3 ft = ? in.

**22.** 1 mi = ? ft

**23.** 3 mi = ? yd

**24.** 18 yd = ? ft

**25.** 18 ft = ? yd

**26.** 3 c = ? pt

**27.** 8 qt = ? gal

**28.** 5 lb = ? oz

**29.** 4 tons = ? lb

# CHAPTER PROJECT

How much water would be wasted if you left the drinking fountain on all night?

1. Find out how many seconds you have to leave the fountain on to collect a liter of water.
2. How long is it from the time your school ends one day to the time it begins the next day? How many seconds is that?
3. How many liters of water would be wasted?
4. If the same amount of water were in your classroom, how deep would it be?

# CHAPTER REVIEW

**Complete.**

1. The length is ? centimeters.
2. The width is ? centimeters.
3. The perimeter is ? centimeters.
4. The area is ? square centimeters.

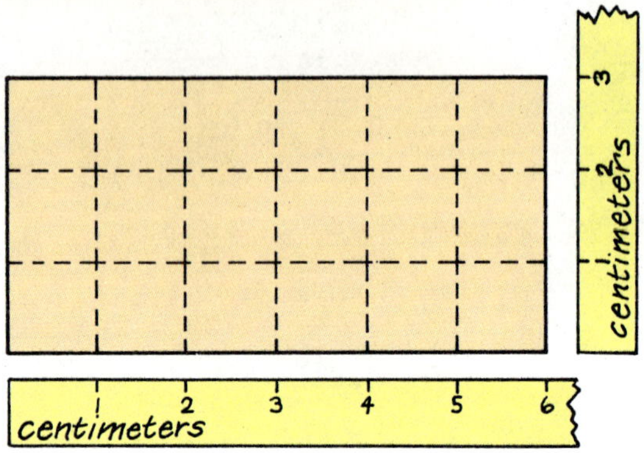


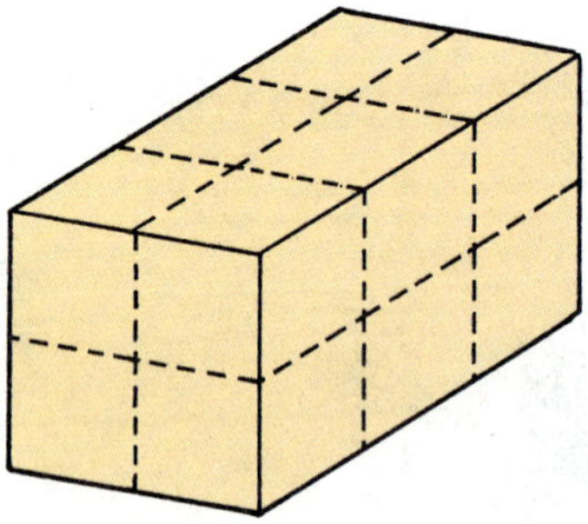

5. The volume is ? cubic centimeters.

**Complete.**

6. 1 centimeter = ? millimeters
7. 1 meter = ? centimeters
8. 1 kilometer = ? meters

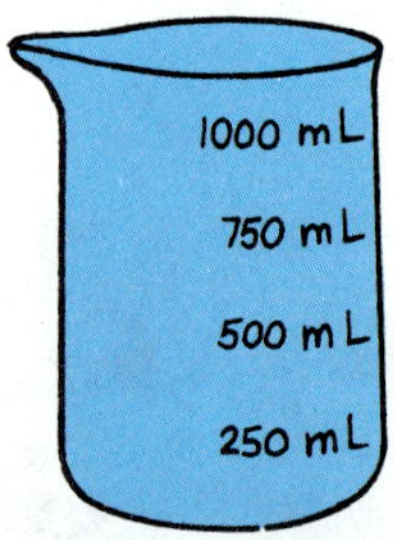


9. 1 liter = ? milliliters
10. 1 kilogram = ? grams

**Measure the length of the pen to the nearest**

11. $\frac{1}{4}$ inch.
12. $\frac{1}{8}$ inch.
13. $\frac{1}{16}$ inch.

**Complete.**

14. 1 foot = ? inches
15. 1 yard = ? feet

# CHAPTER CHALLENGE

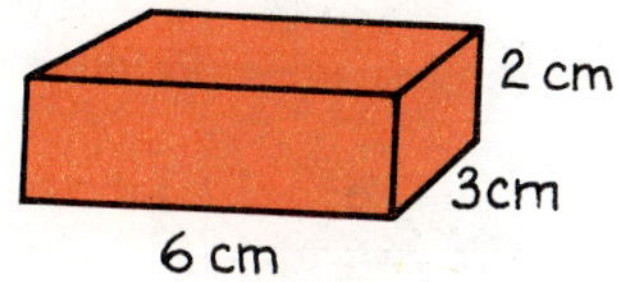


**Look at the red solid.**
**What is the area of the**

1. top?
2. bottom?
3. front?
4. back?
5. right side?
6. left side?

7. What is the surface area of the green solid?

**Give the surface area of each solid.**

8. 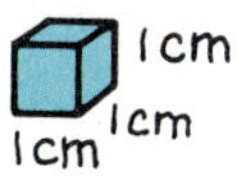


9. 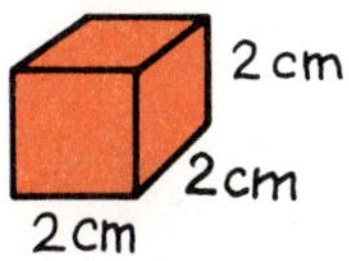


10. 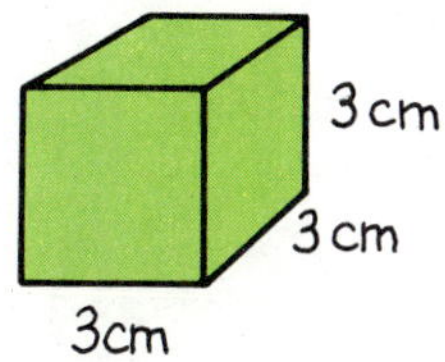


11. 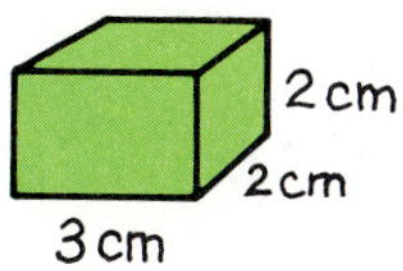


12. 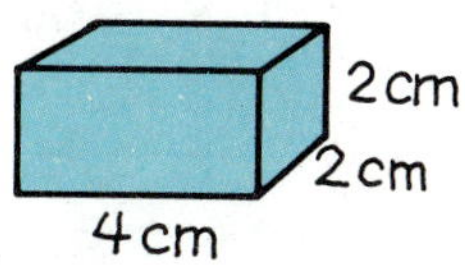


13. 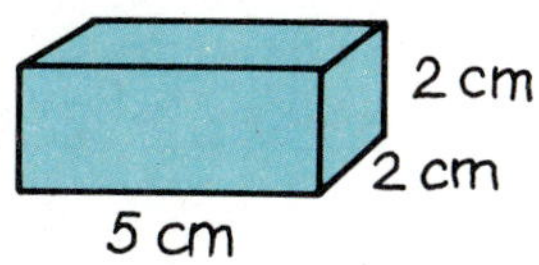


14. 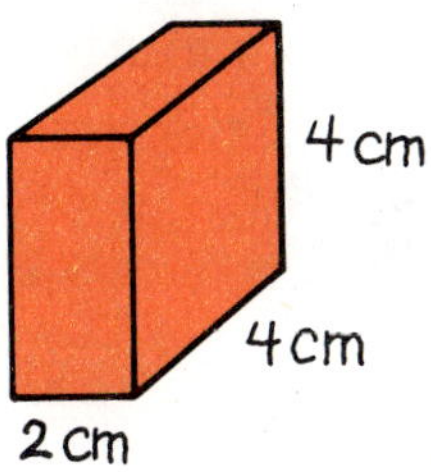


15. 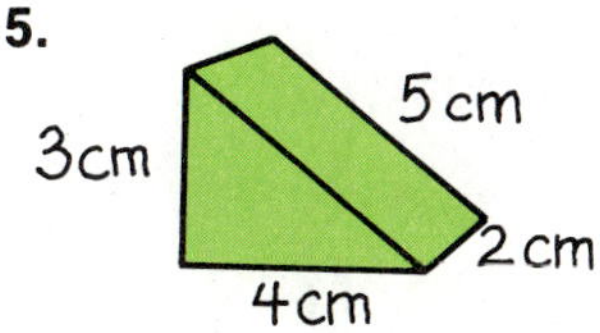


16. 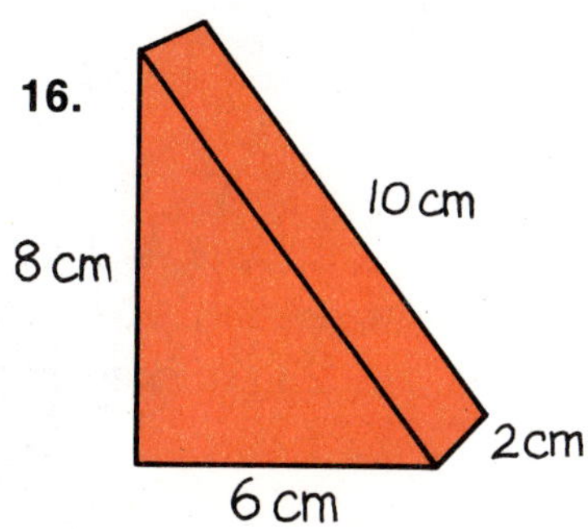

# MAJOR CHECKUP
STANDARDIZED FORMAT

**Choose the correct letter.**

**1.** Multiply.

$$\begin{array}{r} 386 \\ \times 147 \\ \hline \end{array}$$

a. 56,742
b. 56,502
c. 4632
d. none of these

**2.** Divide.

$65\overline{)7839}$

a. 12 R39
b. 120 R39
c. 102 R39
d. none of these

**3.** $\overleftrightarrow{CD}$ is parallel to $\underline{?}$

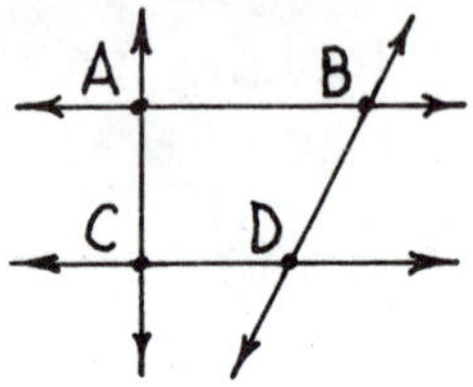


a. $\overleftrightarrow{AC}$
b. $\overleftrightarrow{BD}$
c. $\overleftrightarrow{AB}$
d. none of these

**4.** $\frac{5}{6} - \frac{1}{4} = \underline{?}$

a. $\frac{15}{4}$
b. 2
c. $\frac{1}{6}$
d. none of these

**5.** $\frac{17}{3} = \underline{?}$

a. $5\frac{1}{3}$
b. $6\frac{1}{3}$
c. $5\frac{2}{3}$
d. none of these

**6.** $\frac{3}{8} \times \frac{5}{6} = \underline{?}$

a. $\frac{5}{16}$
b. $\frac{1}{6}$
c. $\frac{15}{24}$
d. none of these

**7.** $85.92 + 6.849 = \underline{?}$

a. 154.41
b. 92.769
c. 81.769
d. none of these

**8.** $50.6 - 3.94 = \underline{?}$

a. 11.2
b. 46.66
c. 47.66
d. none of these

**9.** 27 cm = $\underline{?}$ mm

a. 2.7
b. 27
c. 27,000
d. none of these

**10.** Find the area.

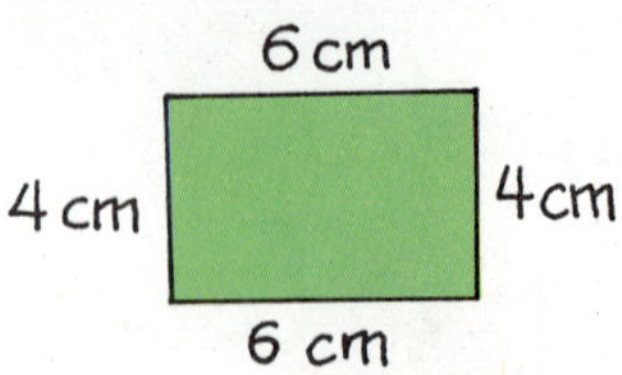


a. 24 cm
b. 23 square cm
c. 20 cm
d. none of these

**11.** Fence costs $5 a meter. How much will it cost to build a fence around this lot?

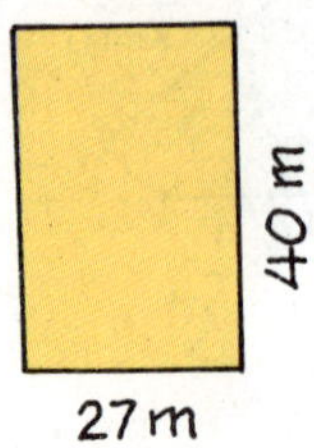


a. 134 m
b. $5400
c. $670
d. none of these

**12.** The plane left at 9:15. The flight took 3 hours 53 minutes. When did the plane arrive?

a. 1:08
b. 2:08
c. 3:08
d. none of these

# Consumer Mathematics

# 10

# Making change

Mary bought a record for $3.44.
She gave the clerk a ten-dollar bill.

Here is how the clerk counted out Mary's change.

three forty-four

three forty-five

three fifty

three seventy-five

4 dollars

5 dollars

10 dollars

Notice that the clerk didn't always say *dollars* and *cents*. How much change did Mary get back?

## EXERCISES

**How much money?**

1. 

2. 

**The cost is shown on the price tag. Suppose that you give the clerk the money shown. List the change that the clerk might count out.**

3. 

4. 

5. 

6. 

7. 

8. 

**Solve.**

9. Robert bought a book for $3.29. He gave the clerk $5. How much change did he get back?

10. Donna bought 2 packages of notebook paper for $1.69 each. She gave the clerk $10. How much change did she get back?

11. Karen bought a book for $.89. She gave the clerk $1. How much change did she get back?

★ 12. David has 2 coins in his pocket. Neither coin is a silver dollar. How much money might he have?

# Computing prices

Suppose that you want to buy 1 ball of twine. Here is how to compute the price.

```
     29 2/3 ¢ each
  3 ) 8 9 ¢
     -6
      29
     -27
       2
```

The exact price is $29\frac{2}{3}$¢. However, you do not have a fraction of a cent. Fractional prices are *rounded up to the next cent.* So the price of 1 ball of twine is 30¢.

**EXERCISES**

**Find the price of one item.**

| | | |
|---|---|---|
| **1.** Paint<br>2 tubes for<br>\$.69 | **2.** Tagboard<br>3 sheets for<br>\$1.25 | **3.** Paint Brushes<br>4 for<br>\$1.35 |
| **4.** Pens<br>2 for<br>\$.97 | **5.** Colored Pencils<br>4 for<br>\$.45 | **6.** Glue<br>3 bottles for<br>\$1.59 |
| **7.** Beads<br>5 for<br>\$.39 | **8.** Wooden Blocks<br>3 for<br>\$.40 | **9.** String<br>2 balls for<br>\$1.23 |
| **10.** Dowel Rods<br>2 for<br>\$.91 | **11.** Paste<br>3 jars for<br>\$1.87 | **12.** Construction Paper<br>5 sheets for<br>\$.67 |

**Give the total price.**

13. 

14. 

15. 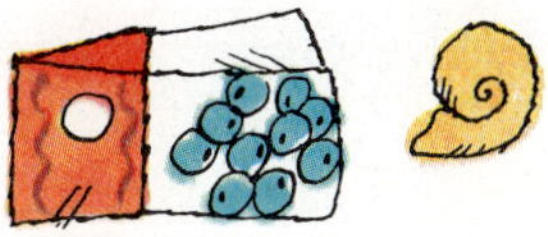

16. 

*Hint:* Add the prices of 2 packages and 1 package.

**Solve.**

17. Find the total price of 2 feathers, 1 package of large beads, and 3 tubes of glue.

18. Find the price of 9 packages of small beads and 4 shells.

19. Which costs more, 3 tubes of glue or 2 packages of large beads? How much more?

★ 20. a. If you bought 1 of each item, what would be the total price?

b. If you gave the clerk a five-dollar bill, how much change would you get?

# More about computing prices

Carol is making a model airplane. She needs 2 tubes of glue. Here is how to compute the price.

**Step 1.** Find the exact price of one tube.

$$
\begin{array}{r}
\$\,.35\tfrac{1}{3} \\
3\overline{)\$1.06} \\
-\ 9\phantom{6} \\
\hline
16 \\
-15 \\
\hline
1
\end{array}
$$

**Step 2.** Multiply the exact price of one tube by 2. To do this, multiply \$.35 by 2 and then $\frac{1}{3}$¢ by 2.

$$\$.35\tfrac{1}{3} \times 2 = \$.70\tfrac{2}{3}$$

**Step 3.** Round up to the next cent.

Two tubes of glue would cost \$.71.

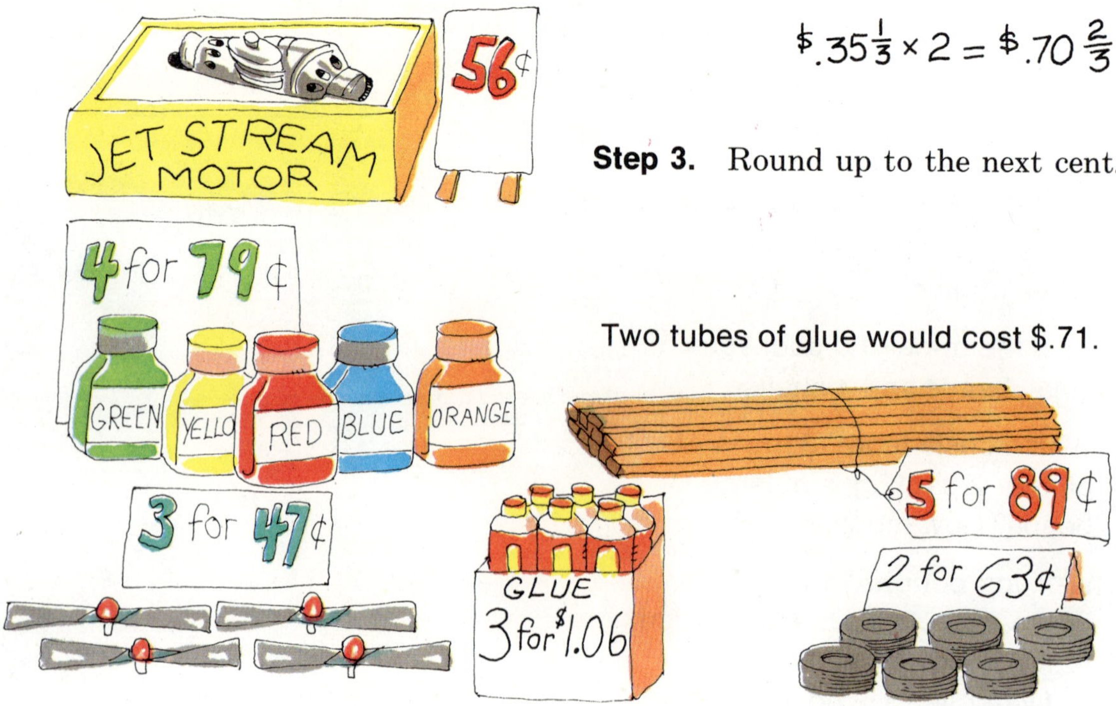

## EXERCISES

**Compute the total price.**

1. 

2. 

3. 

4. 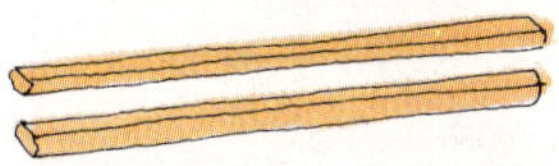

5. 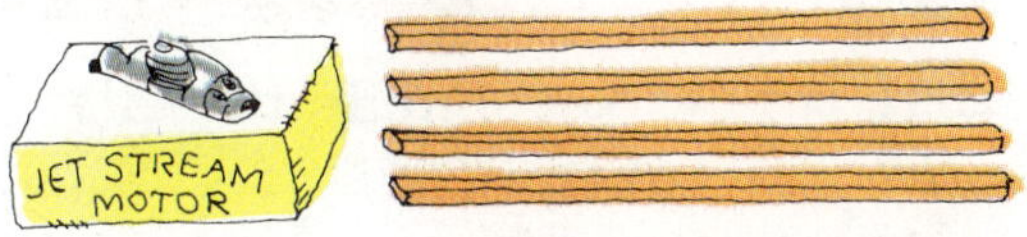


6. 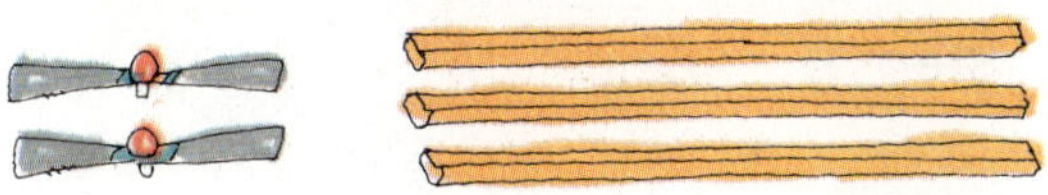

7. 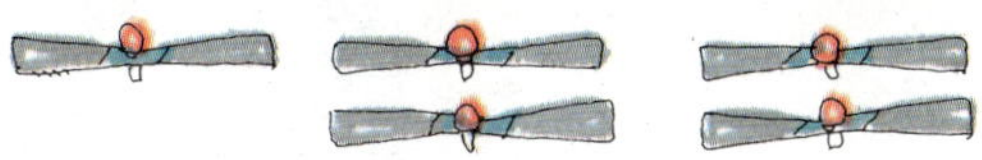

*Hint:* Add the prices of 3  and 2 .

8. 

**Solve.**

9. Model airplane kits are on sale for 3 for $5. Al has $4 and wants to buy 2 kits. How much money will he have left?

10. Janet wants to buy 3 packages of tiles. They are on sale for 2 packages for 79¢. She has 94¢. How much more money does she need?

## KEEPING SKILLS SHARP

1. $125 \times 3$
2. $245 \times 7$
3. $318 \times 6$
4. $729 \times 8$
5. $502 \times 9$
6. $635 \times 5$
7. $423 \times 12$
8. $896 \times 22$
9. $527 \times 40$
10. $744 \times 57$
11. $529 \times 92$

# Computing discounts

Bob Ford owns a bicycle store. He advertises when he has a sale. This gets the customers' attention.

It is a good idea to take advantage of sales. However, you shouldn't buy an item just because it is on sale. You should have a use for it. You should also check the quality of the item on sale.

Here is how to compute the sale price (discount price).

**Step 1.** Divide the regular price by 3.

$$
\begin{array}{r}
\$1.82\frac{2}{3} \\
3\overline{)\$5.48} \\
-3\phantom{.48} \\
\hline
2\,4\phantom{8} \\
-2\,4\phantom{8} \\
\hline
8 \\
-6 \\
\hline
2
\end{array}
$$

**Step 2.** Round to the nearest cent.

$1.83 discount

**Step 3.** Subtract the discount from the regular price.

$$
\begin{array}{r}
\overset{4}{\not{5}}\ \overset{1}{.}48 \\
\$5.48 \\
-1.83 \\
\hline
\$3.65
\end{array}
\leftarrow \text{sale price}
$$

The lock would sell for $3.65.

## EXERCISES

**Compute the amount of discount and the sale price.**

**1.**

$\frac{1}{4}$ off

**2.**

$\frac{1}{2}$ off

**3.**

$\frac{1}{3}$ off

**4.**

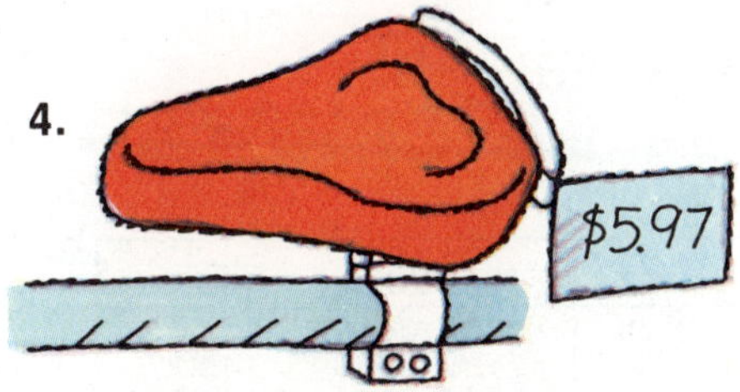


$\frac{1}{5}$ off

**5.**

$\frac{1}{3}$ off

**6.**

$\frac{1}{4}$ off

**7.**

$\frac{1}{5}$ off

**8.**

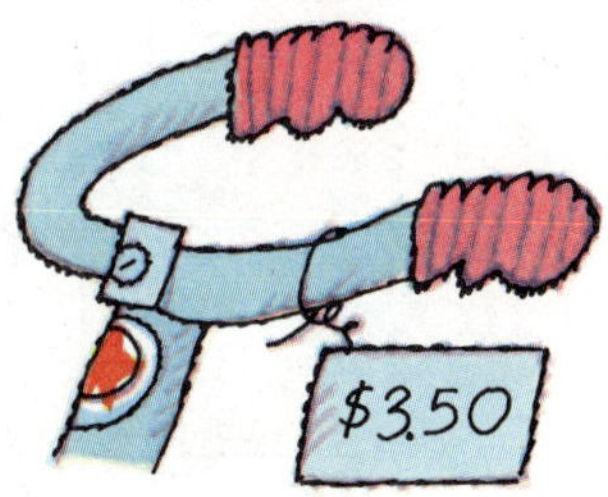


$\frac{1}{3}$ off

### *Challenge!*

**9.** A racing helmet regularly sells for \$48.60. It was on sale for $\frac{1}{3}$ *off* the regular price. Two weeks later the price was reduced to $\frac{3}{4}$ *of* the sale price. What was the price then?

**10.** Racing shoes regularly sell for \$41.40. They were first put on sale for $\frac{1}{4}$ off the regular price. Later they were reduced by $\frac{1}{5}$ off the sale price. What was the price then?

# Sales tax

In most states you pay the merchant a tax (a sales tax) when you buy goods. The merchant then gives the tax to the state. The rate of sales tax depends on the state in which you live.

There are sales-tax tables that a clerk uses to find the amount of tax to charge a customer. The table shown is based on the rate of 5 cents per $1.00 (a rate of 5%, "five percent"). Study the sales receipt.

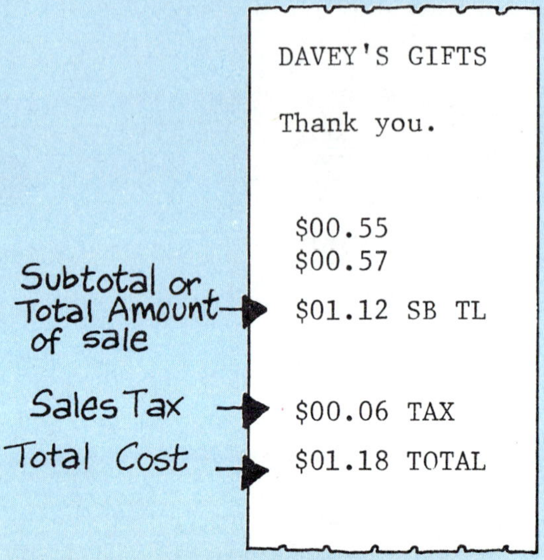


| Sales Tax Table (5%) | | | | | | | |
|---|---|---|---|---|---|---|---|
| **Sale** | **Tax** | **Sale** | **Tax** | **Sale** | **Tax** | **Sale** | **Tax** |
| .01– .09 | .00 | 7.90– 8.09 | .40 | 15.90–16.09 | .80 | 23.90–24.09 | 1.20 |
| .10– .29 | .01 | 8.10– 8.29 | .41 | 16.10–16.29 | .81 | 24.10–24.29 | 1.21 |
| .30– .49 | .02 | 8.30– 8.49 | .42 | 16.30–16.49 | .82 | 24.30–24.49 | 1.22 |
| .50– .69 | .03 | 8.50– 8.69 | .43 | 16.50–16.69 | .83 | 24.50–24.69 | 1.23 |
| .70– .89 | .04 | 8.70– 8.89 | .44 | 16.70–16.89 | .84 | 24.70–24.89 | 1.24 |
| .90– 1.09 | .05 | 8.90– 9.09 | .45 | 16.90–17.09 | .85 | 24.90–25.09 | 1.25 |
| 1.10– 1.29 | .06 | 9.10– 9.29 | .46 | 17.10–17.29 | .86 | 25.10–25.29 | 1.26 |
| 1.30– 1.49 | .07 | 9.30– 9.49 | .47 | 17.30–17.49 | .87 | 25.30–25.49 | 1.27 |
| 1.50– 1.69 | .08 | 9.50– 9.69 | .48 | 17.50–17.69 | .88 | 25.50–25.69 | 1.28 |
| 1.70– 1.89 | .09 | 9.70– 9.89 | .49 | 17.70–17.89 | .89 | 25.70–25.89 | 1.29 |
| 1.90– 2.09 | .10 | 9.90–10.09 | .50 | 17.90–18.09 | .90 | 25.90–26.09 | 1.30 |
| 2.10– 2.29 | .11 | 10.10–10.29 | .51 | 18.10–18.29 | .91 | 26.10–26.29 | 1.31 |
| 2.30– 2.49 | .12 | 10.30–10.49 | .52 | 18.30–18.49 | .92 | 26.30–26.49 | 1.32 |
| 2.50– 2.69 | .13 | 10.50–10.69 | .53 | 18.50–18.69 | .93 | 26.50–26.69 | 1.33 |
| 2.70– 2.89 | .14 | 10.70–10.89 | .54 | 18.70–18.89 | .94 | 26.70–26.89 | 1.34 |
| 2.90– 3.09 | .15 | 10.90–11.09 | .55 | 18.90–19.09 | .95 | 26.90–27.09 | 1.35 |
| 3.10– 3.29 | .16 | 11.10–11.29 | .56 | 19.10–19.29 | .96 | 27.10–27.29 | 1.36 |
| 3.30– 3.49 | .17 | 11.30–11.49 | .57 | 19.30–19.49 | .97 | 27.30–27.49 | 1.37 |
| 3.50– 3.69 | .18 | 11.50–11.69 | .58 | 19.50–19.69 | .98 | 27.50–27.69 | 1.38 |
| 3.70– 3.89 | .19 | 11.70–11.89 | .59 | 19.70–19.89 | .99 | 27.70–27.89 | 1.39 |
| 3.90– 4.09 | .20 | 11.90–12.09 | .60 | 19.90–20.09 | 1.00 | 27.90–28.09 | 1.40 |
| 4.10– 4.29 | .21 | 12.10–12.29 | .61 | 20.10–20.29 | 1.01 | 28.10–28.29 | 1.41 |
| 4.30– 4.49 | .22 | 12.30–12.49 | .62 | 20.30–20.49 | 1.02 | 28.30–28.49 | 1.42 |
| 4.50– 4.69 | .23 | 12.50–12.69 | .63 | 20.50–20.69 | 1.03 | 28.50–28.69 | 1.43 |
| 4.70– 4.89 | .24 | 12.70–12.89 | .64 | 20.70–20.89 | 1.04 | 28.70–28.89 | 1.44 |
| 4.90– 5.09 | .25 | 12.90–13.09 | .65 | 20.90–21.09 | 1.05 | 28.90–29.09 | 1.45 |
| 5.10– 5.29 | .26 | 13.10–13.29 | .66 | 21.10–21.29 | 1.06 | 29.10–29.29 | 1.46 |
| 5.30– 5.49 | .27 | 13.30–13.49 | .67 | 21.30–21.49 | 1.07 | 29.30–29.49 | 1.47 |
| 5.50– 5.69 | .28 | 13.50–13.69 | .68 | 21.50–21.69 | 1.08 | 29.50–29.69 | 1.48 |
| 5.70– 5.89 | .29 | 13.70–13.89 | .69 | 21.70–21.89 | 1.09 | 29.70–29.89 | 1.49 |
| 5.90– 6.09 | .30 | 13.90–14.09 | .70 | 21.90–22.09 | 1.10 | 29.90–30.09 | 1.50 |
| 6.10– 6.29 | .31 | 14.10–14.29 | .71 | 22.10–22.29 | 1.11 | 30.10–30.29 | 1.51 |
| 6.30– 6.49 | .32 | 14.30–14.49 | .72 | 22.30–22.49 | 1.12 | 30.30–30.49 | 1.52 |
| 6.50– 6.69 | .33 | 14.50–14.69 | .73 | 22.50–22.69 | 1.13 | 30.50–30.69 | 1.53 |
| 6.70– 6.89 | .34 | 14.70–14.89 | .74 | 22.70–22.89 | 1.14 | 30.70–30.89 | 1.54 |
| 6.90– 7.09 | .35 | 14.90–15.09 | .75 | 22.90–23.09 | 1.15 | | |
| 7.10– 7.29 | .36 | 15.10–15.29 | .76 | 23.10–23.29 | 1.16 | | |
| 7.30– 7.49 | .37 | 15.30–15.49 | .77 | 23.30–23.49 | 1.17 | | |
| 7.50– 7.69 | .38 | 15.50–15.69 | .78 | 23.50–23.69 | 1.18 | | |
| 7.70– 7.89 | .39 | 15.70–15.89 | .79 | 23.70–23.89 | 1.19 | | |

## EXERCISES

**Give the amount of the sale, the sales tax, and the total cost.**

1.

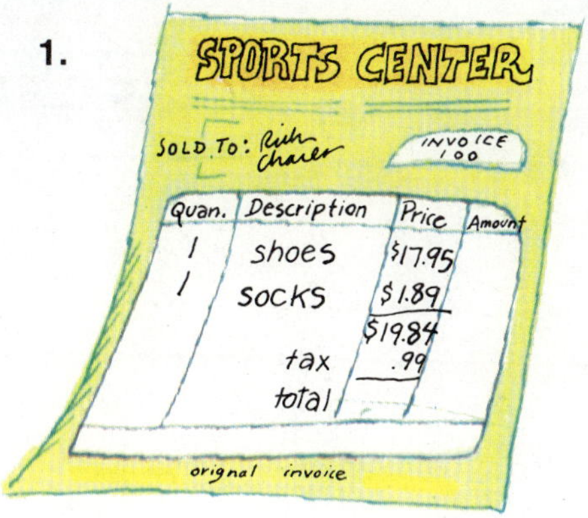


2.

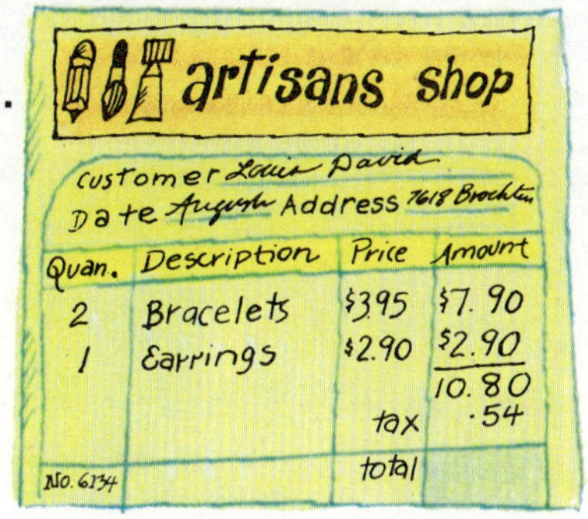


3.

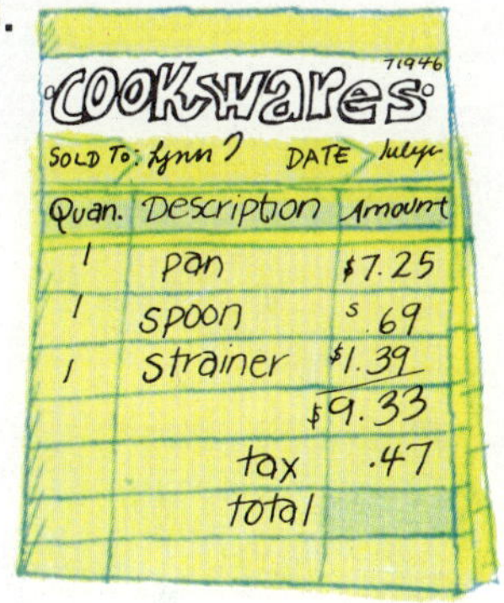


**If the sales tax is 5%, what would the sales tax be for these sales? (*Hint:* Use the table.)**

4.

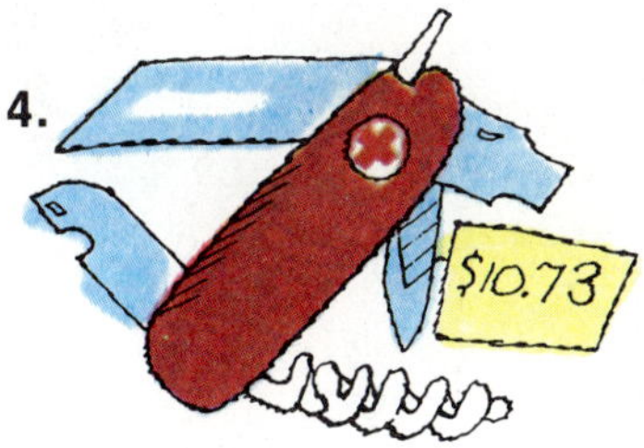


5.

6.

**Solve.**

7. Jack bought a football for $8.95 and a jersey for $3.75. He had to pay a 5% sales tax. What was the total cost?

8. Allison bought some jogging shoes for $21.69 and a pair of socks for $1.89 a pair. What was the total cost if she paid a 5% sales tax?

★ 9. John bought a $15.60 tennis racket that was on sale for $\frac{1}{4}$ off. If he paid a 5% sales tax, what was the total price?

★ 10. Anne bought 2 pairs of tennis socks that were on sale at 3 pairs for $5. She also bought a can of balls for $2.69. What was the total cost if she paid a 5% sales tax?

# Practice exercises

**Compute the total cost. Use the sales tax table on page 294 to include a 5% sales tax.**

1. 

2. 

3. 

4. 

**Solve. Do not include a sales tax.**

5. Andrea had \$4.05. She bought a notebook for \$2.89 and 1 package of paper for \$.97. How much money did she have left?

6. John has \$5.75. He returned 54 soda bottles for 3¢ each. He bought a camping knife for \$5.19. How much money did he have left?

7. The backpack that Sarah wants cost \$12.50. If she waits two weeks, it will be on sale for $\frac{1}{3}$ off. She now has \$6.30. How much more money will she need if she buys the pack on sale?

8. Terry wants some \$21.95 ice skates. The skates are now marked down $\frac{1}{4}$ $\left(\frac{1}{4}\text{ off}\right)$. He can earn \$1.75 an hour raking leaves. He plans to work 6 hours. How much more money will he need?

# Computer service

Many small businesses and professional people do not have their own computers. So they hire a computer service company to do some jobs for them. For example, the Digit Computer Company will prepare monthly bills for any business. It charges \$125 to start the service, \$.17 for each bill prepared, and \$15 to prepare a monthly report. The charge to start the service is paid just one time and the other charges are paid each month.

**EXERCISES**

**Solve.**

1. The Rainbow Paint Company hired the Digit Computer Company to prepare 375 bills each month. What is the total charge the first month?

2. What would the Rainbow Paint Company pay for this service for a year?

3. Dr. Paul Smith estimates that it costs him \$1500 a year to have his office workers prepare 400 patient bills each month. How much could he save the first year by hiring the Digit Computer Company to prepare the bills?

4. Suppose that in the second year of service 450 patient bills have to be prepared each month. If it costs \$1600 for the office workers to prepare these bills, how much would Dr. Smith save by hiring the computer company?

**KEEPING SKILLS SHARP**

1. $4\overline{)284}$
2. $8\overline{)392}$
3. $7\overline{)588}$
4. $6\overline{)906}$
5. $9\overline{)477}$
6. $12\overline{)348}$
7. $23\overline{)596}$
8. $79\overline{)8342}$
9. $68\overline{)5974}$
10. $46\overline{)6835}$

# Sales

Have you noticed that you always get a sales slip when you buy something? The slip gives you a record of the sale. Also, if you should decide to return something, most merchants ask that you return the sales slip with the item.

## EXERCISES

**Look at the sales slip to answer each question.**

1. Where was the purchase made?
2. What was the date of the purchase?
3. How many grocery (GR) items were bought?
4. How much tax was charged?
5. What was the total cost?
6. How much money did the clerk receive?
7. How much change did the clerk give back?

```
Al's Supermarket
        10-08-83

$     4.50 MT
$     1.22 MT
$      .49 GR
$      .99 GR TX
$      .39 GR
$      .30 PR
$      .69 GR TX
$      .25 GR
$      .79 PR

$     9.62 SUB
$      .09 TAX
$     9.71 TOTL

$    20.00 CASH
$    10.29 CHANGE
```

A clerk at Richard's Paint Store filled out this sales slip. The Amount column has not yet been completed.

Use the sales slip to answer.

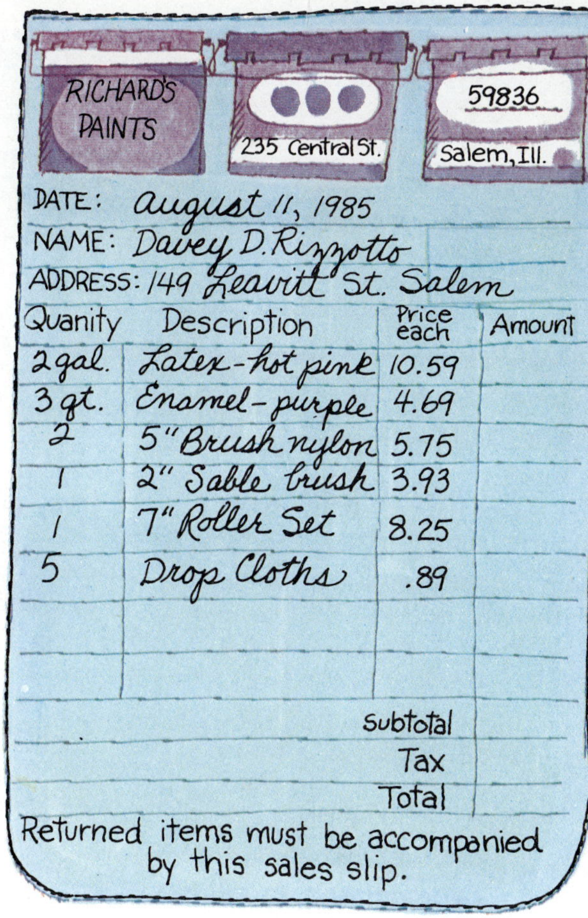

RICHARD'S PAINTS

235 Central St.

59836

Salem, Ill.

DATE: August 11, 1985

NAME: Davey D. Rizzotto

ADDRESS: 149 Leavitt St. Salem

| Quanity | Description | Price each | Amount |
|---|---|---|---|
| 2 gal. | Latex-hot pink | 10.59 | |
| 3 qt. | Enamel-purple | 4.69 | |
| 2 | 5" Brush nylon | 5.75 | |
| 1 | 2" Sable brush | 3.93 | |
| 1 | 7" Roller Set | 8.25 | |
| 5 | Drop Cloths | .89 | |
| | | Subtotal | |
| | | Tax | |
| | | Total | |

Returned items must be accompanied by this sales slip.

8. a. What was the customer's name and address?

   b. On what date was the sale made?

   c. How many gallons of latex paint were ordered?

   d. How many quarts of enamel were ordered?

   e. What was the price of a 7″ roller set?

   f. How many drop cloths were ordered?

   g. What should be the amount charged for the latex paint?

9. What should be the amount charged for these items?

   a. 3 quarts of enamel  b. 2 5-inch brushes  c. 5 drop cloths

10. What is the total amount before tax (the subtotal)?

11. If the sales tax is 5%, how much should be charged for sales tax? (Use the table on page 294.)

12. What is the total cost?

★ 13. Mr. Rizzotto returned 1 quart of the enamel and 2 drop cloths. How much was his refund? (Don't forget to include the sales tax!)

# CHAPTER CHECKUP

1. What is the total price of the boots and stove? [pages 288–289]
2. What is the price of 1 candle? of two candles? [pages 288–291]
3. What is the price of 3 packages of freeze-dried food? of 5 packages? [pages 290–291]
4. The camping store is going to sell the boots for $\frac{1}{3}$ off. How much will the discount be? What will the sale price be? [pages 292–293]
5. A hiking club wants to buy a compass, 2 knives, 1 dozen candles, and 2 packages of freeze-dried foods. What is the total price? If they are charged a 5% sales tax, what will be the total cost? (Use the table on page 294.) [pages 294–296, 298–299]

# CHAPTER PROJECT

1. Get a catalog that has an order form in it.
2. Pretend that a favorite relative gave you $100. Choose from the catalog the things you would buy with the $100.
3. Fill out the order form.

# CHAPTER REVIEW

1. What is the price of 1 sign?
(Remember to round *up*.)

2. What is the price of 2 decals?
(*Hint:* Double the price for
1 decal.)

3. a. What is the discount?

(Remember to round to the
nearest cent.)

b. What is the sale price?

4. a. What is the total price?
b. How much is a 5% sales tax on the total price? (Use the table on page 294.)
c. What would the total cost be, including the 5% sales tax?

# CHAPTER CHALLENGE

You may have a savings account. The most common kind of savings account is called a passbook account. A record of all the transactions is kept in the passbook. The bank (or savings and loan company) pays you **interest** (like rent) for the use of your money.

No. 72357

| Date | Withdrawal | Deposit | Interest | Balance |
|---|---|---|---|---|
| 1/3/85 | | 5.00 | | 88.47 |
| 3/21/85 | | | 1.02 | 89.49 |
| 4/18/85 | 15.00 | | | 74.49 |
| 5/3/85 | 23.00 | | | 51.49 |
| 5/19/85 | | 11.50 | | |

VILLAGE SAVINGS BANK

**Look at the passbook above to answer the following questions.**

1. How much was deposited on 1/3/85 (January 3, 1985)?
2. On what date was interest added (credited) to the account?
3. What was the balance of the account on 4/18/85?
4. How much was taken out of (withdrawn from) the account on 5/3/85?
5. What transaction took place on 4/18/85?
6. How much was in the savings account after the deposit on 5/19/85?
7. If possible, study savings-account passbooks from some local banks.

# MAJOR CHECKUP
## STANDARDIZED FORMAT

b c d 2. a b c d a b c d a b c d 5. a b c d 6. a b c

b c d 8. a b c d 9. a b c d 10. a b c d 11. a b c d 12. a b c

**Choose the correct letter.**

1. Multipy.
$$\begin{array}{r} 586 \\ \times 304 \\ \hline \end{array}$$
   a. 19,924
   b. 19,804
   c. 178,144
   d. none of these

2. Divide.
$36\overline{)8489}$
   a. 235 R29
   b. 225 R29
   c. 236 R29
   d. none of these

3. How many lines of symmetry does a parallelogram have?

   a. 1
   b. 2
   c. 4
   d. none of these

4. $\frac{5}{9} - \frac{1}{3} = \underline{?}$
   a. $\frac{2}{3}$
   b. $\frac{2}{9}$
   c. $\frac{4}{9}$
   d. none of these

5. $\frac{3}{8} \times \frac{2}{3} = \underline{?}$
   a. $\frac{5}{24}$
   b. $\frac{9}{16}$
   c. $\frac{1}{4}$
   d. none of these

6. $5.6 + 3.78 = \underline{?}$
   a. 9.38
   b. 4.34
   c. 43.4
   d. none of these

7. How long is it from 11:20 A.M. to 1:45 P.M.?
   a. $2\frac{1}{2}$ h
   b. 1 h 25 min
   c. 2 h 25 min
   d. none of these

8. 

Volume = ?
   a. 10 cm
   b. 36 square cm
   c. 36 cubic cm
   d. none of these

9. 875 mL = ?
   a. 87.5 liters
   b. 8.75 liters
   c. 0.875 liter
   d. none of these

10. Soup: 2 cans for 57¢. How much for 1 can?
   a. $1.14
   b. 29¢
   c. 28¢
   d. none of these

11. Art paper: 5 sheets for 42¢. How much for 2 sheets?
   a. 21¢
   b. 16¢
   c. 17¢
   d. none of these

12. Eden is 85 miles east of Kent. York is 73 miles east of Kent. How far is Eden from York?
   a. 12 mi
   b. 158 mi
   c. 6205 mi
   d. none of these

# Mixed Numbers— Addition and Subtraction

# 11

## READY OR NOT?

**Give each answer as both a mixed number and a fraction.**

1. How many circles?

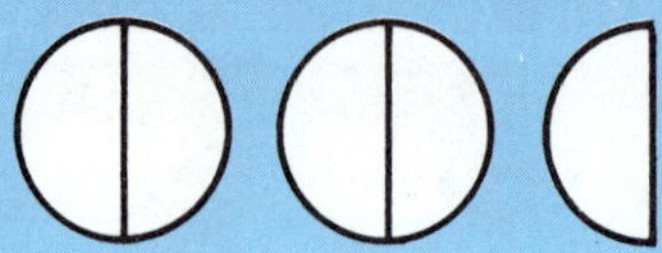

$2\frac{1}{2}, \frac{5}{2}$

2. How many circles?

3. How many circles?

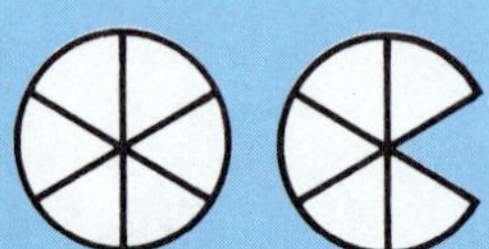

4. How many circles?

# Fractions and mixed numbers

Here is how to change a mixed number to a fraction.

$2\frac{3}{4} = ?$

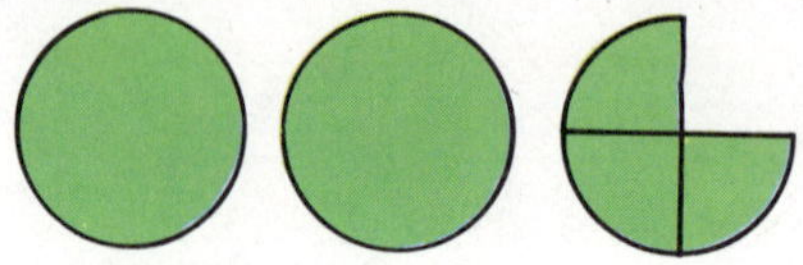

(1) Multiply 2 by 4 to find the number of fourths in 2. Then add 3, the number of fourths in $\frac{3}{4}$.

(2) Write the result, 11, over 4.

$$2\frac{3}{4} = \frac{11}{4}$$

To change a fraction to a mixed number, divide the numerator by the denominator.

$\frac{11}{4} = ?$

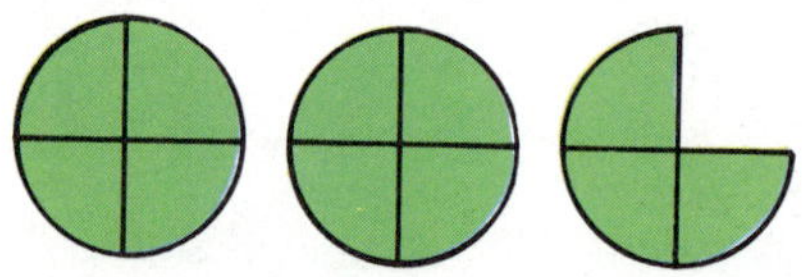

$$\begin{array}{r} 2\frac{3}{4} \\ 4\overline{)11} \\ \underline{-8} \\ 3 \end{array}$$

Number of fourths in one → 4

11 ← Number of fourths

3 ← Number of fourths left over

$$\frac{11}{4} = 2\frac{3}{4}$$

## EXERCISES

Give each length as both a mixed number and a fraction.

1.

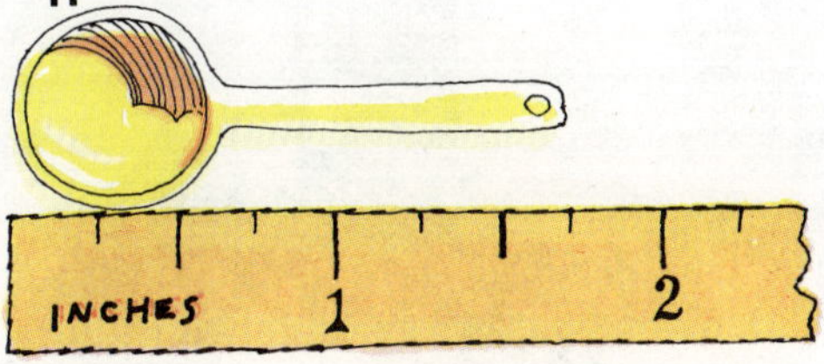


2.

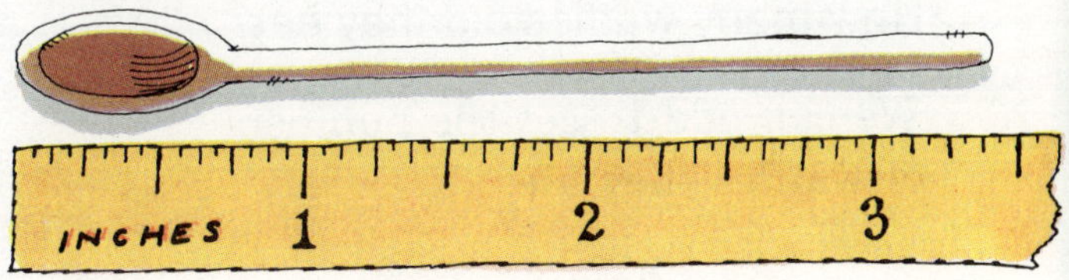


Change to fractions.

3. $1\frac{1}{3}$ 4. $1\frac{1}{2}$ 5. $2\frac{1}{4}$ 6. $3\frac{1}{5}$ 7. $2\frac{1}{6}$ 8. $3\frac{1}{8}$

9. $3\frac{1}{4}$ 10. $2\frac{3}{4}$ 11. $2\frac{1}{3}$ 12. $2\frac{1}{2}$ 13. $4\frac{2}{3}$ 14. $2\frac{3}{8}$

15. $1\frac{7}{8}$ 16. $3\frac{1}{2}$ 17. $3\frac{3}{4}$ 18. $3\frac{5}{8}$ 19. $3\frac{1}{3}$ 20. $4\frac{1}{2}$

21. $5\frac{3}{8}$ 22. $8\frac{5}{6}$ 23. $9\frac{3}{10}$ 24. $12\frac{1}{2}$ 25. $15\frac{2}{3}$ 26. $20\frac{3}{4}$

Write in simplest form.

27. $\frac{3}{2}$ 28. $\frac{4}{3}$ 29. $\frac{5}{2}$ 30. $\frac{9}{4}$ 31. $\frac{6}{5}$ 32. $\frac{9}{5}$

33. $\frac{9}{8}$ 34. $\frac{11}{4}$ 35. $\frac{11}{8}$ 36. $\frac{11}{6}$ 37. $\frac{5}{3}$ 38. $\frac{15}{4}$

39. $\frac{13}{8}$ 40. $\frac{7}{3}$ 41. $\frac{9}{5}$ 42. $\frac{13}{4}$ 43. $\frac{11}{5}$ 44. $\frac{11}{3}$

45. $\frac{12}{5}$ 46. $\frac{26}{3}$ 47. $\frac{39}{4}$ 48. $\frac{43}{8}$ 49. $\frac{61}{2}$ 50. $\frac{73}{5}$

### KEEPING SKILLS SHARP

Give each product in simplest form.

1. $\frac{3}{5} \times \frac{2}{1}$ 2. $\frac{3}{4} \times \frac{2}{5}$ 3. $\frac{5}{8} \times \frac{4}{3}$ 4. $\frac{2}{3} \times \frac{3}{2}$

5. $5 \times \frac{3}{4}$ 6. $4 \times \frac{2}{3}$ 7. $\frac{7}{8} \times 2$ 8. $\frac{5}{9} \times 3$

9. $\frac{5}{9} \times \frac{3}{5}$ 10. $\frac{5}{8} \times \frac{8}{5}$ 11. $\frac{2}{5} \times \frac{10}{3}$ 12. $\frac{3}{2} \times \frac{7}{9}$

# Adding and subtracting mixed numbers

Derek ate $2\frac{1}{4}$ waffles. Ann ate $1\frac{1}{2}$ waffles. How many waffles did they eat?

We can add to get the number of waffles they ate.

$$\begin{array}{r} 2\frac{1}{4} \\ +1\frac{1}{2} \\ \hline \end{array}$$

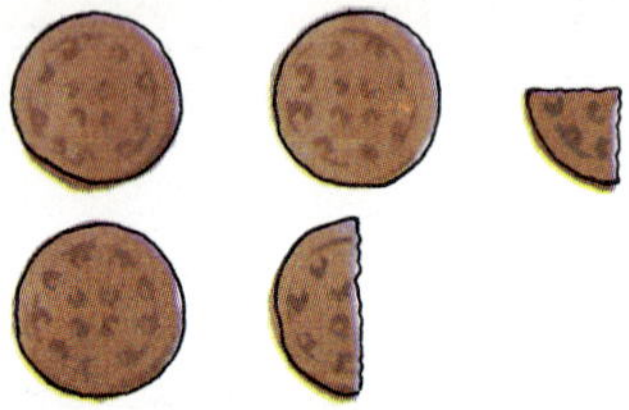

**Step 1.**
Change to fractions with a common denominator.

$$\begin{array}{rl} 2\frac{1}{4} & = 2\frac{1}{4} \\ +1\frac{1}{2} & = 1\frac{2}{4} \\ \hline \end{array}$$

**Step 2.**
Add fractions.

$$\begin{array}{rl} 2\frac{1}{4} & = 2\frac{1}{4} \\ +1\frac{1}{2} & = 1\frac{2}{4} \\ \hline & \quad\frac{3}{4} \end{array}$$

**Step 3.**
Add whole numbers.

$$\begin{array}{rl} 2\frac{1}{4} & = 2\frac{1}{4} \\ +1\frac{1}{2} & = 1\frac{2}{4} \\ \hline & 3\frac{3}{4} \end{array}$$

They ate $3\frac{3}{4}$ waffles.

You can subtract "in columns" too.

**EXAMPLE.**

$$\begin{array}{r} 3\frac{5}{6} \\ -2\frac{1}{6} \\ \hline \end{array}$$

**Step 1.**
Subtract fractions.

$$\begin{array}{r} 3\frac{5}{6} \\ -2\frac{1}{6} \\ \hline \frac{4}{6} \end{array}$$

**Step 2.**
Subtract whole numbers.

$$\begin{array}{r} 3\frac{5}{6} \\ -2\frac{1}{6} \\ \hline 1\frac{4}{6} \end{array}$$

**Step 3.**
Write the difference in simplest form.

$$\begin{array}{r} 3\frac{5}{6} \\ -2\frac{1}{6} \\ \hline 1\frac{4}{6} = 1\frac{2}{3} \end{array}$$

## EXERCISES

**Add. Write the sum in simplest form.**

1. $3\frac{1}{3} + 2\frac{1}{3}$
2. $4\frac{1}{8} + 1\frac{3}{8}$
3. $5\frac{1}{4} + 2\frac{1}{4}$
4. $2\frac{1}{6} + 2\frac{1}{6}$
5. $2\frac{1}{2} + 3\frac{1}{4}$
6. $4\frac{1}{3} + 5\frac{1}{2}$
7. $3\frac{1}{2} + 4\frac{1}{8}$
8. $4\frac{1}{3} + 5\frac{1}{6}$
9. $6\frac{1}{2} + 5$
10. $7\frac{1}{2} + 8\frac{1}{3}$
11. $9\frac{5}{9} + 6\frac{1}{9}$
12. $42\frac{1}{6} + 19\frac{2}{3}$

**Subtract. Write the difference in simplest form.**

13. $8\frac{2}{3} - 5$
14. $9\frac{5}{8} - 3\frac{3}{8}$
15. $8\frac{5}{6} - 4\frac{1}{6}$
16. $5\frac{5}{8} - 1\frac{1}{8}$
17. $7\frac{5}{6} - 2\frac{1}{3}$
18. $5\frac{5}{8} - 2\frac{1}{4}$
19. $8\frac{5}{6} - 4\frac{1}{2}$
20. $8\frac{5}{6} - \frac{1}{4}$
21. $8\frac{7}{10} - 6\frac{1}{5}$
22. $8\frac{5}{6} - 4\frac{1}{4}$
23. $7\frac{2}{3} - 7\frac{2}{5}$
24. $19\frac{4}{5} - 12\frac{3}{10}$

**Solve.**

25. A recipe requires $4\frac{3}{4}$ cups of flour. William has $1\frac{1}{8}$ cups of flour. How much more flour does he need?

26. A recipe requires 4 eggs. Leona has $\frac{3}{4}$ of a dozen eggs. How many eggs will she have left after she makes the recipe?

27. A recipe requires $2\frac{1}{2}$ cups of white sugar and $1\frac{3}{8}$ cups of brown sugar. How much sugar does the recipe require?

28. A recipe requires $2\frac{2}{3}$ cups of milk. Caroline has $5\frac{3}{4}$ cups of milk. How much milk will she have left after she makes the recipe?

# More about adding mixed numbers

When the sum of the fractions is greater than 1, you will need to regroup.

$$\begin{array}{r} 2\frac{3}{4} \\ +1\frac{1}{2} \\ \hline \end{array}$$

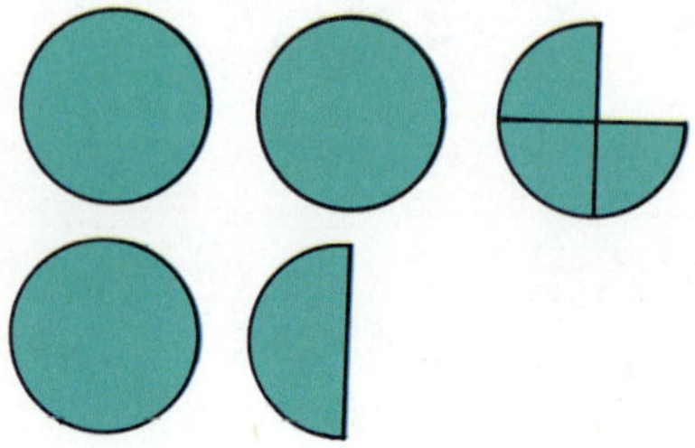

**Step 1.**
Change to a common denominator.

$$\begin{array}{rl} 2\frac{3}{4} = & 2\frac{3}{4} \\ +1\frac{1}{2} = & 1\frac{2}{4} \\ \hline \end{array}$$

**Step 2.**
Add fractions.

$$\begin{array}{rl} 2\frac{3}{4} = & 2\frac{3}{4} \\ +1\frac{1}{2} = & 1\frac{2}{4} \\ \hline & \frac{5}{4} \end{array}$$

**Step 3.**
Add whole numbers.

$$\begin{array}{rl} 2\frac{3}{4} = & 2\frac{3}{4} \\ +1\frac{1}{2} = & 1\frac{2}{4} \\ \hline & 3\frac{5}{4} \end{array}$$

**Step 4.**
Regroup.

$$\begin{array}{rl} 2\frac{3}{4} = & 2\frac{3}{4} \\ +1\frac{1}{2} = & 1\frac{2}{4} \\ \hline & 3\frac{5}{4} = 4\frac{1}{4} \end{array}$$

**Step 1.**
Add and regroup.

$$\begin{array}{r} {}^{1}2\frac{3}{4} \\ +1\frac{1}{2} \\ \hline \frac{1}{4} \end{array} \quad \begin{array}{r} \frac{3}{4} \\ +\frac{2}{4} \\ \hline \frac{5}{4} \end{array}$$

**Step 2.**
Add.

$$\begin{array}{r} {}^{1}2\frac{3}{4} \\ +1\frac{1}{2} \\ \hline 4\frac{1}{4} \end{array}$$

**EXERCISES**

**Regroup the sum.**

1. $$\begin{array}{r} 2\frac{2}{3} \\ +1\frac{1}{3} \\ \hline 3\frac{3}{3} = 4 \end{array}$$

2. $$\begin{array}{rl} 5\frac{1}{2} = & 5\frac{2}{4} \\ +2\frac{3}{4} = & 2\frac{3}{4} \\ \hline & 7\frac{5}{4} = \end{array}$$

3. $$\begin{array}{rl} 4\frac{3}{4} = & 4\frac{6}{8} \\ +2\frac{3}{8} = & 2\frac{3}{8} \\ \hline & 6\frac{9}{8} = \end{array}$$

**Add. Write the sum in simplest form.**

4. $3\frac{1}{9} + 5\frac{4}{9}$
5. $1\frac{3}{5} + 6\frac{2}{5}$
6. $4\frac{3}{8} + 3\frac{1}{4}$
7. $3\frac{2}{3} + 3\frac{1}{6}$
8. $6\frac{1}{2} + 2\frac{1}{3}$
9. $5\frac{1}{10} + 4\frac{7}{10}$

10. $4\frac{1}{2} + 6\frac{1}{2}$
11. $3\frac{1}{4} + 8\frac{3}{4}$
12. $8\frac{3}{4} + 4\frac{5}{8}$
13. $4\frac{2}{3} + 9\frac{1}{2}$
14. $7\frac{3}{4} + 7\frac{2}{3}$
15. $9\frac{7}{8} + 8\frac{1}{4}$

16. $7\frac{5}{9} + 8\frac{2}{3}$
17. $8\frac{2}{3} + 9\frac{5}{6}$
18. $5\frac{3}{4} + 8\frac{3}{8}$
19. $5\frac{7}{10} + 5\frac{9}{10}$
20. $9\frac{7}{8} + 6\frac{3}{4}$
21. $9\frac{5}{6} + 9\frac{2}{3}$

22. $10\frac{2}{3} + 8\frac{3}{4}$
23. $12\frac{1}{2} + 7\frac{2}{3}$
24. $15\frac{1}{4} + 14\frac{7}{8}$
25. $26\frac{5}{8} + 18\frac{3}{4}$
26. $39\frac{5}{6} + 15\frac{3}{8}$
27. $43\frac{5}{6} + 38\frac{3}{4}$

**Find the perimeter of each figure.**

28.

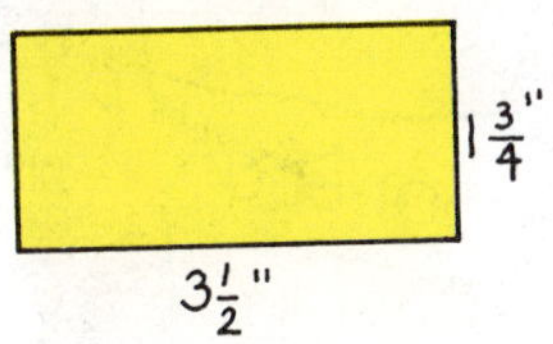


29.

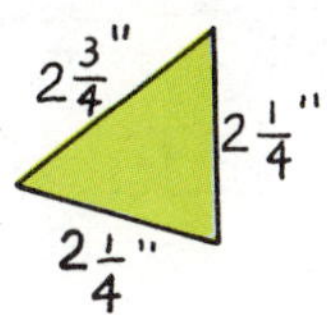


30.

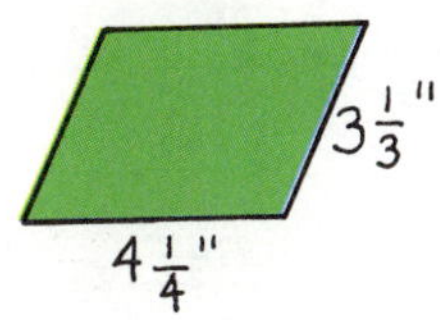


**Solve.**

31. Alison bought $5\frac{2}{3}$ yards of red ribbon and $4\frac{1}{2}$ yards of blue ribbon. How many yards of ribbon did she buy?

32. Cindy rode her bicycle $6\frac{7}{10}$ miles the first hour and $5\frac{1}{2}$ miles the second hour. How many miles did she ride altogether?

33. To get to school, Samuel walks $3\frac{3}{4}$ blocks to the bus stop and then rides the bus $18\frac{1}{2}$ blocks. How far does he travel to get to school?

34. John is $59\frac{3}{4}$ inches tall. Kay is $2\frac{3}{4}$ inches taller than John. Bill is $1\frac{1}{2}$ inches taller than Kay. How tall is Kay?

# More about subtracting mixed numbers

Sometimes you have to regroup before you subtract.

**EXAMPLE.**

$$\begin{array}{r} 8\frac{1}{4} \\ -3\frac{3}{8} \\ \hline \end{array}$$

**Step 1.**
Change to a common denominator.

$$\begin{array}{rl} 8\frac{1}{4} & = 8\frac{2}{8} \\ -3\frac{3}{8} & = 3\frac{3}{8} \\ \hline \end{array}$$

**Step 2.**
Since $\frac{2}{8}$ is less than $\frac{3}{8}$, regroup 1 as $\frac{8}{8}$.

$$\begin{array}{rll} 8\frac{1}{4} & = 8\frac{2}{8} & = 7\frac{10}{8} \\ -3\frac{3}{8} & = 3\frac{3}{8} & = 3\frac{3}{8} \\ \hline \end{array}$$

**Step 3.**
Subtract.

$$\begin{array}{rll} 8\frac{1}{4} & = 8\frac{2}{8} & = 7\frac{10}{8} \\ -3\frac{3}{8} & = 3\frac{3}{8} & = 3\frac{3}{8} \\ \hline & & 4\frac{7}{8} \end{array}$$

$$\begin{array}{rl} 8\frac{1}{4} & = \not{8}\frac{\not{2}}{8} \;\; (7\frac{10}{8}) \\ -3\frac{3}{8} & = 3\frac{3}{8} \\ \hline & 4\frac{7}{8} \end{array}$$

**EXERCISES**

**Subtract. Write the difference in simplest form.**

1. $\begin{array}{rll} 5\frac{1}{4} & = 5\frac{1}{4} & = 4\frac{5}{4} \\ -2\frac{1}{2} & = 2\frac{2}{4} & = 2\frac{2}{4} \\ \hline \end{array}$

2. $\begin{array}{rll} 6\frac{1}{8} & = 6\frac{1}{8} & = 5\frac{9}{8} \\ -3\frac{3}{4} & = 3\frac{6}{8} & = 3\frac{6}{8} \\ \hline \end{array}$

3. $\begin{array}{rll} 7\frac{2}{3} & = 7\frac{8}{12} & = 6\frac{20}{12} \\ -4\frac{3}{4} & = 4\frac{9}{12} & = 4\frac{9}{12} \\ \hline \end{array}$

4. $\begin{array}{rl} 6 & = 5\frac{4}{4} \\ -2\frac{1}{4} & = 2\frac{1}{4} \\ \hline \end{array}$

5. $\begin{array}{rl} 9 & = 8\frac{3}{3} \\ -5\frac{2}{3} & = 5\frac{2}{3} \\ \hline \end{array}$

6. $\begin{array}{rl} 7 & = 6\frac{8}{8} \\ -3\frac{5}{8} & = 3\frac{5}{8} \\ \hline \end{array}$

7. $9\frac{5}{9} - 4\frac{1}{9}$

8. $8\frac{4}{5} - 2\frac{3}{5}$

9. $5\frac{3}{4} - 1\frac{1}{2}$

10. $9\frac{5}{6} - 4\frac{2}{3}$

11. $7\frac{1}{2} - 2\frac{1}{3}$

12. $9\frac{3}{4} - 4\frac{2}{3}$

13. $8 - 3\frac{1}{2}$

14. $6 - 4\frac{3}{5}$

15. $12 - 3\frac{7}{10}$

16. $11 - 5\frac{3}{4}$

17. $15 - 6\frac{3}{8}$

18. $16 - 9\frac{4}{5}$

19. $9\frac{1}{4} - 3\frac{1}{2}$

20. $8\frac{1}{2} - 4\frac{3}{4}$

21. $7\frac{3}{8} - 2\frac{3}{4}$

22. $10\frac{2}{3} - 8\frac{5}{6}$

23. $11\frac{1}{8} - 6\frac{3}{4}$

24. $12\frac{2}{3} - 8\frac{5}{6}$

**Solve.**

25. Diane is $8\frac{3}{4}$ inches taller than her little brother. If Diane is $58\frac{1}{2}$ inches tall, how tall is her brother?

26. One week, David watched $6\frac{1}{4}$ hours of television. The next week he watched 8 hours of television. How much more did he watch the second week?

27. In basketball, the rim of the basket is 10 feet above the floor. Stan can jump up and reach $7\frac{1}{4}$ feet above the floor. How much higher must he jump to touch the basket rim?

28. A share of certain stock sold for $32\frac{5}{8}$ dollars. Two weeks later it sold for $34\frac{1}{8}$ dollars a share. How much did the value of the stock increase?

## KEEPING SKILLS SHARP

**Add or subtract.**

1. $8.4 + 3.29$

2. $7.34 - 2.6$

3. $13 + 89.2$

4. $87 - 9.6$

5. $0.67 + 5.8$

6. $9 - 2.57$

7. $37 + 8.6$

8. $37 - 8.6$

9. $2.6 + 0.57$

10. $2.6 - 0.57$

# Practice

**Add. Write the sum in simplest form.**

**1.** $5\frac{1}{3}+2\frac{1}{3}$ **2.** $8\frac{1}{9}+2\frac{1}{9}$ **3.** $4\frac{1}{3}+3\frac{1}{3}$

**4.** $8\frac{1}{4}+2\frac{1}{2}$ **5.** $3\frac{1}{8}+6\frac{1}{4}$ **6.** $9\frac{2}{5}+4\frac{1}{10}$

**7.** $8\frac{2}{3}+4\frac{1}{2}$ **8.** $6\frac{4}{5}+9\frac{1}{2}$ **9.** $7\frac{2}{3}+7\frac{5}{6}$

**10.** $5\frac{2}{3}+6\frac{3}{4}$ **11.** $9\frac{4}{5}+8\frac{9}{10}$ **12.** $6\frac{5}{8}+4\frac{3}{4}$

**Subtract. Write the difference in simplest form.**

**13.** $8\frac{5}{6}-3\frac{1}{6}$ **14.** $4\frac{1}{2}-2\frac{1}{2}$ **15.** $7\frac{3}{4}-5$ **16.** $9\frac{5}{6}-4\frac{2}{3}$

**17.** $9-3\frac{1}{2}$ **18.** $12-4\frac{5}{8}$ **19.** $10-5\frac{3}{4}$ **20.** $12-6\frac{3}{5}$

**21.** $12-9\frac{1}{2}$ **22.** $13\frac{3}{8}-6\frac{3}{4}$ **23.** $16\frac{4}{9}-12\frac{2}{3}$ **24.** $18\frac{1}{8}-4\frac{5}{6}$

**Solve.**

**25.**

How many dozen doughnuts?

**26.**

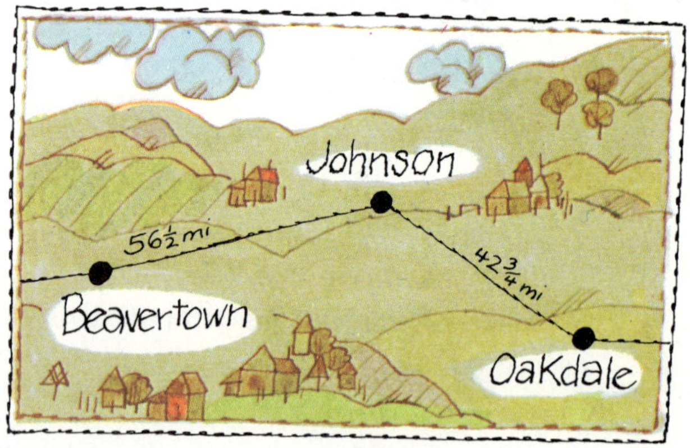


**a.** How many miles from Beavertown to Oakdale?

**b.** Johnson is how much farther from Beavertown than from Oakdale?

**27.** On her first try, Mary long-jumped $12\frac{5}{6}$ feet. On her second try, she jumped $13\frac{1}{4}$ feet. How much farther did she jump the second time?

**28.** A rectangular garden is $10\frac{1}{2}$ feet wide and $14\frac{1}{4}$ feet long. How much fencing is needed to go around the garden?

## SKILL GAME

Lisa missed the target number 5 by $2\frac{19}{30}$.

**How close did these players come to the target number?**

1. $\left(3\frac{2}{5}+4\frac{4}{6}\right)-2\frac{1}{8} \Rightarrow 5$

2. $\left(2\frac{4}{5}+4\frac{3}{6}\right)-1\frac{2}{8} \Rightarrow 5$

3. $\left(5\frac{2}{4}+8\frac{6}{2}\right)-3\frac{1}{4} \Rightarrow 5$

4. $\left(8\frac{6}{4}+3\frac{2}{2}\right)-5\frac{1}{4} \Rightarrow 5$

**Play the game.**

1. Make a digit card for each of the digits 1 through 9.
2. Choose a game leader.
3. Draw a table like this:

$$\left(\square\frac{\square}{\square}+\square\frac{\square}{\square}\right)-\square\frac{\square}{\square} \Rightarrow 5$$

4. Without looking, the leader picks a card. Each player writes the digit in his or her table.
5. The leader replaces the card in the deck, and step 4 is repeated until the tables have been filled in.
6. The player who gets closest to the target number 5 wins the game.

# Problem solving

A bird-watching club decided to make some bluebird houses from the plan shown. Here are some problems that they solved.

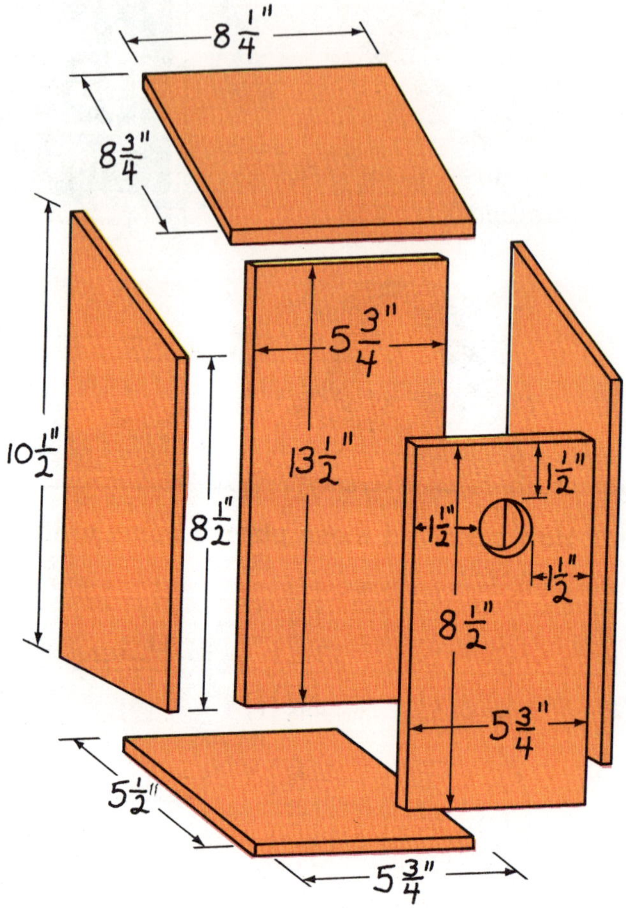


1. How wide is the top of the birdhouse?

2. How tall is the back piece?

3. How much taller is the back piece than the front piece?

4. The club ordered a 4′ × 8′ piece of plywood. The plywood was 62¢ per square foot. What was the total price?

5. A front was cut from a piece of plywood that was $6\frac{3}{8}$ inches wide. How wide was the unused piece that was cut off?

6. Two sides were to be cut from a rectangular piece of plywood $5\frac{1}{2}''$ wide. How long must the piece be if a saw cut (a kerf) uses $\frac{1}{8}''$?

7. Dennis had a piece of plywood that was 6 inches wide and 13 inches long. How many bottoms could be cut from it?

8. Each birdhouse had 2 hinges. The hinges cost $.65 each. How much did the club spend for hinges for one birdhouse?

9. Carla drilled the hole in each front.
   a. What was the diameter of the hole?
   b. How far from the side was the center of the hole?

10. The club could buy brown paint for $4.29 a pint or $7.65 a quart. They needed 3 pints.
    a. What was the cheapest way to buy the paint?
    b. How much did the club spend for paint?

11. The club bought $2\frac{1}{4}$ pounds of number 4 common nails and $1\frac{3}{4}$ pounds of number 6 common nails. How many pounds of nails did they buy? If nails cost 60¢ a pound, what was the total price of the nails?

12. They bought $3\frac{1}{2}$ dozen screws for 28¢ a dozen. What was the total price of the screws?

13. Each birdhouse was to be placed on a wooden post $4\frac{3}{4}$ feet above the ground. If the posts were $7\frac{1}{2}$ feet long, how deep did the holes have to be?

14. This record was kept of the number of hours that each member worked on the project.

| Name | Hours |
|---|---|
| Bill | $3\frac{1}{2}$ |
| Carla | 4 |
| Dennis | $2\frac{3}{4}$ |
| Howie | $4\frac{1}{2}$ |
| John | $3\frac{1}{4}$ |
| Sue | 5 |

How many hours were spent on the project?

# Practice

**Add or subtract. Write the answer in simplest form.**

1. $2\frac{1}{2} + 1\frac{3}{4}$
2. $3\frac{1}{4} + 2\frac{5}{8}$
3. $2\frac{1}{3} - 1\frac{1}{6}$
4. $4\frac{3}{8} - 2\frac{1}{2}$
5. $3\frac{5}{6} - 2$
6. $4\frac{3}{8} + 1\frac{3}{8}$
7. $6\frac{5}{7} - \frac{3}{7}$
8. $8\frac{3}{4} + 7\frac{1}{4}$
9. $5\frac{1}{3} - 2\frac{5}{6}$
10. $3\frac{1}{3} - 2\frac{1}{2}$
11. $3\frac{1}{3} + 1\frac{1}{3}$
12. $5\frac{7}{8} - 4\frac{1}{8}$
13. $3\frac{1}{5} - \frac{1}{4}$
14. $2\frac{3}{5} + 4\frac{2}{3}$
15. $4\frac{5}{6} + 2\frac{3}{8}$
16. $13\frac{4}{5} - 8\frac{1}{4}$
17. $7\frac{5}{8} - 4\frac{1}{8}$
18. $3\frac{3}{10} + 6\frac{5}{6}$
19. $6\frac{3}{4} - 1\frac{7}{8}$
20. $8 - 5\frac{2}{3}$
21. $7\frac{3}{5} + 5$
22. $6\frac{3}{5} + 2\frac{3}{4}$
23. $15\frac{2}{9} - 8\frac{5}{6}$
24. $21\frac{4}{5} + 2\frac{5}{6}$

## *Challenge!*

**Find the missing input or output.**

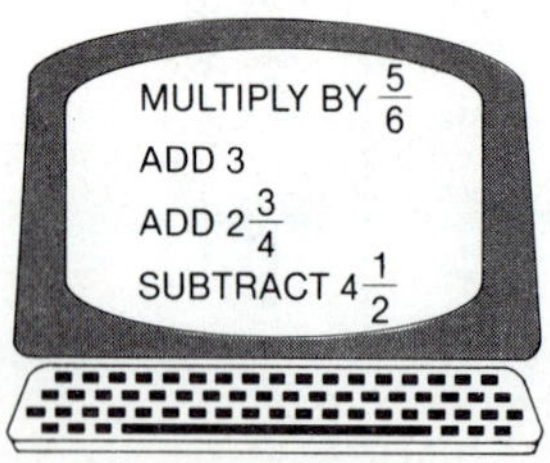


| | Input | Output |
|---|---|---|
| | $\frac{5}{6}$ | $1\frac{17}{18}$ |
| 25. | 1 | ? |
| 26. | $\frac{1}{2}$ | ? |
| 27. | ? | $1\frac{9}{10}$ |

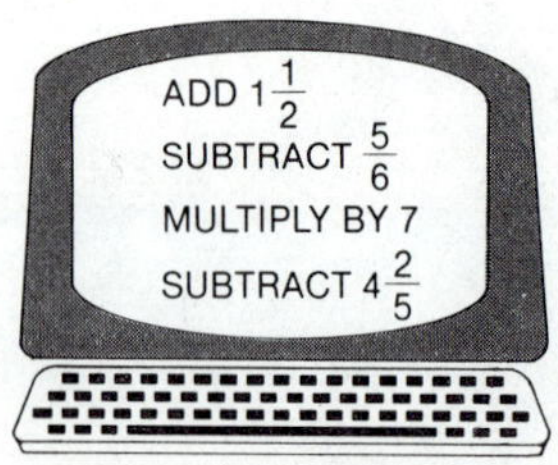


| | Input | Output |
|---|---|---|
| | $\frac{1}{3}$ | $2\frac{3}{5}$ |
| 28. | 1 | ? |
| 29. | $\frac{1}{2}$ | ? |
| 30. | ? | 4 |

# Problem solving

**Solve.**

1. A hiking club had \$45. The club spent \$38.53 for supplies. How much money was left?

2. A total of 336 pounds had to be carried. There were 12 hikers to share the weight. How many pounds did each hiker carry?

3. The base of the mountain was 1309 feet above sea level. The peak of the mountain was 6159 feet above sea level. How high was the mountain?

4. The hikers rested for $\frac{1}{2}$ hour at one place and for $\frac{3}{4}$ hour at another place. How long did they rest?

5. Each of the 12 hikers carried $\frac{3}{4}$ quart of water. How many quarts of water were carried?

6. They hiked $3\frac{1}{2}$ hours before lunch and $2\frac{3}{4}$ hours after lunch. How many hours did they hike?

7. For lunch they made $1\frac{1}{2}$ gallons of soup. They ate $1\frac{1}{4}$ gallons. How much soup was left?

8. For dinner they used 8 packages of chili. Each package was to be added to $\frac{3}{4}$ quart of water. How much water was used to make the chili?

9. They went to bed at 10:30 P.M. and got up at 6:00 A.M. How long did they sleep?

10. On the first day they hiked $6\frac{1}{2}$ miles. On the second day they hiked $8\frac{2}{3}$ miles. How many miles did they hike?

11. The trail they took to go up the mountain was $6\frac{1}{6}$ miles. The trail they took to come down the mountain was $9\frac{3}{4}$ miles. How much longer was the trail back?

# CHAPTER CHECKUP

**Change to a fraction.** [pages 306–307]

1. $2\frac{1}{2}$ 2. $4\frac{1}{3}$ 3. $2\frac{2}{3}$ 4. $1\frac{3}{4}$ 5. $5\frac{5}{6}$ 6. $4\frac{5}{8}$

**Change to a mixed number.** [pages 306–307]

7. $\frac{9}{2}$ 8. $\frac{10}{3}$ 9. $\frac{11}{4}$ 10. $\frac{12}{5}$ 11. $\frac{15}{8}$ 12. $\frac{15}{6}$

**Add. Write the sum in simplest form.** [pages 308–311, 314–315, 318]

13. $4\frac{1}{2} + 3$
14. $6\frac{2}{3} + 2$
15. $5\frac{1}{4} + 2\frac{1}{4}$
16. $7\frac{3}{4} + 8\frac{1}{8}$
17. $9\frac{1}{2} + 3\frac{1}{3}$
18. $4\frac{2}{3} + 6\frac{1}{4}$
19. $8\frac{2}{3} + 7\frac{1}{2}$
20. $9\frac{3}{8} + 6\frac{3}{4}$
21. $15\frac{5}{9} + 8\frac{2}{3}$
22. $23\frac{7}{8} + 17\frac{3}{4}$
23. $35\frac{5}{6} + 9\frac{5}{8}$
24. $38\frac{7}{10} + 26\frac{9}{10}$

**Subtract. Write the difference in simplest form.** [pages 308–309, 312–315, 318]

25. $9\frac{5}{8} - 2$
26. $6\frac{3}{5} - 4$
27. $10\frac{5}{9} - 2\frac{2}{9}$
28. $12\frac{5}{8} - 5\frac{1}{4}$
29. $13\frac{2}{3} - 6\frac{1}{2}$
30. $15\frac{3}{4} - 7\frac{2}{3}$
31. $6 - 4\frac{2}{3}$
32. $5 - 3\frac{7}{8}$
33. $18\frac{1}{4} - 9\frac{3}{5}$
34. $26\frac{1}{3} - 8\frac{5}{6}$
35. $19\frac{3}{8} - 7\frac{2}{3}$
36. $34\frac{1}{6} - 17\frac{5}{8}$

**Solve.** [pages 316–317, 319]

| Name | Denise | Ella | Frank | George |
|---|---|---|---|---|
| Pounds of Paper Collected | $10\frac{3}{4}$ | $8\frac{1}{8}$ | $7\frac{1}{2}$ | $12\frac{5}{8}$ |

37. Together Denise and Ella collected how many pounds?

38. Who collected $3\frac{1}{4}$ pounds less than Denise?

# CHAPTER PROJECT

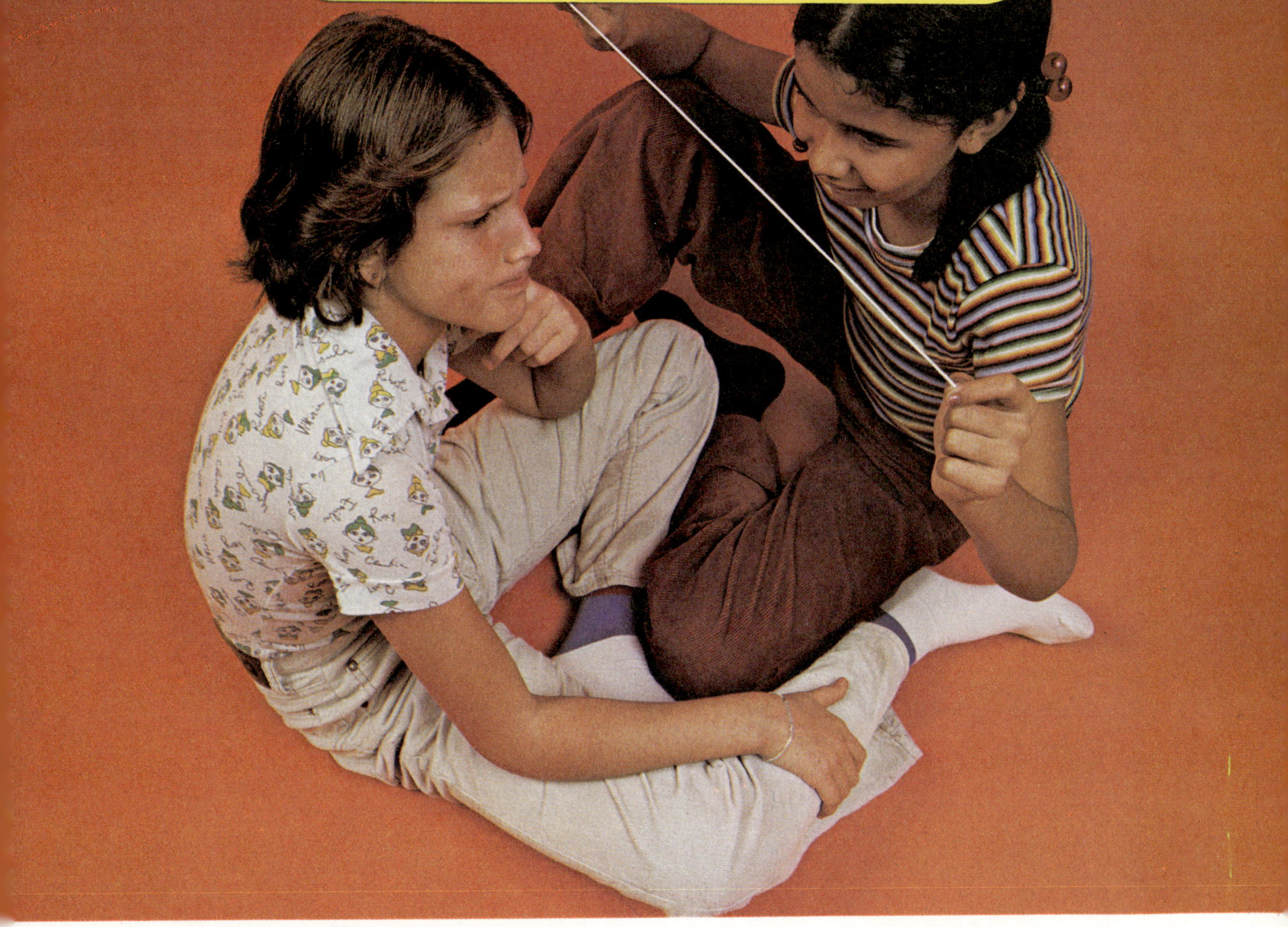

1. *Without measuring,* cut off a piece of string that is somewhere between 10 and 20 inches long.
2. Ask your classmates to estimate the length of the string, and keep a record of the estimates like this:

| Name | Estimate of Length | Actual Length | Error |
|---|---|---|---|
| Denise | $16\frac{1}{2}$" | | |

3. Measure the string to the nearest $\frac{1}{4}$ inch and compute the error of each estimate.
4. Who made the smallest error? The greatest error?
5. Make a bar graph of the errors.

# CHAPTER REVIEW

$3\frac{2}{5} = \frac{17}{5}$

**Change to a fraction.**

1. $1\frac{1}{2}$ 2. $2\frac{1}{3}$ 3. $2\frac{3}{4}$ 4. $5\frac{2}{3}$

5. $6\frac{1}{4}$ 6. $3\frac{1}{7}$ 7. $9\frac{2}{5}$ 8. $4\frac{3}{8}$

$\frac{9}{2} = 4\frac{1}{2}$

$$\begin{array}{r} 4\frac{1}{2} \\ 2\overline{)9} \\ -8 \\ \hline 1 \end{array}$$

**Change to a mixed number.**

9. $\frac{7}{2}$ 10. $\frac{7}{4}$ 11. $\frac{14}{3}$ 12. $\frac{18}{5}$

13. $\frac{16}{3}$ 14. $\frac{27}{8}$ 15. $\frac{33}{4}$ 16. $\frac{68}{9}$

$$\begin{array}{rcl} 4\frac{1}{3} & = & 4\frac{4}{12} \\ +2\frac{1}{4} & = & 2\frac{3}{12} \\ \hline & & 6\frac{7}{12} \end{array}$$

$$\begin{array}{rcl} 8\frac{3}{4} & = & 8\frac{3}{4} \\ +2\frac{1}{2} & = & 2\frac{2}{4} \\ \hline & & 10\frac{5}{4} = 11\frac{1}{4} \end{array}$$

**Add. Write the sum in simplest form.**

17. $2\frac{1}{5} + 1$ 18. $5\frac{5}{9} + 2\frac{2}{9}$ 19. $6\frac{3}{4} + 3\frac{1}{2}$ 20. $8\frac{3}{5} + 6\frac{1}{4}$

21. $5\frac{1}{2} + 8\frac{1}{2}$ 22. $7\frac{3}{4} + 9\frac{5}{8}$ 23. $12\frac{2}{3} + 5\frac{5}{6}$ 24. $9\frac{3}{10} + 14\frac{4}{5}$

$$\begin{array}{rcl} 5\frac{2}{3} & = & 5\frac{4}{6} \\ -1\frac{1}{2} & = & 1\frac{3}{6} \\ \hline & & 4\frac{1}{6} \end{array}$$

$$\begin{array}{rcl} 9\frac{1}{4} & = & 9\frac{3}{12} = 8\frac{15}{12} \\ -2\frac{2}{3} & = & 2\frac{8}{12} = 2\frac{8}{12} \\ \hline & & 6\frac{7}{12} \end{array}$$

**Subtract. Write the difference in simplest form.**

25. $7\frac{1}{3} - 3$ 26. $5\frac{3}{4} - 1\frac{1}{4}$ 27. $8\frac{5}{6} - 4\frac{2}{3}$ 28. $12\frac{5}{8} - 3\frac{1}{2}$

29. $12 - 8\frac{5}{8}$ 30. $15\frac{1}{4} - 6\frac{1}{2}$ 31. $17\frac{3}{8} - 8\frac{2}{3}$ 32. $13\frac{1}{6} - 5\frac{3}{4}$

# CHAPTER CHALLENGE

How much you weigh depends on the force of gravity. You would not weigh the same on another planet as you weigh on the planet Earth, because the force of gravity is different on each planet. You can use a weight multiplier to find your weight on another planet.

| Planet | Weight multiplier |
|---|---|
| Earth | 1 |
| Jupiter | $2\frac{5}{8}$ |
| Mars | $\frac{3}{8}$ |
| Mercury | $\frac{3}{10}$ |
| Neptune | $1\frac{1}{10}$ |
| Pluto | $\frac{7}{10}$ |
| Saturn | $1\frac{1}{5}$ |
| Uranus | $\frac{9}{10}$ |
| Venus | $\frac{7}{8}$ |

**EXAMPLE.**

Michael weighs 73 pounds on Earth. How much would he weigh on Jupiter?

Multiply Michael's weight on Earth by the weight multiplier for Jupiter shown in the table.

$$73 \times 2\frac{5}{8} = 73 \times \frac{21}{8}$$

Change to a fraction.

$$= \frac{1533}{8}$$

$$= 191\frac{5}{8}$$

Michael would weigh $191\frac{5}{8}$ pounds on Jupiter.

**Give each answer in simplest form.**

1. What would a 10-pound radio weigh on Mercury? On Neptune?
2. What would a 60-pound dog weigh on Pluto? On Uranus?
3. What would a 24-pound bicycle weigh on Mars? On Jupiter?
4. What would a 40-pound television set weigh on Venus? On Saturn?
5. On which planet would you weigh the most? The least?
6. Compute your weight on each of the planets.

b c d  2. a b c d  a b c d  a b c d  5. a b c d  6. a b c

b c d  8. a b c d  9. a b c d  10. a b c d  11. a b c d  12. a b c

# MAJOR CHECKUP

## STANDARDIZED FORMAT

**Choose the correct letter.**

1. A right angle measures
   - a. 100°
   - b. 50°
   - c. 25°
   - d. none of these

2. The triangles are congruent. $\overline{BC}$ is congruent to

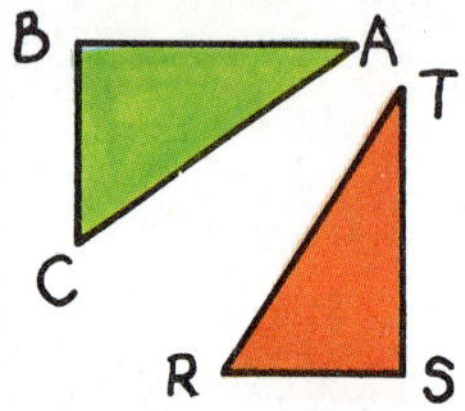


   - a. $\overline{RT}$
   - b. $\overline{SR}$
   - c. $\overline{ST}$
   - d. none of these

3. $\frac{5}{6} \times \frac{2}{9} = \underline{?}$
   - a. $\frac{5}{27}$
   - b. $\frac{5}{9}$
   - c. $\frac{5}{18}$
   - d. none of these

4. $74.3 + 113.82 = \underline{?}$
   - a. 121.25
   - b. 1212.5
   - c. 188.12
   - d. none of these

5. $56.4 - 4.35 = \underline{?}$
   - a. 52.05
   - b. 1.29
   - c. 12.9
   - d. none of these

6. 1 km = $\underline{?}$ m
   - a. 10
   - b. 1000
   - c. 100
   - d. none of these

7. Area = $\underline{?}$

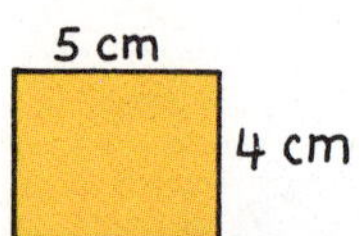


   - a. 20 square cm
   - b. 18 square cm
   - c. 9 square cm
   - d. none of these

8. $\frac{1}{3}$ off marked price. What is the sale price?

   - a. $8.93
   - b. $4.46
   - c. $17.85
   - d. none of these

9. $3\frac{3}{8} + 4\frac{7}{8} = \underline{?}$
   - a. $7\frac{1}{4}$
   - b. $8\frac{1}{4}$
   - c. $1\frac{1}{2}$
   - d. none of these

10. $5\frac{1}{4} - 2\frac{1}{2} = \underline{?}$
    - a. $2\frac{3}{4}$
    - b. $3\frac{3}{4}$
    - c. $3\frac{1}{4}$
    - d. none of these

11. Pencils: 3 for 59¢ Erasers: 2 for 49¢ How much for a pencil and an eraser?
    - a. 22¢
    - b. 45¢
    - c. $1.08
    - d. none of these

12. Vanessa works 8 hours a day, 5 days a week. She is paid $3.75 an hour. How much does she earn in a day?
    - a. $30
    - b. $75
    - c. $150
    - d. none of these

# Decimals—Multiplication and Division

# 12

## READY OR NOT?

**Give a decimal for the number of shaded unit squares.**

1.  0.30

2. 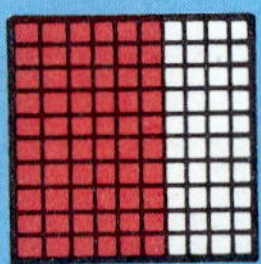

3. 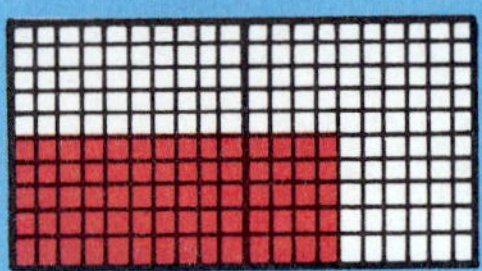

4. 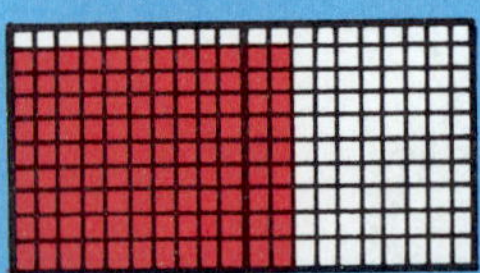

5. 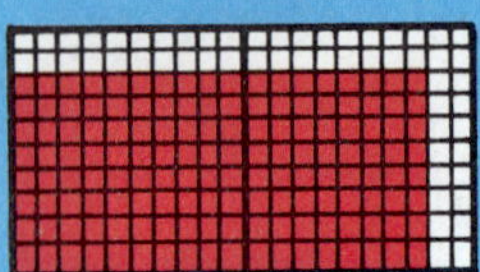



# Multiplying decimals

If we multiply, we get the number of unit squares that are yellow.

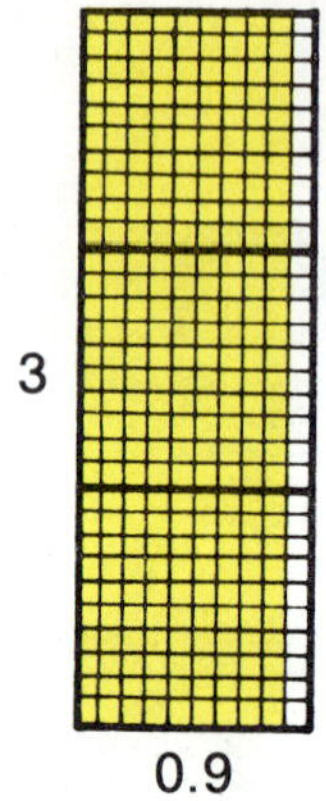


$$\begin{array}{r} 3 \\ \times 0.9 \\ \hline 2.7 \end{array}$$

2.7 unit squares are yellow.

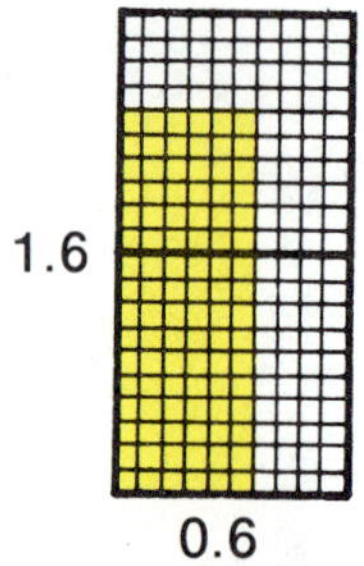


$$\begin{array}{r} 1.6 \\ \times 0.6 \\ \hline 0.96 \end{array}$$

0.96 of a unit square is yellow.

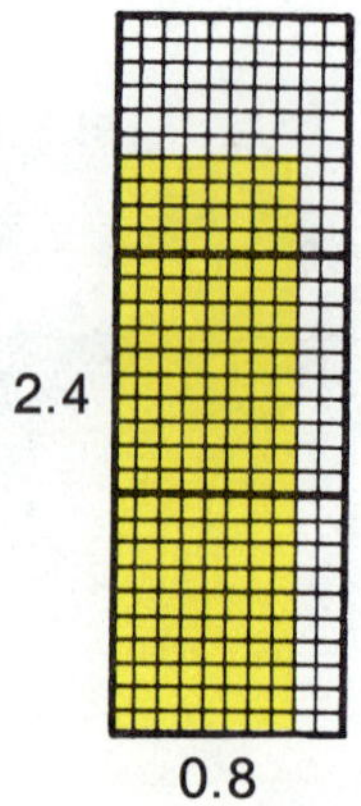


$$\begin{array}{r} 2.4 \\ \times 0.8 \\ \hline 1.92 \end{array}$$

1.92 unit squares are yellow.

## EXERCISES

First study the picture. Then give the product.

1\.
0.8 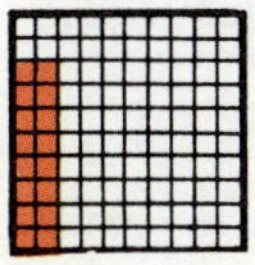
0.2

$\begin{array}{r} 0.8 \\ \times 0.2 \\ \hline \end{array}$

2\.
0.9 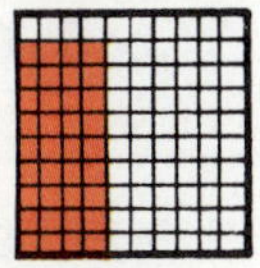
0.4

$\begin{array}{r} 0.9 \\ \times 0.4 \\ \hline \end{array}$

3\.
0.7 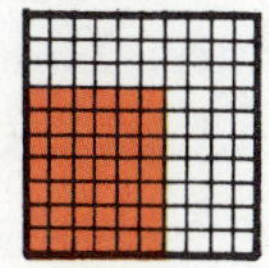
0.6

$\begin{array}{r} 0.7 \\ \times 0.6 \\ \hline \end{array}$

4\.
2 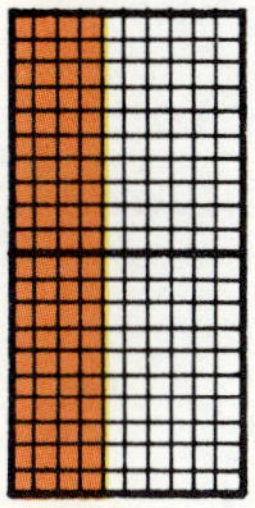
0.4

$\begin{array}{r} 2 \\ \times 0.4 \\ \hline \end{array}$

5\.
1.5 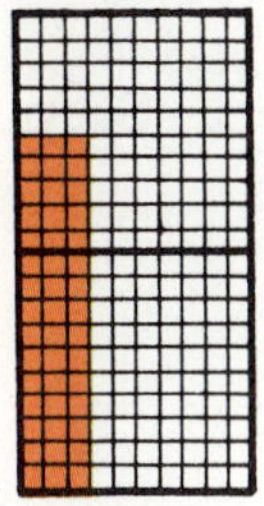
0.3

$\begin{array}{r} 1.5 \\ \times 0.3 \\ \hline \end{array}$

6\.
1.8 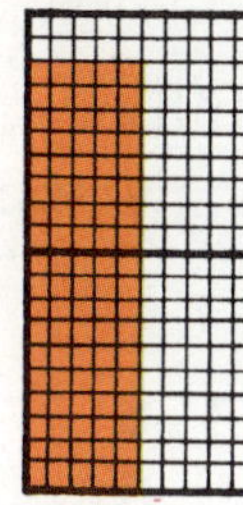
0.5

$\begin{array}{r} 1.8 \\ \times 0.5 \\ \hline \end{array}$

Give the product. Then check your answer by looking at the picture.

7\.
3 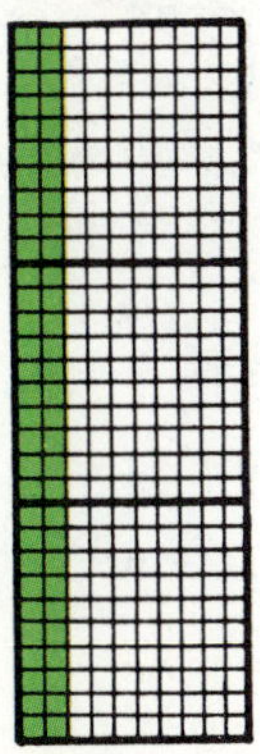
0.2

$\begin{array}{r} 3 \\ \times 0.2 \\ \hline \end{array}$

8\.
2.2 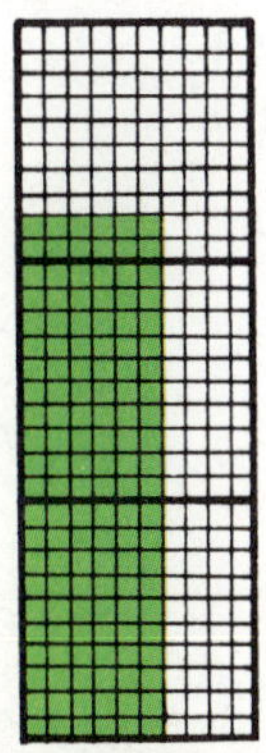
0.6

$\begin{array}{r} 2.2 \\ \times 0.6 \\ \hline \end{array}$

9\.
2.8 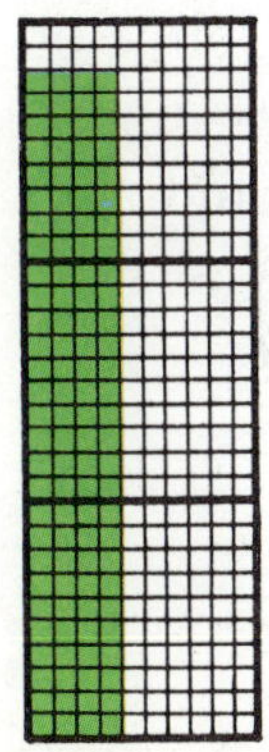
0.4

$\begin{array}{r} 2.8 \\ \times 0.4 \\ \hline \end{array}$

### KEEPING SKILLS SHARP

Give each sum in simplest form.

1\. $\begin{array}{r} \frac{2}{5} \\ +\frac{2}{5} \\ \hline \end{array}$
2\. $\begin{array}{r} \frac{3}{8} \\ +\frac{1}{8} \\ \hline \end{array}$
3\. $\begin{array}{r} \frac{1}{6} \\ +\frac{1}{3} \\ \hline \end{array}$
4\. $\begin{array}{r} \frac{3}{8} \\ +\frac{1}{4} \\ \hline \end{array}$
5\. $\begin{array}{r} \frac{1}{4} \\ +\frac{1}{2} \\ \hline \end{array}$
6\. $\begin{array}{r} \frac{1}{9} \\ +\frac{5}{6} \\ \hline \end{array}$

7\. $\begin{array}{r} \frac{1}{3} \\ +\frac{1}{2} \\ \hline \end{array}$
8\. $\begin{array}{r} \frac{1}{8} \\ +\frac{3}{4} \\ \hline \end{array}$
9\. $\begin{array}{r} \frac{1}{4} \\ +\frac{2}{3} \\ \hline \end{array}$
10\. $\begin{array}{r} \frac{1}{9} \\ +\frac{2}{3} \\ \hline \end{array}$
11\. $\begin{array}{r} \frac{1}{6} \\ +\frac{3}{8} \\ \hline \end{array}$
12\. $\begin{array}{r} \frac{3}{5} \\ +\frac{3}{10} \\ \hline \end{array}$

# Multiplying decimals—a shortcut

You can find the product by looking at the picture. You can also use this shortcut to find the product.

$$\begin{array}{r} 1.8 \\ \times 0.6 \\ \hline \end{array}$$

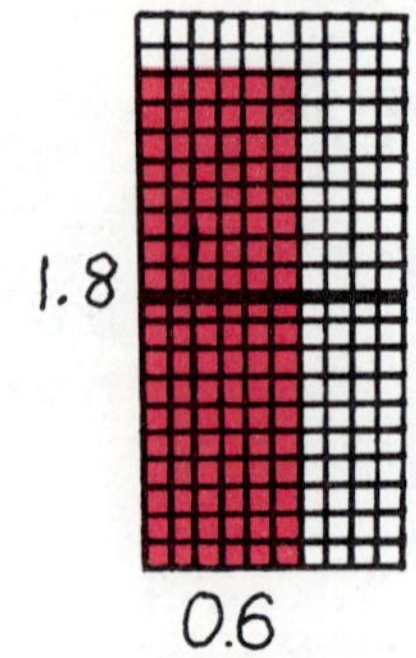


**Step 1.** Multiply the numbers as whole numbers.

$$\begin{array}{r} 1.8 \\ \times 0.6 \\ \hline 108 \end{array}$$

**Step 2.** Count the digits to the right of the decimal points in the factors.

$$\begin{array}{rl} 1.8 & \leftarrow 1 \text{ digit} \\ \times 0.6 & \leftarrow 1 \text{ digit} \\ \hline 108 & \quad 2 \text{ digits} \end{array}$$

**Step 3.** Count off the same number of digits to place the decimal point in the product.

$$\begin{array}{r} 1.8 \\ \times 0.6 \\ \hline 1.08 \end{array}$$

Study these examples.

$$\begin{array}{rl} 3.25 & \leftarrow 2 \text{ digits} \\ \times 0.6 & \leftarrow 1 \text{ digit} \\ \hline 1.950 & \quad 3 \text{ digits} \end{array} \qquad \begin{array}{r} 25.2 \\ \times 6.4 \\ \hline 1008 \\ 15120 \\ \hline 161.28 \end{array} \qquad \begin{array}{r} 53.8 \\ \times 2.14 \\ \hline 2152 \\ 5380 \\ 107600 \\ \hline 115.132 \end{array}$$

## EXERCISES

**Multiply.**

1. $28 \times 0.5$
2. $56 \times 0.3$
3. $135 \times 0.6$
4. $2.78 \times 8$
5. $39.5 \times 3$
6. $32.1 \times 0.3$
7. $72.4 \times 0.8$
8. $2.95 \times 0.6$
9. $6.08 \times 0.4$
10. $59.9 \times 0.7$
11. $8.21 \times 1.5$
12. $74.8 \times 0.08$
13. $5.93 \times 3.7$
14. $36.9 \times 0.09$
15. $57.4 \times 8.5$
16. $42.6 \times 16.2$
17. $3.78 \times 2.57$
18. $29.7 \times 34.8$
19. $54.6 \times 5.62$
20. $97.8 \times 2.09$

$$\begin{array}{r} 3.86 \\ \times 10 \\ \hline 0\,00 \\ 3860 \\ \hline 38.60 \end{array}$$

**Use the shortcut to find these products.**

21. $35.2 \times 10$
22. $2.96 \times 10$
23. $0.789 \times 10$
24. $74.1 \times 10$

**See if you can find a shortcut for multiplying by 100.**

25. $1.56 \times 100$
26. $0.65 \times 100$
27. $9.34 \times 100$
28. $0.878 \times 100$

**Solve.**

29. The gasoline tank on a certain car will hold 22.4 gallons.
    a. If gasoline costs \$1.35 a gallon, how much will it cost to fill the tank?
    b. If the car averages 21.2 miles per gallon, how far can the car travel on a tank of gas?

30. If a car averages 52.5 miles per hour, how far will it travel in 2 hours and 45 minutes? (*Hint:* Change $2\frac{3}{4}$ to a decimal.)

★ 31. It is 11:30 A.M. You want to arrive at Disney World by 1:45 P.M. You are 115 miles from Disney World. If you average 48.5 miles per hour, will you be late?

# More about multiplying decimals

Sometimes you will need to write one or more zeros in the product before you can place the decimal point.

$$\begin{array}{r} 0.3 \\ \times 0.2 \\ \hline 0.06 \end{array}$$

2 digits

$$\begin{array}{r} 0.04 \\ \times 0.2 \\ \hline 0.008 \end{array}$$

3 digits

**EXERCISES**

**Multiply.**

1. $0.2 \times 3$
2. $0.2 \times 0.3$
3. $0.02 \times 3$
4. $0.4 \times 0.2$
5. $0.3 \times 0.1$
6. $0.5 \times 0.3$
7. $0.02 \times 0.4$
8. $0.02 \times 0.1$
9. $23 \times 0.02$
10. $4 \times 0.2$
11. $13.2 \times 0.06$
12. $24.3 \times 0.05$
13. $1.14 \times 4.2$
14. $527 \times 0.26$
15. $39.5 \times 1.8$
16. $86.4 \times 0.25$
17. $29.1 \times 3.8$
18. $74.3 \times 14.2$
19. $651 \times 1.56$
20. $507 \times 35.4$

**Use shortcuts to find these products.**

21. $3.82 \times 10$
22. $70.8 \times 10$
23. $56.3 \times 10$
24. $4.26 \times 10$
25. $0.915 \times 10$
26. $35.9 \times 10$
27. $2.63 \times 100$
28. $3.952 \times 100$
29. $35.48 \times 100$
30. $46.17 \times 100$
31. $6.18 \times 100$
32. $394.5 \times 100$
33. $7.826 \times 1000$
34. $5.37 \times 1000$
35. $0.653 \times 1000$
36. $8.291 \times 1000$
37. $742.83 \times 1000$
38. $561.94 \times 1000$

**Complete.**

39. To change kilometers to meters, multiply by _?_.

40. To change meters to centimeters, multiply by _?_.

41. To change liters to milliliters, multiply by _?_.

42. To change kilograms to grams, multiply by _?_.

**Complete.**

43. 3.825 km = _?_ m

44. 5.23 km = _?_ m

45. 8.6 km = _?_ m

46. 4.65 m = _?_ cm

47. 9.06 m = _?_ cm

48. 12.1 m = _?_ cm

49. 5.18 L = _?_ mL

50. 4.3 L = _?_ mL

51. 21.62 L = _?_ mL

52. 7.456 kg = _?_ g

53. 12.4 kg = _?_ g

54. 20.36 kg = _?_ g

**Complete.**

55. Longest Bicycle

21.946 m, or _?_ cm

56. Heaviest Turtle

408 kg, or _?_ g

57. Biggest Paper Cup

2.154 kL, or _?_ L

1 kiloliter (kL) = 1000 L

58. Longest Snake

11.43 m, or _?_ cm

# Dividing a decimal

You divide a decimal the same way you divide a whole number.

**EXAMPLE.** $3\overline{)2.58}$

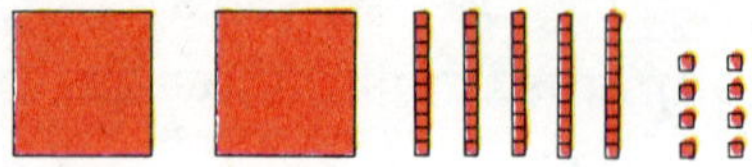

**Step 1.** Not enough ones. Regroup 2 ones for 20 tenths.

$3\overline{)2.58}$

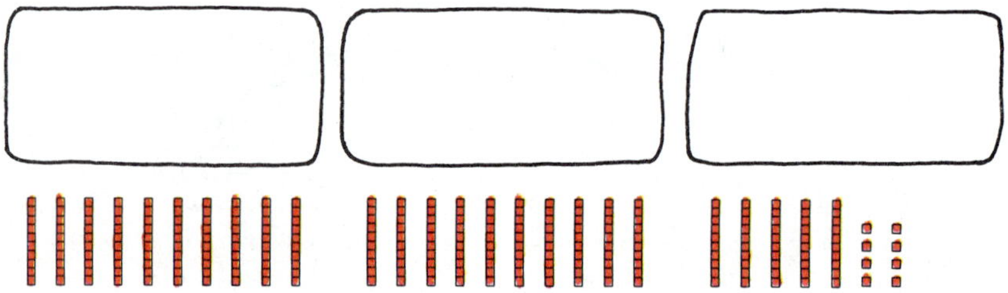

**Step 2.** Divide tenths. Subtract.

$$\begin{array}{r} 0.8\phantom{8} \\ 3\overline{)2.58} \\ -2\,4\phantom{8} \\ \hline 1\phantom{8} \end{array}$$

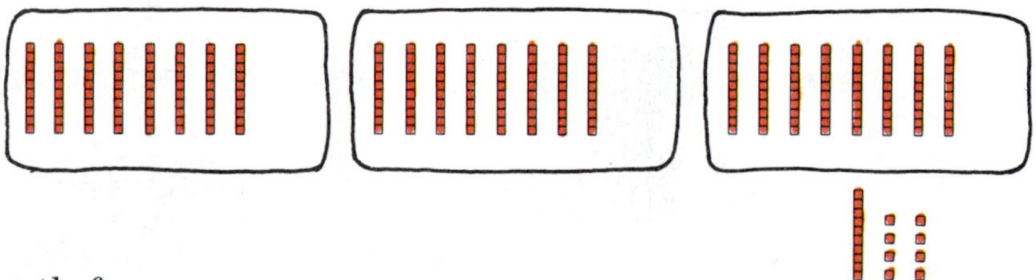

**Step 3.** Regroup 1 tenth for 10 hundredths.

$$\begin{array}{r} 0.8\phantom{8} \\ 3\overline{)2.58} \\ -2\,4\phantom{8} \\ \hline 1\,8 \end{array}$$

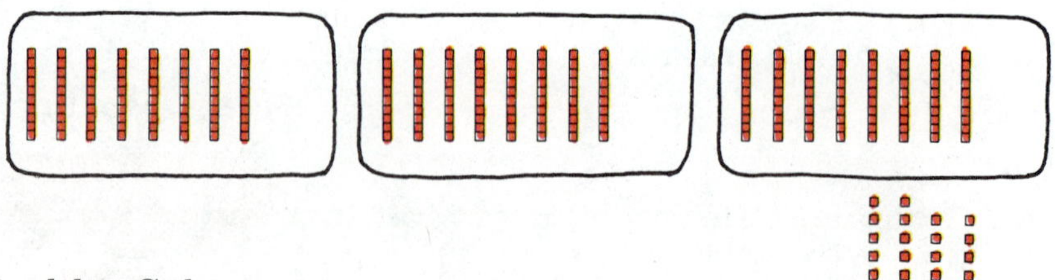

**Step 4.** Divide hundredths. Subtract.

$$\begin{array}{r} 0.86 \\ 3\overline{)2.58} \\ -2\,4\phantom{8} \\ \hline 1\,8 \\ -1\,8 \\ \hline 0 \end{array}$$

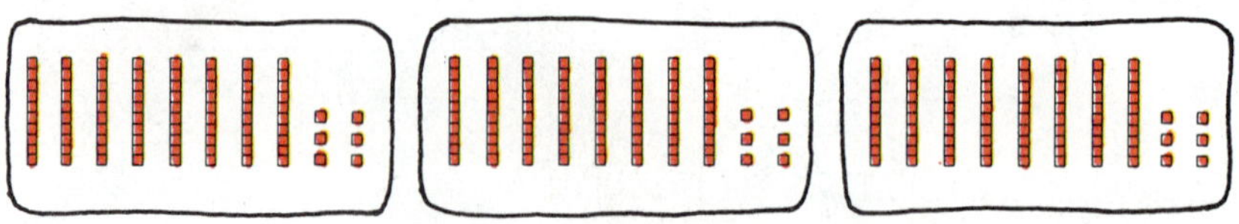

Study these examples.

$$\begin{array}{r} 0.124 \\ 8\overline{)0.992} \\ -8\phantom{92} \\ \hline 19\phantom{2} \\ -16\phantom{2} \\ \hline 32 \\ -32 \\ \hline 0 \end{array} \qquad \begin{array}{r} 5.34 \\ 12\overline{)64.08} \\ -60\phantom{.08} \\ \hline 40\phantom{8} \\ -36\phantom{8} \\ \hline 48 \\ -48 \\ \hline 0 \end{array} \qquad \begin{array}{r} 0.461 \\ 243\overline{)112.023} \\ -972\phantom{23} \\ \hline 1482\phantom{3} \\ -1458\phantom{3} \\ \hline 243 \\ -243 \\ \hline 0 \end{array}$$

**EXERCISES**

**Divide.**

| | | | | |
|---|---|---|---|---|
| **1.** $3\overline{)52.5}$ | **2.** $5\overline{)6.85}$ | **3.** $7\overline{)3.92}$ | **4.** $8\overline{)0.360}$ | **5.** $2\overline{)76.4}$ |
| **6.** $4\overline{)5.96}$ | **7.** $6\overline{)29.4}$ | **8.** $9\overline{)0.783}$ | **9.** $4\overline{)70.8}$ | **10.** $8\overline{)0.896}$ |
| **11.** $7\overline{)7.56}$ | **12.** $3\overline{)92.1}$ | **13.** $5\overline{)8.55}$ | **14.** $9\overline{)0.666}$ | **15.** $6\overline{)92.4}$ |
| **16.** $14\overline{)33.74}$ | **17.** $25\overline{)975.50}$ | **18.** $39\overline{)95.94}$ | **19.** $56\overline{)565.6}$ | **20.** $45\overline{)5.670}$ |

**Solve.**

**21.** 

Weight: 2 kilograms
Price: $12.58
What is the price per kilogram?

**22.** 

What is the total cost? (Give the answer to the nearest cent.)

**23.** Mr. Johnson drove his car 283.6 kilometers in 4 hours. How many kilometers per hour did he average?

**24.** Barbara earned $25.90 for working 14 hours. How much did she earn per hour?

**25.** During one week, Sarah jogged an average of 4.2 kilometers each day. How far did she jog during that week?

**26.** Arthur jogs 2.75 kilometers each school day and 4.5 kilometers each day of the weekend. How many kilometers does he average each day?

# More about dividing decimals

Study the example. Notice the zeros written in the dividend.

**EXAMPLE.** $4\overline{)3.1}$

**Step 1.** Divide.

$$\begin{array}{r} 0.7 \\ 4\overline{)3.1} \\ -2\,8 \\ \hline 3 \end{array}$$

**Step 2.** Regroup 3 tenths for 30 hundredths.

$$\begin{array}{r} 0.7\phantom{0} \\ 4\overline{)3.10} \\ -2\,8\phantom{0} \\ \hline 30 \end{array}$$

**Step 3.** Divide.

$$\begin{array}{r} 0.77 \\ 4\overline{)3.10} \\ -2\,8\phantom{0} \\ \hline 30 \\ -28 \\ \hline 2 \end{array}$$

**Step 4.** Regroup 2 hundredths for 20 thousandths.

$$\begin{array}{r} 0.77\phantom{0} \\ 4\overline{)3.100} \\ -2\,8\phantom{00} \\ \hline 30\phantom{0} \\ -28\phantom{0} \\ \hline 20 \end{array}$$

**Step 5.** Divide.

$$\begin{array}{r} 0.775 \\ 4\overline{)3.100} \\ -2\,8\phantom{00} \\ \hline 30\phantom{0} \\ -28\phantom{0} \\ \hline 20 \\ -20 \\ \hline 0 \end{array}$$

Other examples.

$$\begin{array}{r} 0.125 \\ 8\overline{)1.000} \\ -\,8\phantom{00} \\ \hline 20\phantom{0} \\ -16\phantom{0} \\ \hline 40 \\ -40 \\ \hline 0 \end{array}$$

$$\begin{array}{r} 1.375 \\ 16\overline{)22.000} \\ 16\phantom{.000} \\ \hline 6\,0\phantom{00} \\ -4\,8\phantom{00} \\ \hline 1\,2\,0\phantom{0} \\ -1\,1\,2\phantom{0} \\ \hline 80 \\ -80 \\ \hline 0 \end{array}$$

## EXERCISES

**Divide.**

1. $5\overline{)2.7}$
2. $4\overline{)3.5}$
3. $2\overline{)9.1}$
4. $8\overline{)5.2}$
5. $4\overline{)9.6}$
6. $8\overline{)19}$
7. $5\overline{)24}$
8. $2\overline{)27}$
9. $4\overline{)19}$
10. $8\overline{)23}$
11. $24\overline{)9}$
12. $24\overline{)15}$
13. $10\overline{)62.4}$
14. $14\overline{)2.8}$
15. $32\overline{)20}$
16. $84\overline{)39.48}$
17. $48\overline{)3.408}$
18. $92\overline{)7.636}$
19. $65\overline{)2.47}$
20. $38\overline{)1311}$

**Use the shortcut to find these quotients.**

21. $62.4 \div 10$
22. $345 \div 10$
23. $8.95 \div 10$
24. $0.6 \div 10$

**See if you can find a shortcut for dividing by 100.**

25. $42.8 \div 100$
26. $725 \div 100$
27. $5.9 \div 100$
28. $0.8 \div 100$

## KEEPING SKILLS SHARP

**Give each difference in simplest form.**

1. $\frac{5}{9} - \frac{1}{9}$
2. $\frac{5}{8} - \frac{3}{8}$
3. $\frac{1}{2} - \frac{1}{4}$
4. $\frac{3}{4} - \frac{1}{2}$
5. $\frac{7}{8} - \frac{1}{4}$
6. $\frac{1}{2} - \frac{3}{8}$
7. $\frac{7}{8} - \frac{5}{6}$
8. $\frac{3}{4} - \frac{5}{8}$
9. $\frac{5}{6} - \frac{2}{3}$
10. $\frac{2}{3} - \frac{1}{4}$
11. $\frac{2}{3} - \frac{5}{8}$
12. $\frac{3}{4} - \frac{1}{6}$

# Estimating

If you learn to estimate before you calculate, you will know whether your answer makes sense.

Remember to work inside the grouping symbols first.

**EXERCISES**

**First estimate, then calculate.**

**1.** (5.8 + 4.2) − 3.9

**2.** 5.8 + (4.2 − 3.9)

**3.** (3.9 × 2.8) + 4.2

**4.** 3.9 × (2.8 + 4.2)

**5.** (19.7 − 3.1) × 2.8

**6.** 19.7 − (3.1 × 2.8)

**7.** (6.24 ÷ 3) × 2

**8.** 6.24 ÷ (3 × 2)

**9.** (16.8 − 8.1) − 2.9

**10.** 16.8 − (8.1 − 2.9)

## *Challenge!*

**Put decimal points as needed to get the answer shown.**

**11.** (41 × 202) ÷ 2 — 41.41

**12.** 41 × (202 ÷ 2) — 4.141

**13.** (252 × 39) − 203 — 96.25

**14.** 252 × (39 − 203) — 47.124

**15.** (436 + 189) × 417 — 260.625

**16.** 436 + (189 × 417) — 51.4813

1. Calculate the problem shown above.
2. How close was Team A's estimate?
   How close was Team B's estimate?
3. Which estimate was closer?

**Play the game.**

1. Divide the class into two teams.
2. A player from each team goes to the chalkboard and faces the class.
3. The teacher writes a decimal problem.
4. On a signal, the two players look at the problem and have 15 seconds to write their estimates.
5. The player with the closer estimate earns 1 point for his or her team.
6. Continue to play until each player has had at least 1 turn.
7. The team with the greater score wins.

# Problem solving—more than one way to solve

1. Study and understand.
2. Plan and do.
3. Answer and check.

Sandy took in 4 negatives to Ken's Photo Shop. She ordered one 5-inch by 7-inch enlargement for \$1.99 and one 8-inch by 10-inch enlargement for \$3.89 from each of them. What was the total cost of the enlargements?

**Brad's way to solve.**

**Step 1.**
Multiply to find the cost of the 5-inch by 7-inch enlargements.

$$\$1.99 \times 4 = \$7.96$$

**Step 2.**
Multiply to find the cost of the 8-inch by 10-inch enlargements.

$$\$3.89 \times 4 = \$15.56$$

**Step 3.**
Add to find the total cost of the enlargements.

$$\$7.96 + 15.56 = \$23.52$$

The total cost of the enlargements is \$23.52.

**Diana's way to solve.**

**Step 1.**
Add to find the cost of the enlargements for each negative.

$$\$1.99 + 3.89 = \$5.88$$

**Step 2.**
Multiply to find the total cost of the enlargements.

$$\$5.88 \times 4 = \$23.52$$

The total cost of the enlargements is \$23.52.

## EXERCISES

**Solve.**

1. Richard bought a roll of film with 36 exposures and a roll of film with 20 exposures. Then he took 8 pictures. How many more pictures could he take?

2. Nine of the pictures on a 36-exposure roll of film had Anne in them. Five of the pictures on a 20-exposure roll of film had Anne in them. How many of the pictures on the 2 rolls did not have Anne in them?

3. A flash cube can be used to take 4 indoor pictures. There are 3 flash cubes in a package. Christine bought 5 packages. How many indoor pictures could the cubes be used for?

4. Fred had $10. He bought a package of flash cubes for $1.59 and a roll of 12-exposure film for $2.79. How much money did he have left?

5. A 36-exposure roll of film that usually sells for $6.29 is on sale for $5.69. How much would Dorothy save if she bought 3 rolls?

6. Leon wants to buy a camera that costs $139.95. He has $146.18 and can trade in his old camera for $25. If he buys the new camera and trades in his old camera, how much money will he have left?

7. A 64-page album and a 96-page album can each hold three 3-inch by 5-inch pictures on a page. How many 3-inch by 5-inch pictures can the 2 albums hold altogether?

8. Antonio bought a camera for $27.99 and some film for $6.29. He was charged a 5% sales tax. What was the total tax? See the table on page 294.

# CHAPTER CHECKUP

**Multiply.** [pages 326–331, 336–337]

1. $0.8 \times 0.3$
2. $1.4 \times 0.5$
3. $37 \times 0.8$
4. $19.6 \times 4$
5. $2.14 \times 0.7$
6. $4.2 \times 1.2$
7. $53 \times 0.32$
8. $7.8 \times 6.1$
9. $42.6 \times 0.48$
10. $5.37 \times 5.6$
11. $52.3 \times 10.3$
12. $60.4 \times 251$
13. $49.3 \times 3.25$
14. $8.26 \times 472$
15. $7.14 \times 56.8$

**Divide.** [pages 332–337]

16. $2\overline{)43.6}$
17. $4\overline{)9.12}$
18. $6\overline{)5.88}$
19. $9\overline{)2.52}$
20. $7\overline{)38.5}$
21. $8\overline{)3}$
22. $4\overline{)3}$
23. $8\overline{)15}$
24. $5\overline{)31}$
25. $4\overline{)53}$
26. $19\overline{)124.07}$
27. $28\overline{)229.88}$
28. $32\overline{)203.2}$
29. $57\overline{)62.13}$
30. $75\overline{)126}$

**Solve.** [pages 338–339]

31. 

How much will 3.5 kilograms of bananas cost?

32. 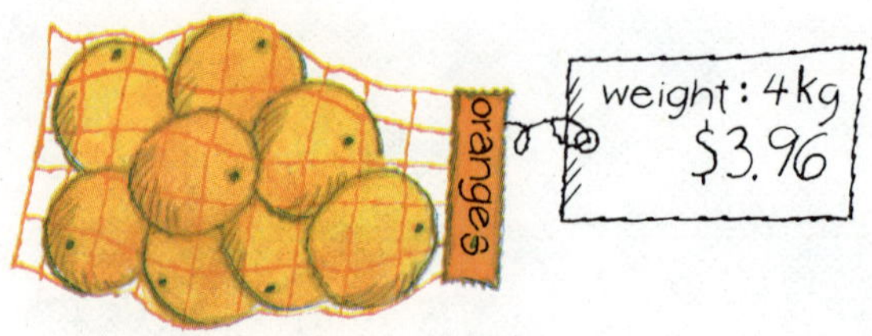


How much will 1 kilogram of oranges cost?

33. A rectangular lawn is 52.4 meters by 30.5 meters. What is the area of the lawn?

34. An airplane flew 2619 kilometers in 4 hours. How many kilometers did it average per hour?

# CHAPTER PROJECT

To find out how much one paper clip weighs, you can first weigh 10 paper clips to the nearest gram and then divide by 10.

**Use the same method to "weigh" these objects.**

**1.** a larger paper clip

**2.** a straight pin

**3.** a thumbtack

**4.** a brass paper fastener

# CHAPTER REVIEW

**Multiply.**

28.6 ← 1 digit
× 0.24 ← 2 digits
1144
5720
6.864 ← 3 digits

1. $\begin{array}{r} 3.5 \\ \times 0.7 \\ \hline \end{array}$
2. $\begin{array}{r} 7.5 \\ \times 0.8 \\ \hline \end{array}$
3. $\begin{array}{r} 6.45 \\ \times 0.6 \\ \hline \end{array}$
4. $\begin{array}{r} 12.4 \\ \times 3.5 \\ \hline \end{array}$
5. $\begin{array}{r} 52.6 \\ \times 4.3 \\ \hline \end{array}$
6. $\begin{array}{r} 82.5 \\ \times 0.47 \\ \hline \end{array}$

**Multiply.**

0.02
× 0.3
0.006

7. $\begin{array}{r} 0.4 \\ \times 0.2 \\ \hline \end{array}$
8. $\begin{array}{r} 0.6 \\ \times 0.1 \\ \hline \end{array}$
9. $\begin{array}{r} 0.03 \\ \times 0.3 \\ \hline \end{array}$
10. $\begin{array}{r} 1.2 \\ \times 0.04 \\ \hline \end{array}$
11. $\begin{array}{r} 2.6 \\ \times 0.06 \\ \hline \end{array}$
12. $\begin{array}{r} 0.21 \\ \times 0.2 \\ \hline \end{array}$

$\begin{array}{r} 0.59 \\ 4\overline{)2.36} \\ -2\,0 \\ \hline 36 \\ -36 \\ \hline 0 \end{array}$

**Divide.**

13. $6\overline{)16.2}$
14. $8\overline{)3.52}$
15. $9\overline{)0.162}$
16. $12\overline{)1.44}$
17. $24\overline{)84.48}$
18. $63\overline{)17.829}$

$\begin{array}{r} 0.375 \\ 16\overline{)6.000} \\ -4\,8 \\ \hline 1\,20 \\ -1\,12 \\ \hline 80 \\ -80 \\ \hline 0 \end{array}$

**Divide.**

19. $4\overline{)3}$
20. $8\overline{)3}$
21. $2\overline{)3}$
22. $16\overline{)50}$
23. $36\overline{)54}$
24. $25\overline{)105}$

# CHAPTER CHALLENGE

**Copy and complete.**

**1.** Across: + Down: +

| 3.64 | ? | 8.52 |
|---|---|---|
| 4.18 | ? | 7.09 |
| ? | ? | ? |

**2.** Across: × Down: ×

| 3 | 1.51 | ? |
|---|---|---|
| 1.2 | 0.8 | ? |
| ? | ? | ? |

**3.** Across: + Down: −

| 9.34 | 3.86 | ? |
|---|---|---|
| ? | ? | ? |
| 5.16 | 2.05 | ? |

**4.** Across: × Down: ÷

| 3 | 1 | ? |
|---|---|---|
| 4 | 2 | ? |
| ? | ? | ? |

**5.** Across: × Down: ÷

| 6.4 | 1.5 | ? |
|---|---|---|
| 2 | ? | 6 |
| ? | ? | ? |

**6.** Across: + Down: −

| ? | 6.23 | ? |
|---|---|---|
| ? | ? | 5.14 |
| ? | 2.26 | 4.45 |

**7.** Across: + Down: −

| 9.6 | 2.4 | ? |
|---|---|---|
| ? | ? | 1.8 |
| 8.3 | ? | ? |

**8.** Across: × Down: ÷

| 7.2 | 4.4 | ? |
|---|---|---|
| 6 | ? | 12 |
| ? | ? | ? |

# MAJOR CHECKUP
## STANDARDIZED FORMAT

1. $\overleftrightarrow{AB}$ is parallel to

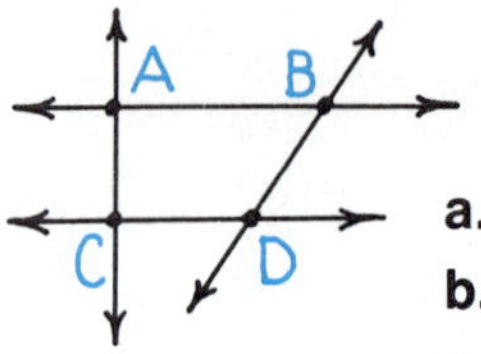


a. $\overleftrightarrow{CD}$
b. $\overleftrightarrow{DB}$
c. $\overleftrightarrow{CA}$
d. none of these

2. $\frac{5}{6} - \frac{1}{4} = \underline{?}$

a. 2
b. $\frac{7}{12}$
c. $\frac{1}{3}$
d. none of these

3. $\frac{5}{9} \times \frac{3}{2} = \underline{?}$

a. $\frac{5}{6}$
b. $\frac{4}{9}$
c. $\frac{15}{9}$
d. none of these

4. $49.6 - 3.82 = \underline{?}$

a. 45.78
b. 11.4
c. 1.14
d. none of these

5. The perimeter of this triangle is

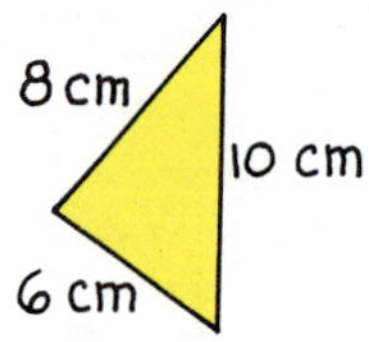


a. 48 cm
b. 24 cm
c. 14 cm
d. none of these

6. What is the price of 1?

a. 22¢
b. 23¢
c. 24¢
d. none of these

7. Add.

$$\begin{array}{r} 4\frac{1}{2} \\ +3\frac{5}{8} \\ \hline \end{array}$$

a. $7\frac{1}{8}$
b. $8\frac{1}{8}$
c. $7\frac{7}{8}$
d. none of these

8. Subtract.

$$\begin{array}{r} 6\frac{1}{3} \\ -2\frac{3}{4} \\ \hline \end{array}$$

a. $4\frac{5}{12}$
b. $4\frac{7}{12}$
c. $3\frac{7}{12}$
d. none of these

9. Multiply.

$$\begin{array}{r} 6.21 \\ \times 0.8 \\ \hline \end{array}$$

a. 49.68
b. 4.968
c. 0.4968
d. none of these

10. Divide.

$7\overline{)36.61}$

a. 5 R161
b. 52.9
c. 5.29
d. none of these

11. Alicia skated for $3\frac{1}{2}$ hours. She began at 12:40. When did she finish?

a. 3:10
b. 3:50
c. 4:10
d. none of these

12. A record costs \$5. A tape costs \$9. Dan has \$12. How much more does he need to buy one of each?

a. \$4
b. \$14
c. \$26
d. none of these

# RESOURCES


GETTING READY FOR COMPUTER PROGRAMMING 346

SKILL TEST 356

EXTRA PRACTICE 360

EXTRA PROBLEM SOLVING 379

GLOSSARY 390

# GETTING READY FOR COMPUTER PROGRAMMING

## Using commands in programs

You write a **program** to tell the helicopter where to fly. The directions you write in the program are called **commands.**

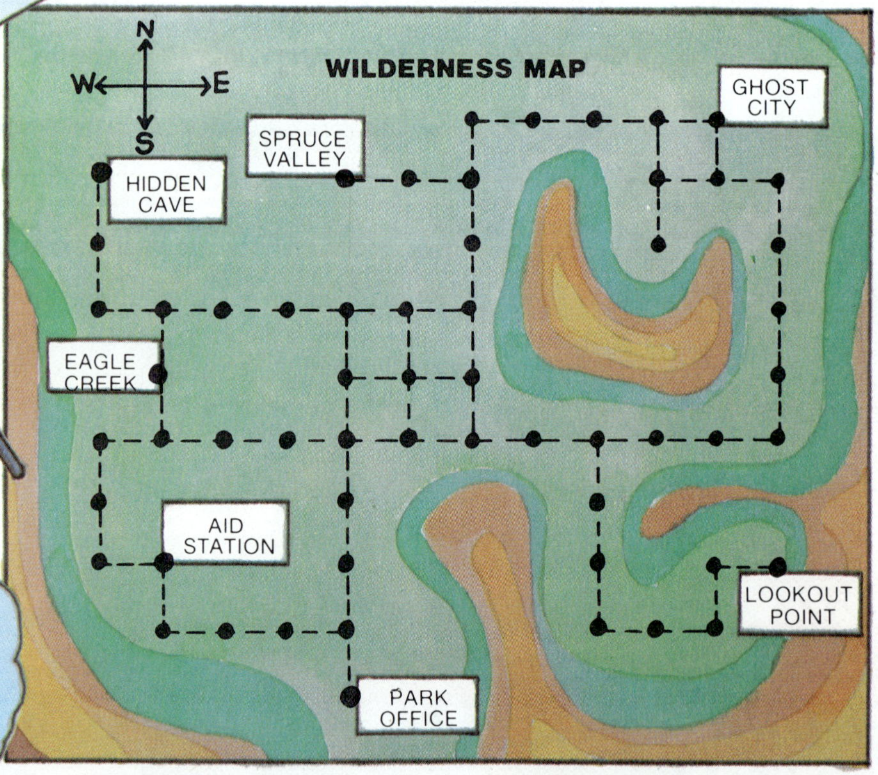


**Follow the path on the map.**

1. PROGRAM PARK/SPRUCE
   BEGIN PARK OFFICE
   N6
   E2
   N2
   W2
   END SPRUCE VALLEY

   This command tells the helicopter to fly north 6 spaces.

   a. What is the name of the program?
   b. Where did you begin?
   c. Where did you end?

**Complete each program.**

2. PROGRAM PARK/ ___a___
BEGIN PARK OFFICE
N4
E4
S3
E2
N1
E1
END ___b___

3. PROGRAM GHOST/ ___a___
BEGIN GHOST CITY
W4
S5
W6
S2
E1
END ___b___

**There is a mistake in each program.**
**Rewrite each program correctly.**

4. PROGRAM HIDDEN/AID
BEGIN HIDDEN CAVE
S2
E1
S2
W1
S2
W1
END AID STATION

5. PROGRAM EAGLE/LOOKOUT
BEGIN EAGLE CREEK
S1
E7
S3
E2
N2
E1
END LOOKOUT POINT

**Complete each program.**

6. PROGRAM SPRUCE/AID
BEGIN SPRUCE VALLEY
__a__
S4
W6
__b__
__c__
END ___d___

★7. PROGRAM ___a___ /PARK
BEGIN ___b___
W1
S1
W2
N3
W4
S4
END PARK OFFICE

# New commands—TAKE and LEAVE

You can pick up a hiker by using the command TAKE HIKER. You can leave the hiker by using the command LEAVE HIKER.

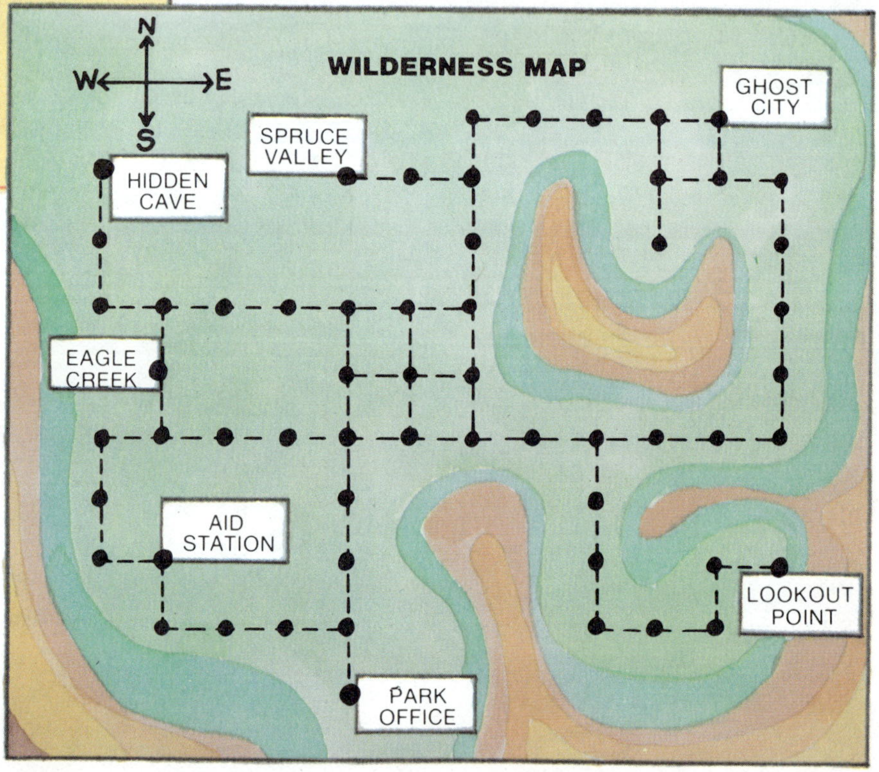


**Follow the path on the map.**

1. PROGRAM PARK/EAGLE/AID
   BEGIN PARK OFFICE
   N4
   W3
   N1
   TAKE HIKER
   S1
   W1
   S2
   E1
   LEAVE HIKER
   END AID STATION

Use this command to take or pick up a hiker.

Use this command to leave a hiker.

a. What is the name of the program?
b. Where did you begin?
c. Where did you pick up the hiker?
d. Where did you leave the hiker?
e. Where did you end?

**Complete each program.**

2. PROGRAM PARK/SPRUCE/____a____
BEGIN PARK OFFICE
N6
E2
N2
W2
TAKE HIKER
E2
S4
W6
S2
E1
LEAVE HIKER
END ____b____

3. PROGRAM PARK/GHOST/____a____
BEGIN PARK OFFICE
N6
E2
N3
E4
TAKE HIKER
W4
S3
W2
S6
LEAVE HIKER
END ____b____

4. PROGRAM PARK/GHOST/AID
BEGIN PARK OFFICE
N4
__a__
__b__
E4
TAKE HIKER
W4
S5
W6
__c__
__d__
LEAVE HIKER
END ____e____

5. PROGRAM PARK/SPRUCE/PARK
BEGIN PARK OFFICE
N6
__a__
__b__
__c__
TAKE HIKER
__d__
S4
__e__
__f__
____g____
END PARK OFFICE

**Write the program that uses the fewest commands.**

6. Begin at the Park Office, pick up a hiker at Hidden Cave, and leave the hiker at the Park Office.

7. Begin at the Park Office, pick up a hiker at Lookout Point, and leave the hiker at the Aid Station.

# Procedures

A **procedure** is a list of commands that is used often. Each procedure has a name. You can use the name of a procedure as a command in a program.

**Follow the path for the refueling procedure on the map.**

1. PROCEDURE REFUEL
   BEGIN PARK OFFICE
   S2
   E1
   FILL TANKS
   W1
   N2

   a. What is the name of the procedure?
   b. Where did you begin?
   c. Where were you when the command FILL TANKS was given?
   d. Where are you now?

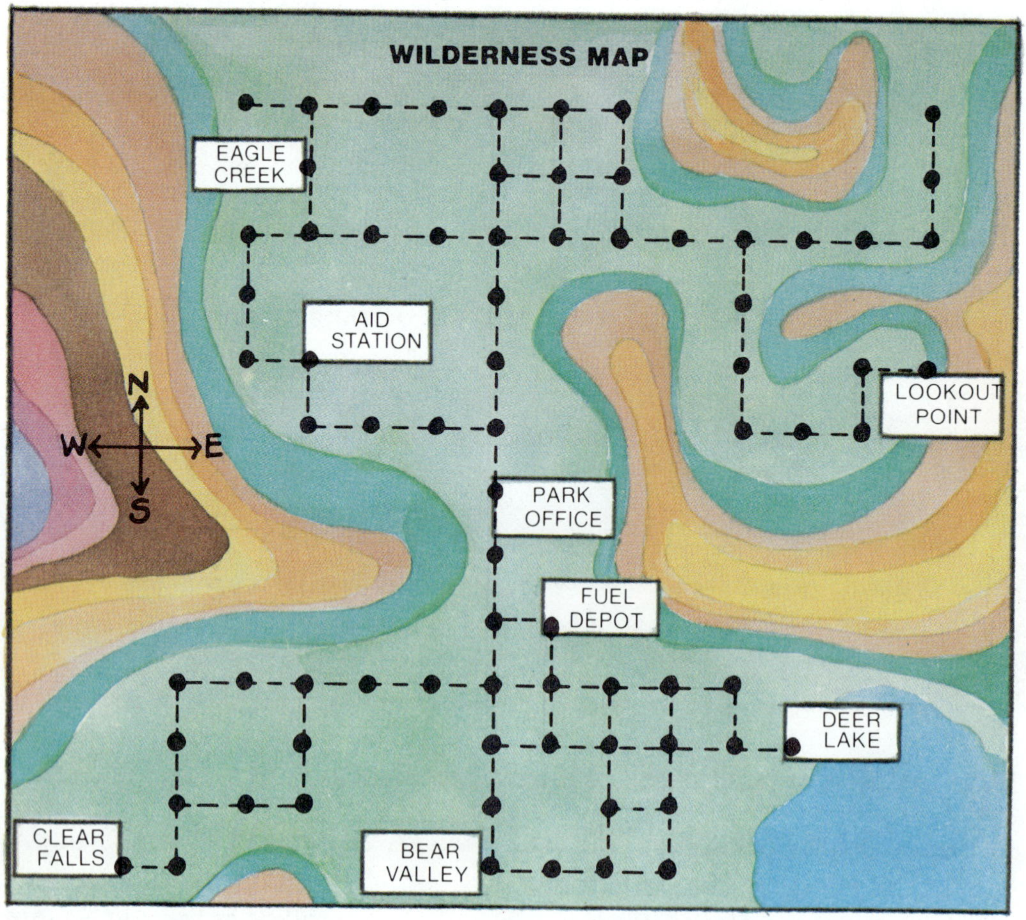

**Complete each procedure.**

2. PROCEDURE AID/PARK
   BEGIN AID STATION
   ____a____
   ____b____
   ____c____

3. PROCEDURE PARK/CLEAR
   BEGIN PARK OFFICE
   ____a____
   W5
   ____b____
   ____c____

The path for a rescue must go from the Park Office to the hiker, then to the Aid Station. After waiting for the hiker to be checked, the helicopter must take the hiker to the Park Office, refuel, and then return to the Park Office.

**Complete each program.**

4. PROGRAM RESCUE DEER
   BEGIN PARK OFFICE
   ____a____
   ____b____
   TAKE HIKER
   ____c____
   N5
   W3
   ____d____
   LEAVE HIKER
   WAIT
   TAKE HIKER
   AID/PARK
   LEAVE HIKER
   REFUEL
   END ____e____

5. PROGRAM RESCUE BEAR
   BEGIN PARK OFFICE
   ____a____
   TAKE HIKER
   ____b____
   ____c____
   ____d____
   LEAVE HIKER
   WAIT
   TAKE HIKER
   ____e____
   ____f____
   ____g____
   END PARK OFFICE

**Write the program that uses the fewest commands.**

★6.

# More procedures

In the last lesson, you used two procedures, REFUEL and AID/PARK, in programs. You also wrote procedure PARK/CLEAR. In this lesson, you will write some more procedures and then use them in programs.

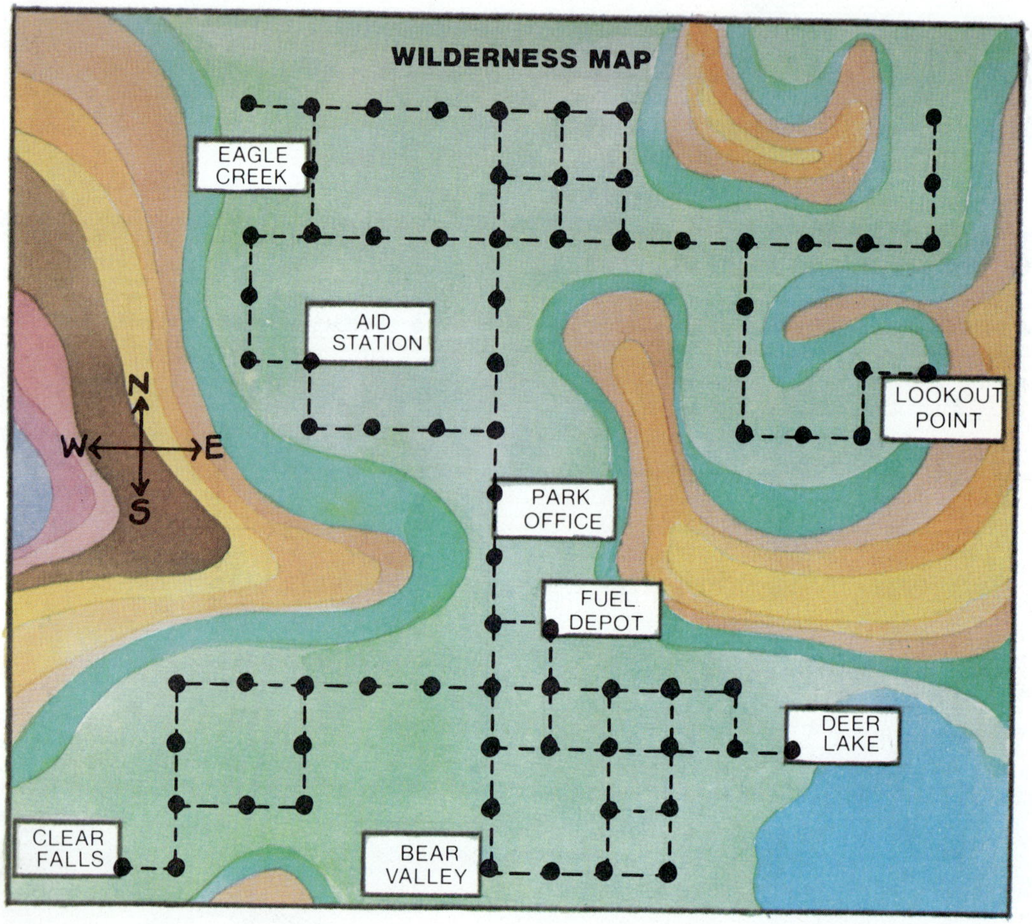


**Complete each procedure.**

1. PROCEDURE PARK/EAGLE
   BEGIN PARK OFFICE
   N4
   __a__
   __b__

2. PROCEDURE EAGLE/AID
   BEGIN ___a___
   __b__
   __c__
   __d__
   __e__

3. PROCEDURE CLEAR/AID
BEGIN ____a____
E1
N3
____b____
____c____
____d____
____e____

In these programs, rescued hikers are taken to the Aid Station and then to the Park Office. Remember that the helicopter always refuels after a rescue.

**Complete each program.**

4. PROGRAM RESCUE EAGLE
BEGIN PARK OFFICE
PARK/EAGLE
TAKE HIKER
____a____
LEAVE HIKER
WAIT
TAKE HIKER
AID/PARK
LEAVE HIKER
____b____
END PARK OFFICE

5. PROGRAM RESCUE CLEAR
BEGIN PARK OFFICE
PARK/CLEAR
TAKE HIKER
____a____
LEAVE HIKER
WAIT
____b____
____c____
LEAVE HIKER
____d____
END ____e____

From now on, assume that you have procedures that will take the helicopter from any labeled point on the map to any other labeled point on the map.

**Write the program that uses the fewest commands.**

6. The helicopter goes to Deer Lake, picks up a hiker, takes the hiker to Clear Falls, leaves the hiker there. Then the helicopter refuels and returns to the Park Office.

7. The helicopter leaves the Park Office, picks up a hiker at Clear Falls, lets the hiker off at the Park Office, and then refuels.

# Decisions

The helicopter always begins at the Park Office. When a rescue call comes in, it picks up the hiker. The pilot decides whether the hiker is injured. If the hiker is injured, the helicopter flies directly to the Aid Station. If not, it flies to the Park Office.

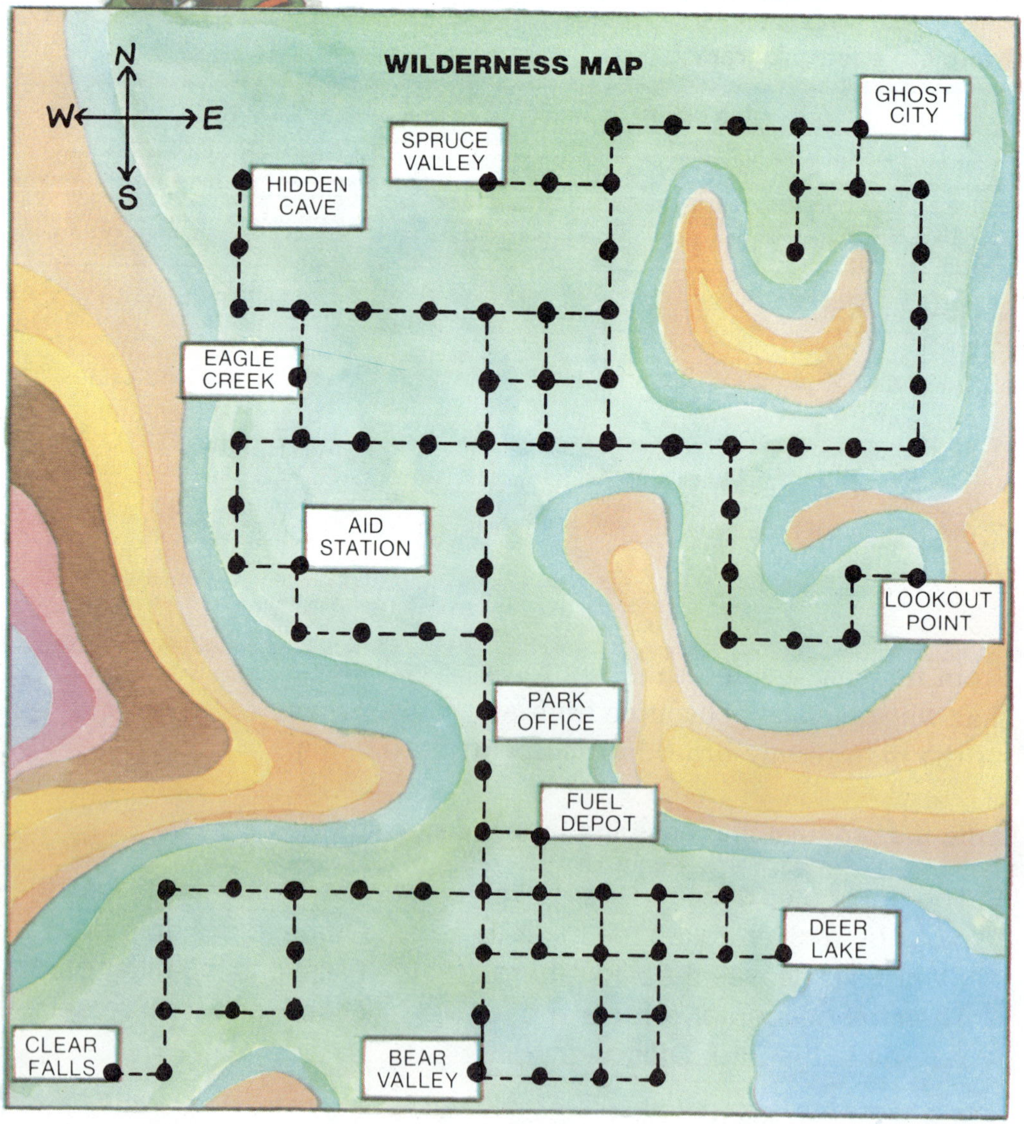

**Study this program.**

1.
```
PROGRAM RESCUE GHOST
BEGIN PARK OFFICE
PARK/GHOST
TAKE HIKER
IF INJURED
   THEN GHOST/AID
        LEAVE HIKER
        WAIT
        TAKE HIKER
        AID/PARK
   OTHERWISE GHOST/PARK
LEAVE HIKER
REFUEL
END PARK OFFICE
```

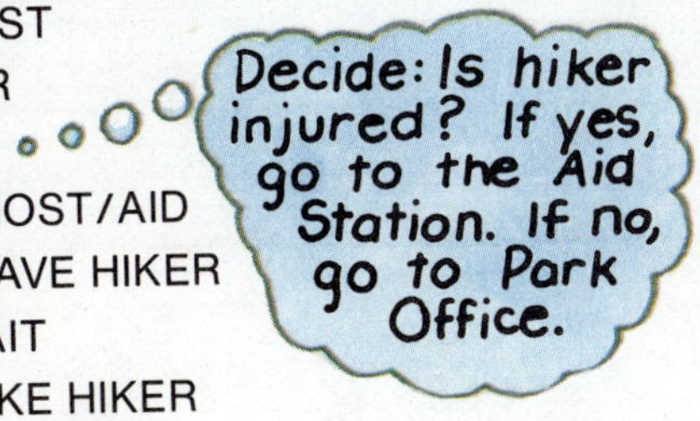


a. Where did you begin?
b. Where did you first pick up the hiker?
c. If injured, where was the hiker taken?
d. After being treated, where was the injured hiker taken?
e. If not injured, where was the hiker taken?

**Complete each program. Don't forget to refuel.**

2.
```
PROGRAM RESCUE SPRUCE
BEGIN PARK OFFICE
___a___
TAKE HIKER
IF INJURED
   THEN ___b___
        LEAVE HIKER
        WAIT
        TAKE HIKER
        ___c___
   OTHERWISE ___d___
LEAVE HIKER
___e___
END PARK OFFICE
```

3.
```
PROGRAM RESCUE EAGLE
BEGIN PARK OFFICE
___a___
___b___
IF INJURED
   THEN ___c___
        LEAVE HIKER
        ___d___
        TAKE HIKER
        ___e___
   OTHERWISE ___f___
___g___
___h___
END PARK OFFICE
```

**Write the program that uses the fewest commands.**

4. PROGRAM RESCUE CLEAR

5. PROGRAM RESCUE LOOKOUT

6. PROGRAM RESCUE DEER

7. PROGRAM RESCUE HIDDEN

# SKILL TEST

| | | | | | | |
|---|---|---|---|---|---|---|
| 1 | Basic addition facts, sums through 18 | 3<br>+4 | 9<br>+0 | 6<br>+7 | 9<br>+8 | 8<br>+7 |
| 2 | Addition without or with regrouping | 42<br>+23 | 32<br>+48 | 37<br>+26 | 58<br>+79 | |
| 3 | Addition with regrouping | 528<br>+146 | 359<br>+670 | 546<br>+278 | 945<br>+697 | |
| 4 | Addition with regrouping | 3789<br>+1546 | 84,962<br>+57,698 | 348,716<br>+655,029 | | |
| 5 | Addition of three or more numbers | 74<br>29<br>+63 | 185<br>76<br>640<br>+57 | 3921<br>1823<br>4765<br>+5283 | 86,945<br>6,913<br>7,428<br>+5,317 | |
| 6 | Basic subtraction facts, sums through 18 | 8<br>−8 | 12<br>−5 | 18<br>−9 | 16<br>−7 | |
| 7 | Subtraction without or with regrouping | 58<br>−12 | 70<br>−35 | 93<br>−27 | 64<br>−56 | |
| 8 | Subtraction with regrouping | 750<br>−429 | 621<br>−385 | 803<br>−256 | 900<br>−664 | |
| 9 | Subtraction with regrouping | 4218<br>−1072 | 68,102<br>−34,529 | 726,300<br>−459,275 | | |

| | | | | | | |
|---|---|---|---|---|---|---|
| 10 | Basic multiplication facts, products through 9 × 9 | 8<br>×0 | 3<br>×7 | 9<br>×4 | 6<br>×8 | 7<br>×9 |
| 11 | Multiplication without and with regrouping | 32<br>×2 | 24<br>×4 | 65<br>×8 | 89<br>×7 | |
| 12 | Multiplication with regrouping | 151<br>×5 | 278<br>×3 | 578<br>×6 | 758<br>×9 | |
| 13 | Multiplication with regrouping | 7283<br>×4 | 5096<br>×6 | 4179<br>×8 | | |
| 14 | Multiplying by 10 and 100 | 258<br>×10 | 3192<br>×10 | 764<br>×100 | 5978<br>×100 | |
| 15 | Multiplying by 10's and 100's | 78<br>×30 | 526<br>×60 | 885<br>×400 | 942<br>×700 | |
| 16 | Multiplication with or without regrouping | 42<br>×21 | 68<br>×16 | 325<br>×48 | 674<br>×53 | |
| 17 | Multiplication with regrouping | 328<br>×142 | 536<br>×240 | 755<br>×526 | 889<br>×703 | |
| 18 | Division facts, quotients through 81 ÷ 9 | $9\overline{)36}$ | $5\overline{)0}$ | $6\overline{)42}$ | $9\overline{)54}$ | |
| 19 | Division with or without regrouping | $2\overline{)86}$ | $6\overline{)72}$ | $4\overline{)68}$ | $3\overline{)87}$ | |

| | | | | | | |
|---|---|---|---|---|---|---|
| 20 | Division with remainder | $5\overline{)73}$ | $4\overline{)91}$ | $3\overline{)74}$ | $6\overline{)99}$ | |
| 21 | Division with regrouping | $7\overline{)945}$ | $5\overline{)745}$ | $8\overline{)289}$ | $9\overline{)593}$ | |
| 22 | Division with regrouping | $16\overline{)688}$ | $28\overline{)952}$ | $42\overline{)978}$ | $33\overline{)809}$ | |
| 23 | Division with regrouping | $68\overline{)2516}$ | $74\overline{)95{,}634}$ | $53\overline{)72{,}974}$ | | |
| 24 | Changing to lowest terms | $\frac{5}{10} = \underline{?}$ | $\frac{6}{8} = \underline{?}$ | $\frac{6}{9} = \underline{?}$ | $\frac{8}{12} = \underline{?}$ | |
| 25 | Adding fractions | $\frac{4}{9} + \frac{1}{9}$ | $\frac{3}{8} + \frac{1}{8}$ | $\frac{1}{2} + \frac{1}{4}$ | $\frac{1}{6} + \frac{5}{8}$ | $\frac{2}{3} + \frac{1}{4}$ |
| 26 | Subtracting fractions | $\frac{8}{9} - \frac{3}{9}$ | $\frac{5}{6} - \frac{1}{6}$ | $\frac{3}{4} - \frac{1}{2}$ | $\frac{7}{9} - \frac{1}{2}$ | $\frac{5}{6} - \frac{3}{4}$ |
| 27 | Changing a mixed number to a fraction | $1\frac{1}{2} = \underline{?}$ | $3\frac{1}{4} = \underline{?}$ | $2\frac{2}{3} = \underline{?}$ | $3\frac{4}{5} = \underline{?}$ | |
| 28 | Changing a fraction to a mixed number | $\frac{4}{3} = \underline{?}$ | $\frac{5}{2} = \underline{?}$ | $\frac{10}{3} = \underline{?}$ | $\frac{15}{4} = \underline{?}$ | |

| | | | | | |
|---|---|---|---|---|---|
| 29 | Fraction of a number | $\frac{1}{4}$ of 20 = ? | $\frac{3}{4}$ of 24 = ? | $\frac{2}{3}$ of 18 = ? | |
| 30 | Multiplying fractions | $\frac{2}{3} \times \frac{4}{5}$ | $\frac{5}{8} \times \frac{4}{3}$ | $\frac{5}{6} \times \frac{6}{5}$ | |
| 31 | Adding decimals | $3.9 + 2.6$ | $7.85 + 3.96$ | $2.786 + 9.374$ | $69.78 + 56.95$ |
| 32 | Subtracting decimals | $3.4 - 2.6$ | $7.03 - 2.56$ | $5.078 - 2.193$ | $82.14 - 19.57$ |
| 33 | Adding mixed numbers | $5\frac{4}{9} + 2\frac{1}{9}$ | $3\frac{5}{8} + 6\frac{1}{4}$ | $6\frac{1}{2} + 5\frac{3}{4}$ | $8\frac{2}{3} + 6\frac{3}{4}$ |
| 34 | Subtracting mixed numbers | $9\frac{3}{5} - 4\frac{1}{5}$ | $8\frac{2}{3} - 1\frac{1}{6}$ | $6\frac{1}{4} - 2\frac{1}{2}$ | $7\frac{2}{9} - 3\frac{5}{6}$ |
| 35 | Multiplying decimals | $1.3 \times 0.2$ | $4.2 \times 3.4$ | $6.05 \times 5.3$ | $68.5 \times 5.37$ |
| 36 | Dividing decimals | $8\overline{)7.92}$ | $42\overline{)235.2}$ | $38\overline{)28.12}$ | |

# EXTRA PRACTICE

**Set 1  Round to the nearest thousand.**

| | | | |
|---|---|---|---|
| 1. 7926 *8000* | 2. 8034 | 3. 2799 | 4. 6500 |
| 5. 74,824 | 6. 52,009 | 7. 82,643 | 8. 18,500 |
| 9. 365,294 | 10. 381,666 | 11. 592,500 | 12. 763,004 |
| 13. 859,600 | 14. 537,499 | 15. 399,832 | 16. 799,500 |
| 17. 428,950 | 18. 609,085 | 19. 299,406 | 20. 639,500 |

**Set 2  Give the standard numeral.**

1. sixteen million *16,000,000*
2. one hundred forty-six million
3. thirty-two million, three hundred two thousand, two hundred sixty-five
4. five hundred thirty-nine million, eight hundred ninety-seven thousand, four hundred two
5. eight hundred fifty-one million, three hundred thirty-five
6. ninety-five million, fifty-two thousand
7. seventeen million, six thousand, forty-eight
8. two hundred million, fifty-nine

**Set 3  Add.**

| | | | | | | |
|---|---|---|---|---|---|---|
| 1. $3 + 5 = 8$ | 2. $9 + 0$ | 3. $4 + 8$ | 4. $4 + 6$ | 5. $6 + 6$ | 6. $5 + 7$ | 7. $9 + 4$ |
| 8. $6 + 8$ | 9. $3 + 8$ | 10. $8 + 8$ | 11. $5 + 6$ | 12. $9 + 9$ | 13. $3 + 9$ | 14. $6 + 7$ |
| 15. $5 + 8$ | 16. $9 + 8$ | 17. $2 + 7$ | 18. $9 + 6$ | 19. $9 + 7$ | 20. $9 + 5$ | 21. $8 + 4$ |
| 22. $2 + 9$ | 23. $3 + 7$ | 24. $7 + 7$ | 25. $8 + 2$ | 26. $4 + 7$ | 27. $7 + 8$ | 28. $5 + 5$ |

**Set 4 Add.**

| | | | | | | |
|---|---|---|---|---|---|---|
| 1. 5<br>2<br>+5<br>12 | 2. 3<br>7<br>+4 | 3. 5<br>6<br>+4 | 4. 3<br>6<br>+7 | 5. 2<br>8<br>+8 | 6. 4<br>1<br>+9 | 7. 5<br>5<br>+6 |
| 8. 3<br>6<br>5<br>+4 | 9. 7<br>8<br>3<br>+2 | 10. 5<br>3<br>6<br>+5 | 11. 4<br>3<br>6<br>+1 | 12. 7<br>4<br>5<br>+3 | 13. 9<br>4<br>1<br>+6 | 14. 3<br>4<br>6<br>+8 |
| 15. 5<br>3<br>5<br>+3 | 16. 4<br>2<br>3<br>+7 | 17. 2<br>4<br>1<br>+8 | 18. 6<br>3<br>4<br>+2 | 19. 3<br>1<br>7<br>+6 | 20. 3<br>2<br>4<br>+8 | 21. 5<br>3<br>4<br>+5 |

**Set 5 Add.**

| | | | | | |
|---|---|---|---|---|---|
| 1. 1<br>28<br>+44<br>72 | 2. 32<br>+27 | 3. 62<br>+51 | 4. 28<br>+37 | 5. 46<br>+38 | 6. 58<br>+19 |
| 7. 58<br>+36 | 8. 86<br>+53 | 9. 65<br>+87 | 10. 94<br>+93 | 11. 44<br>+50 | 12. 37<br>+93 |
| 13. 92<br>+38 | 14. 73<br>+56 | 15. 68<br>+68 | 16. 35<br>+86 | 17. 90<br>+63 | 18. 77<br>+58 |
| 19. 73<br>+99 | 20. 59<br>+59 | 21. 84<br>+70 | 22. 76<br>+88 | 23. 88<br>+93 | 24. 69<br>+48 |

**Set 6 Add.**

| | | | | |
|---|---|---|---|---|
| 1. 11<br>298<br>+436<br>734 | 2. 275<br>+105 | 3. 645<br>+428 | 4. 736<br>+319 | 5. 827<br>+216 |
| 6. 506<br>+709 | 7. 721<br>+192 | 8. 365<br>+294 | 9. 638<br>+270 | 10. 376<br>+898 |
| 11. 357<br>+296 | 12. 842<br>+158 | 13. 592<br>+321 | 14. 807<br>+511 | 15. 653<br>+328 |
| 16. 429<br>+496 | 17. 688<br>+742 | 18. 275<br>+395 | 19. 378<br>+628 | 20. 553<br>+308 |

**Set 7** Give each sum.

1. 5298 + 2486 = 7784 (carries: 1 1)
2. 3174 + 5834
3. 2653 + 1789
4. 8521 + 6524
5. 72,538 + 29,164
6. 93,825 + 61,943
7. 42,351 + 16,089
8. 51,783 + 29,468
9. 321,748 + 596,317
10. 629,634 + 789,519
11. 639,768 + 417,956
12. 399,999 + 427,835
13. 234,821 + 92,528
14. 566,097 + 483,465
15. 438,291 + 73,856
16. 756,815 + 52,234

**Set 8** Give each sum.

1. 335 + 558 + 219 = 1112 (carries: 1 2)
2. 315 + 218 + 476
3. 598 + 421 + 782
4. 495 + 342 + 687
5. 627 + 258 + 846
6. 3088 + 7809 + 2546
7. 5005 + 9637 + 2824
8. 9057 + 4706 + 3958
9. 6787 + 2953 + 3069
10. 4308 + 7295 + 6578
11. 238 + 567 + 429 + 531
12. 698 + 42 + 385 + 756
13. 8276 + 3755 + 9284 + 3615
14. 4038 + 192 + 78 + 5964
15. 36,175 + 8,930 + 65,274 + 5,971

**Set 9** Subtract.

1. 9 − 4 = 5
2. 10 − 7
3. 12 − 9
4. 14 − 5
5. 11 − 8
6. 13 − 5
7. 15 − 6
8. 16 − 8
9. 13 − 6
10. 11 − 9
11. 18 − 9
12. 12 − 8
13. 11 − 4
14. 12 − 7
15. 12 − 6
16. 14 − 6
17. 17 − 9
18. 13 − 4
19. 15 − 7
20. 14 − 7
21. 16 − 9
22. 13 − 8
23. 15 − 9
24. 12 − 4
25. 14 − 8
26. 16 − 7
27. 15 − 8
28. 10 − 8

## Set 10 Subtract.

| | | | | | |
|---|---|---|---|---|---|
| 1. $\begin{array}{r} \scriptstyle 7\ 1 \\ \not{8}4 \\ -36 \\ \hline 48 \end{array}$ | 2. $\begin{array}{r} 68 \\ -13 \\ \hline \end{array}$ | 3. $\begin{array}{r} 82 \\ -55 \\ \hline \end{array}$ | 4. $\begin{array}{r} 54 \\ -29 \\ \hline \end{array}$ | 5. $\begin{array}{r} 90 \\ -38 \\ \hline \end{array}$ | 6. $\begin{array}{r} 64 \\ -38 \\ \hline \end{array}$ |
| 7. $\begin{array}{r} 62 \\ -58 \\ \hline \end{array}$ | 8. $\begin{array}{r} 56 \\ -49 \\ \hline \end{array}$ | 9. $\begin{array}{r} 87 \\ -48 \\ \hline \end{array}$ | 10. $\begin{array}{r} 62 \\ -46 \\ \hline \end{array}$ | 11. $\begin{array}{r} 55 \\ -35 \\ \hline \end{array}$ | 12. $\begin{array}{r} 87 \\ -87 \\ \hline \end{array}$ |
| 13. $\begin{array}{r} 50 \\ -42 \\ \hline \end{array}$ | 14. $\begin{array}{r} 92 \\ -43 \\ \hline \end{array}$ | 15. $\begin{array}{r} 81 \\ -27 \\ \hline \end{array}$ | 16. $\begin{array}{r} 73 \\ -16 \\ \hline \end{array}$ | 17. $\begin{array}{r} 64 \\ -48 \\ \hline \end{array}$ | 18. $\begin{array}{r} 76 \\ -39 \\ \hline \end{array}$ |
| 19. $\begin{array}{r} 81 \\ -47 \\ \hline \end{array}$ | 20. $\begin{array}{r} 64 \\ -57 \\ \hline \end{array}$ | 21. $\begin{array}{r} 70 \\ -36 \\ \hline \end{array}$ | 22. $\begin{array}{r} 84 \\ -58 \\ \hline \end{array}$ | 23. $\begin{array}{r} 93 \\ -27 \\ \hline \end{array}$ | 24. $\begin{array}{r} 52 \\ -19 \\ \hline \end{array}$ |

## Set 11 Subtract.

| | | | | |
|---|---|---|---|---|
| 1. $\begin{array}{r} \scriptstyle 6\ 17\ 1 \\ \not{7}\not{8}4 \\ -198 \\ \hline 586 \end{array}$ | 2. $\begin{array}{r} 328 \\ -136 \\ \hline \end{array}$ | 3. $\begin{array}{r} 334 \\ -77 \\ \hline \end{array}$ | 4. $\begin{array}{r} 963 \\ -479 \\ \hline \end{array}$ | 5. $\begin{array}{r} 946 \\ -268 \\ \hline \end{array}$ |
| 6. $\begin{array}{r} 587 \\ -398 \\ \hline \end{array}$ | 7. $\begin{array}{r} 458 \\ -157 \\ \hline \end{array}$ | 8. $\begin{array}{r} 970 \\ -375 \\ \hline \end{array}$ | 9. $\begin{array}{r} 452 \\ -175 \\ \hline \end{array}$ | 10. $\begin{array}{r} 715 \\ -149 \\ \hline \end{array}$ |
| 11. $\begin{array}{r} 701 \\ -346 \\ \hline \end{array}$ | 12. $\begin{array}{r} 604 \\ -429 \\ \hline \end{array}$ | 13. $\begin{array}{r} 803 \\ -275 \\ \hline \end{array}$ | 14. $\begin{array}{r} 700 \\ -358 \\ \hline \end{array}$ | 15. $\begin{array}{r} 900 \\ -736 \\ \hline \end{array}$ |
| 16. $\begin{array}{r} 503 \\ -229 \\ \hline \end{array}$ | 17. $\begin{array}{r} 700 \\ -352 \\ \hline \end{array}$ | 18. $\begin{array}{r} 602 \\ -438 \\ \hline \end{array}$ | 19. $\begin{array}{r} 906 \\ -698 \\ \hline \end{array}$ | 20. $\begin{array}{r} 800 \\ -177 \\ \hline \end{array}$ |

## Set 12 Give each difference.

| | | | |
|---|---|---|---|
| 1. $\begin{array}{r} \scriptstyle 7\ 1 \\ 52\not{8}7 \\ -2158 \\ \hline 3129 \end{array}$ | 2. $\begin{array}{r} 3964 \\ -1479 \\ \hline \end{array}$ | 3. $\begin{array}{r} 5203 \\ -3425 \\ \hline \end{array}$ | 4. $\begin{array}{r} 7812 \\ -1699 \\ \hline \end{array}$ |
| 5. $\begin{array}{r} 78{,}134 \\ -12{,}965 \\ \hline \end{array}$ | 6. $\begin{array}{r} 92{,}514 \\ -30{,}427 \\ \hline \end{array}$ | 7. $\begin{array}{r} 63{,}810 \\ -26{,}593 \\ \hline \end{array}$ | 8. $\begin{array}{r} 52{,}004 \\ -24{,}685 \\ \hline \end{array}$ |
| 9. $\begin{array}{r} 365{,}128 \\ -159{,}675 \\ \hline \end{array}$ | 10. $\begin{array}{r} 965{,}437 \\ -374{,}882 \\ \hline \end{array}$ | 11. $\begin{array}{r} 805{,}604 \\ -527{,}607 \\ \hline \end{array}$ | 12. $\begin{array}{r} 920{,}653 \\ -819{,}299 \\ \hline \end{array}$ |
| 13. $\begin{array}{r} 436{,}173 \\ -271{,}821 \\ \hline \end{array}$ | 14. $\begin{array}{r} 733{,}586 \\ -257{,}329 \\ \hline \end{array}$ | 15. $\begin{array}{r} 628{,}000 \\ -113{,}521 \\ \hline \end{array}$ | 16. $\begin{array}{r} 530{,}560 \\ -453{,}846 \\ \hline \end{array}$ |

**Set 13** Multiply.

| | | | | | | |
|---|---|---|---|---|---|---|
| 1. $5 \times 5$ = 25 | 2. $6 \times 8$ | 3. $7 \times 4$ | 4. $6 \times 5$ | 5. $7 \times 7$ | 6. $8 \times 5$ | 7. $8 \times 3$ |
| 8. $9 \times 5$ | 9. $9 \times 3$ | 10. $8 \times 8$ | 11. $4 \times 9$ | 12. $4 \times 8$ | 13. $5 \times 7$ | 14. $6 \times 9$ |
| 15. $7 \times 9$ | 16. $6 \times 6$ | 17. $7 \times 8$ | 18. $9 \times 9$ | 19. $6 \times 7$ | 20. $8 \times 9$ | 21. $5 \times 9$ |
| 22. $4 \times 6$ | 23. $7 \times 3$ | 24. $4 \times 4$ | 25. $9 \times 2$ | 26. $6 \times 3$ | 27. $5 \times 4$ | 28. $8 \times 6$ |

**Set 14** Multiply.

| | | | | | |
|---|---|---|---|---|---|
| 1. $52 \times 5$ = 260 | 2. $76 \times 5$ | 3. $45 \times 8$ | 4. $29 \times 6$ | 5. $53 \times 6$ | 6. $63 \times 7$ |
| 7. $28 \times 9$ | 8. $36 \times 6$ | 9. $47 \times 9$ | 10. $44 \times 8$ | 11. $78 \times 5$ | 12. $86 \times 9$ |
| 13. $57 \times 6$ | 14. $48 \times 8$ | 15. $93 \times 5$ | 16. $65 \times 9$ | 17. $74 \times 7$ | 18. $83 \times 4$ |
| 19. $75 \times 9$ | 20. $86 \times 7$ | 21. $67 \times 8$ | 22. $93 \times 6$ | 23. $77 \times 4$ | 24. $58 \times 5$ |

**Set 15** Multiply.

| | | | | |
|---|---|---|---|---|
| 1. $386 \times 2$ = 772 | 2. $620 \times 4$ | 3. $456 \times 3$ | 4. $578 \times 9$ | 5. $956 \times 6$ |
| 6. $827 \times 9$ | 7. $578 \times 7$ | 8. $742 \times 5$ | 9. $609 \times 5$ | 10. $693 \times 2$ |
| 11. $827 \times 4$ | 12. $755 \times 8$ | 13. $819 \times 7$ | 14. $990 \times 6$ | 15. $748 \times 5$ |
| 16. $628 \times 5$ | 17. $534 \times 9$ | 18. $729 \times 7$ | 19. $806 \times 6$ | 20. $923 \times 8$ |

**Set 16 Multiply.**

1. 5246 × 4 = 20,984 (worked example; carries 1 2 written above)
2. 3805 × 3
3. 9174 × 5
4. 2636 × 8
5. 52,813 × 6
6. 74,159 × 5
7. 63,820 × 4
8. 50,073 × 7
9. 31,526 × 4
10. 40,360 × 6
11. 50,035 × 5
12. 62,903 × 8
13. 491,432 × 8
14. 381,764 × 6
15. 928,314 × 9
16. 743,448 × 7

**Set 17 Multiply.**

1. 248 × 10 = 2480 (worked example)
2. 326 × 10
3. 534 × 100
4. 291 × 100
5. 753 × 100
6. 819 × 10
7. 650 × 100
8. 748 × 10
9. 847 × 100
10. 950 × 10
11. 966 × 10
12. 300 × 10
13. 853 × 10
14. 700 × 100
15. 542 × 100
16. 823 × 100
17. 467 × 10
18. 519 × 100
19. 388 × 10
20. 704 × 100

**Set 18 Multiply.**

1. 142 × 20 = 2840 (worked example)
2. 453 × 300
3. 836 × 70
4. 742 × 400
5. 558 × 60
6. 945 × 90
7. 627 × 500
8. 234 × 30
9. 915 × 700
10. 747 × 80
11. 412 × 600
12. 900 × 50
13. 640 × 800
14. 823 × 900
15. 370 × 40
16. 325 × 200
17. 638 × 80
18. 518 × 400
19. 674 × 50
20. 406 × 700

**Set 19** Give each product.

1. 63 ×27 (worked: 441, 1260, 1701)
2. 62 ×45
3. 86 ×19
4. 67 ×56
5. 99 ×18
6. 76 ×32
7. 605 ×26
8. 567 ×33
9. 558 ×72
10. 288 ×34
11. 693 ×40
12. 873 ×54
13. 475 ×35
14. 320 ×62
15. 781 ×58
16. 625 ×73
17. 914 ×86
18. 846 ×91
19. 635 ×28
20. 827 ×92
21. 507 ×87
22. 843 ×56
23. 921 ×71

**Set 20** Multiply.

1. 572 ×121 (worked: 572, 1144, 572, 69,212)
2. 384 ×235
3. 754 ×352
4. 848 ×218
5. 369 ×205
6. 621 ×243
7. 519 ×309
8. 832 ×642
9. 950 ×538
10. 783 ×436
11. 257 ×103
12. 846 ×208
13. 435 ×236
14. 364 ×316
15. 818 ×514
16. 503 ×254
17. 264 ×178
18. 831 ×603
19. 765 ×480

**Set 21** Divide.

1. $4\overline{)16}$ (worked: 4)
2. $6\overline{)24}$
3. $3\overline{)21}$
4. $6\overline{)18}$
5. $3\overline{)27}$
6. $4\overline{)20}$
7. $8\overline{)32}$
8. $4\overline{)36}$
9. $6\overline{)48}$
10. $6\overline{)36}$
11. $5\overline{)40}$
12. $7\overline{)56}$
13. $5\overline{)45}$
14. $4\overline{)28}$
15. $8\overline{)72}$
16. $9\overline{)54}$
17. $6\overline{)42}$
18. $9\overline{)81}$
19. $7\overline{)49}$
20. $8\overline{)64}$
21. $6\overline{)24}$
22. $9\overline{)63}$
23. $7\overline{)42}$
24. $9\overline{)36}$
25. $9\overline{)45}$
26. $8\overline{)48}$
27. $6\overline{)54}$
28. $9\overline{)72}$
29. $8\overline{)56}$
30. $7\overline{)63}$

**Set 22 Divide.**

| | | | | |
|---|---|---|---|---|
| 1. $4\overline{)48}$ (12) | 2. $3\overline{)39}$ | 3. $2\overline{)48}$ | 4. $5\overline{)55}$ | 5. $2\overline{)68}$ |
| 6. $6\overline{)60}$ | 7. $2\overline{)84}$ | 8. $4\overline{)44}$ | 9. $2\overline{)44}$ | 10. $3\overline{)93}$ |
| 11. $3\overline{)69}$ | 12. $4\overline{)40}$ | 13. $5\overline{)50}$ | 14. $3\overline{)90}$ | 15. $4\overline{)84}$ |
| 16. $2\overline{)86}$ | 17. $2\overline{)64}$ | 18. $3\overline{)99}$ | 19. $3\overline{)63}$ | 20. $4\overline{)80}$ |
| 21. $6\overline{)66}$ | 22. $3\overline{)96}$ | 23. $7\overline{)77}$ | 24. $4\overline{)88}$ | 25. $9\overline{)90}$ |
| 26. $8\overline{)88}$ | 27. $3\overline{)60}$ | 28. $8\overline{)80}$ | 29. $9\overline{)99}$ | 30. $7\overline{)70}$ |

**Set 23 Divide.**

| | | | | |
|---|---|---|---|---|
| 1. $5\overline{)63}$ (12 R3: −5, 13, −10, 3) | 2. $3\overline{)87}$ | 3. $5\overline{)78}$ | 4. $8\overline{)96}$ | 5. $5\overline{)75}$ |
| | 6. $6\overline{)79}$ | 7. $6\overline{)84}$ | 8. $8\overline{)99}$ | 9. $6\overline{)96}$ |
| | 10. $3\overline{)44}$ | 11. $3\overline{)80}$ | 12. $7\overline{)91}$ | 13. $2\overline{)64}$ |
| 14. $5\overline{)94}$ | 15. $4\overline{)56}$ | 16. $6\overline{)90}$ | 17. $4\overline{)63}$ | 18. $8\overline{)91}$ |
| 19. $3\overline{)62}$ | 20. $6\overline{)88}$ | 21. $2\overline{)76}$ | 22. $3\overline{)56}$ | 23. $5\overline{)83}$ |
| 24. $4\overline{)60}$ | 25. $2\overline{)93}$ | 26. $4\overline{)72}$ | 27. $3\overline{)95}$ | 28. $7\overline{)86}$ |

**Set 24 Divide.**

| | | | | |
|---|---|---|---|---|
| 1. $8\overline{)692}$ (86 R4: −64, 52, −48, 4) | 2. $9\overline{)793}$ | 3. $5\overline{)167}$ | 4. $7\overline{)789}$ | 5. $6\overline{)564}$ |
| | 6. $3\overline{)457}$ | 7. $8\overline{)548}$ | 8. $7\overline{)944}$ | 9. $5\overline{)858}$ |
| | 10. $4\overline{)959}$ | 11. $5\overline{)3818}$ | 12. $7\overline{)5000}$ | 13. $8\overline{)4141}$ |
| 14. $6\overline{)5036}$ | 15. $9\overline{)8000}$ | 16. $5\overline{)6458}$ | 17. $6\overline{)4966}$ | 18. $8\overline{)2930}$ |
| 19. $9\overline{)4240}$ | 20. $6\overline{)4355}$ | 21. $7\overline{)6207}$ | 22. $9\overline{)6147}$ | 23. $8\overline{)3330}$ |
| 24. $3\overline{)2579}$ | 25. $5\overline{)2092}$ | 26. $4\overline{)2552}$ | 27. $9\overline{)5899}$ | 28. $6\overline{)5248}$ |

**Set 25 Divide.**

1. 4)429 (107 R1)
2. 3)168
3. 4)436
4. 2)804
5. 6)368
6. 5)4055
7. 9)7471
8. 8)8363
9. 9)5568
10. 6)3053
11. 7)2168
12. 6)3705
13. 4)4096
14. 5)5603
15. 7)6020
16. 8)6217
17. 7)3425
18. 4)3062
19. 9)6348
20. 5)2906
21. 3)5111
22. 6)7823
23. 4)1239
24. 6)5628
25. 5)3546
26. 4)3708
27. 5)3006
28. 8)2974

**Set 26 Use the given multiplication facts to help you divide.**

| 50 | 50 | 50 | 50 | 50 | 50 | 50 | 50 | 50 | 50 |
|---|---|---|---|---|---|---|---|---|---|
| ×0 | ×1 | ×2 | ×3 | ×4 | ×5 | ×6 | ×7 | ×8 | ×9 |
| 0 | 50 | 100 | 150 | 200 | 250 | 300 | 350 | 400 | 450 |

1. 50)7960 (159 R10)
2. 50)3874
3. 50)2968
4. 50)9600
5. 50)8348
6. 50)6207
7. 50)7039
8. 50)3662
9. 50)9256
10. 50)6735
11. 50)5553
12. 50)4086
13. 50)4763
14. 50)5807
15. 50)8719
16. 50)3182
17. 50)6010
18. 50)4357

**Set 27 Divide.**

1. 13)846 (65 R1)
2. 23)690
3. 15)542
4. 28)943
5. 16)706
6. 25)929
7. 34)756
8. 23)758
9. 42)653
10. 39)802
11. 43)829
12. 26)693
13. 37)874
14. 30)658
15. 44)942
16. 46)3621
17. 63)4729
18. 52)2700
19. 39)6377
20. 72)8524
21. 60)7115
22. 58)3894
23. 28)3885
24. 35)8293
25. 49)5743
26. 68)4296
27. 53)6218
28. 43)6271

**Set 28** Divide.

1. $23\overline{)7821}$ → $340$ R1 ($-69$, $92$, $-92$, $1$, $-0$, $1$)  2. $42\overline{)5906}$  3. $18\overline{)3748}$  4. $29\overline{)9261}$  5. $46\overline{)7403}$

6. $65\overline{)8005}$  7. $37\overline{)8916}$  8. $49\overline{)6413}$  9. $58\overline{)5921}$

10. $70\overline{)6508}$  11. $53\overline{)8934}$  12. $62\overline{)2617}$  13. $40\overline{)8311}$

14. $39\overline{)6520}$  15. $56\overline{)7000}$  16. $83\overline{)32,164}$  17. $92\overline{)59,708}$  18. $84\overline{)35,214}$

19. $76\overline{)60,058}$  20. $95\overline{)93,774}$  21. $46\overline{)35,291}$  22. $33\overline{)74,258}$  23. $70\overline{)63,592}$

24. $51\overline{)84,420}$  25. $63\overline{)25,781}$  26. $85\overline{)50,000}$  27. $58\overline{)81,300}$  28. $93\overline{)65,290}$

**Set 29** Write each fraction in lowest terms.

1. $\frac{2}{4}$ $\frac{1}{2}$  2. $\frac{3}{9}$  3. $\frac{6}{8}$  4. $\frac{6}{16}$  5. $\frac{2}{8}$  6. $\frac{10}{15}$  7. $\frac{4}{6}$

8. $\frac{3}{12}$  9. $\frac{8}{10}$  10. $\frac{2}{6}$  11. $\frac{3}{6}$  12. $\frac{6}{9}$  13. $\frac{6}{8}$  14. $\frac{4}{12}$

15. $\frac{5}{10}$  16. $\frac{8}{12}$  17. $\frac{10}{16}$  18. $\frac{5}{15}$  19. $\frac{6}{12}$  20. $\frac{4}{8}$  21. $\frac{9}{12}$

22. $\frac{9}{6}$  23. $\frac{6}{10}$  24. $\frac{10}{8}$  25. $\frac{20}{16}$  26. $\frac{15}{6}$  27. $\frac{12}{8}$  28. $\frac{2}{10}$

29. $\frac{15}{24}$  30. $\frac{20}{30}$  31. $\frac{12}{16}$  32. $\frac{24}{18}$  33. $\frac{18}{30}$  34. $\frac{8}{16}$  35. $\frac{10}{6}$

36. $\frac{4}{10}$  37. $\frac{18}{8}$  38. $\frac{21}{6}$  39. $\frac{12}{30}$  40. $\frac{14}{16}$  41. $\frac{20}{8}$  42. $\frac{24}{16}$

**Set 30** < or >?

1. $\frac{3}{5}$ (>) $\frac{2}{5}$  2. $\frac{4}{9}$ ● $\frac{5}{9}$  3. $\frac{3}{8}$ ● $\frac{1}{8}$  4. $\frac{5}{6}$ ● $\frac{1}{6}$

5. $\frac{1}{2}$ ● $\frac{1}{4}$  6. $\frac{3}{4}$ ● $\frac{1}{2}$  7. $\frac{5}{8}$ ● $\frac{3}{4}$  8. $\frac{3}{4}$ ● $\frac{7}{8}$

9. $\frac{1}{2}$ ● $\frac{1}{3}$  10. $\frac{2}{3}$ ● $\frac{3}{4}$  11. $\frac{3}{2}$ ● $\frac{2}{3}$  12. $\frac{7}{9}$ ● $\frac{5}{6}$

13. $\frac{5}{8}$ ● $\frac{7}{8}$  14. $\frac{1}{2}$ ● $\frac{2}{3}$  15. $\frac{3}{10}$ ● $\frac{1}{4}$  16. $\frac{2}{3}$ ● $\frac{5}{8}$

17. $\frac{5}{12}$ ● $\frac{1}{3}$  18. $\frac{2}{3}$ ● $\frac{7}{12}$  19. $\frac{7}{8}$ ● $\frac{13}{16}$  20. $\frac{5}{12}$ ● $\frac{5}{8}$

21. $\frac{9}{10}$ ● $\frac{7}{8}$  22. $\frac{5}{6}$ ● $\frac{11}{12}$  23. $\frac{9}{16}$ ● $\frac{3}{4}$  24. $\frac{2}{3}$ ● $\frac{3}{5}$

**Set 31** **Add. Give answers in lowest terms.**

1. $\frac{1}{5} + \frac{3}{5} = \frac{4}{5}$
2. $\frac{1}{6} + \frac{1}{6}$
3. $\frac{5}{8} + \frac{1}{8}$
4. $\frac{1}{2} + \frac{0}{4}$
5. $\frac{3}{10} + \frac{3}{5}$
6. $\frac{5}{8} + \frac{1}{4}$
7. $\frac{1}{2} + \frac{1}{12}$
8. $\frac{3}{7} + \frac{3}{7}$
9. $\frac{1}{4} + \frac{5}{12}$
10. $\frac{1}{6} + \frac{1}{3}$
11. $\frac{3}{10} + \frac{1}{2}$
12. $\frac{1}{6} + \frac{7}{12}$
13. $\frac{2}{9} + \frac{1}{3}$
14. $\frac{1}{3} + \frac{5}{12}$
15. $\frac{1}{2} + \frac{1}{6}$
16. $\frac{3}{8} + \frac{1}{2}$
17. $\frac{5}{16} + \frac{1}{4}$
18. $\frac{3}{10} + \frac{1}{5}$
19. $\frac{3}{16} + \frac{3}{8}$
20. $\frac{5}{12} + \frac{1}{2}$
21. $\frac{2}{3} + \frac{1}{15}$
22. $\frac{5}{12} + \frac{1}{4}$
23. $\frac{3}{4} + \frac{1}{12}$
24. $\frac{3}{16} + \frac{1}{2}$

**Set 32** **Add. Give answers in lowest terms.**

1. $\frac{1}{3} + \frac{1}{2} = \frac{2}{6} + \frac{3}{6}$
   $= \frac{5}{6}$
2. $\frac{1}{4} + \frac{1}{3}$
3. $\frac{1}{4} + \frac{1}{5}$
4. $\frac{2}{5} + \frac{3}{10}$
5. $\frac{1}{3} + \frac{5}{8}$
6. $\frac{5}{9} + \frac{1}{9}$
7. $\frac{1}{3} + \frac{4}{9}$
8. $\frac{1}{6} + \frac{1}{4}$
9. $\frac{2}{5} + \frac{1}{3}$
10. $\frac{1}{4} + \frac{3}{10}$
11. $\frac{1}{4} + \frac{1}{2}$
12. $\frac{1}{2} + \frac{2}{5}$
13. $\frac{4}{9} + \frac{1}{6}$
14. $\frac{2}{3} + \frac{1}{6}$
15. $\frac{1}{10} + \frac{7}{10}$
16. $\frac{1}{4} + \frac{3}{10}$
17. $\frac{1}{3} + \frac{3}{5}$
18. $\frac{5}{12} + \frac{3}{8}$
19. $\frac{5}{8} + \frac{5}{16}$
20. $\frac{7}{10} + \frac{1}{15}$
21. $\frac{2}{5} + \frac{1}{6}$
22. $\frac{7}{16} + \frac{1}{2}$
23. $\frac{1}{6} + \frac{3}{4}$

**Set 33** **Subtract. Give answers in lowest terms.**

1. $\frac{3}{5} - \frac{1}{5} = \frac{2}{5}$
2. $\frac{5}{7} - \frac{1}{7}$
3. $\frac{5}{8} - \frac{3}{8}$
4. $\frac{1}{2} - \frac{1}{4}$
5. $\frac{3}{8} - \frac{1}{4}$
6. $\frac{5}{6} - \frac{1}{3}$
7. $\frac{2}{3} - \frac{5}{9}$
8. $\frac{5}{6} - \frac{2}{3}$
9. $\frac{7}{8} - \frac{0}{2}$
10. $\frac{9}{10} - \frac{3}{10}$
11. $\frac{3}{4} - \frac{5}{8}$
12. $\frac{7}{10} - \frac{1}{5}$
13. $\frac{2}{3} - \frac{1}{6}$
14. $\frac{5}{8} - \frac{1}{2}$
15. $\frac{8}{9} - \frac{2}{3}$
16. $\frac{1}{2} - \frac{1}{8}$
17. $\frac{5}{12} - \frac{1}{6}$
18. $\frac{3}{4} - \frac{5}{12}$
19. $\frac{7}{8} - \frac{5}{16}$
20. $\frac{7}{10} - \frac{1}{2}$
21. $\frac{11}{12} - \frac{2}{3}$
22. $\frac{11}{16} - \frac{1}{2}$
23. $\frac{11}{12} - \frac{3}{4}$
24. $\frac{9}{16} - \frac{1}{4}$

## Set 34 Subtract. Give answers in lowest terms.

1. $\frac{1}{2} - \frac{1}{3} = \frac{3}{6} - \frac{2}{6} = \frac{1}{6}$
2. $\frac{1}{4} - \frac{1}{6}$
3. $\frac{3}{4} - \frac{1}{3}$
4. $\frac{4}{5} - \frac{1}{2}$
5. $\frac{7}{8} - \frac{0}{3}$
6. $\frac{2}{5} - \frac{1}{6}$
7. $\frac{7}{10} - \frac{1}{4}$
8. $\frac{2}{3} - \frac{1}{2}$
9. $\frac{2}{3} - \frac{4}{9}$
10. $\frac{3}{4} - \frac{2}{3}$
11. $\frac{7}{8} - \frac{2}{3}$
12. $\frac{5}{6} - \frac{3}{4}$
13. $\frac{5}{8} - \frac{1}{3}$
14. $\frac{2}{3} - \frac{3}{10}$
15. $\frac{1}{2} - \frac{3}{10}$
16. $\frac{5}{6} - \frac{2}{9}$
17. $\frac{5}{12} - \frac{1}{6}$
18. $\frac{3}{5} - \frac{1}{2}$
19. $\frac{13}{16} - \frac{3}{8}$
20. $\frac{8}{9} - \frac{5}{6}$
21. $\frac{15}{16} - \frac{3}{4}$
22. $\frac{11}{12} - \frac{3}{8}$
23. $\frac{9}{10} - \frac{2}{5}$

## Set 35 Write as a fraction.

1. $1\frac{1}{2}$ $\frac{3}{2}$
2. $2\frac{1}{2}$
3. $3\frac{2}{3}$
4. $1\frac{1}{3}$
5. $3\frac{1}{4}$
6. $1\frac{2}{3}$
7. $4\frac{1}{3}$
8. $2\frac{7}{8}$
9. $1\frac{3}{4}$
10. $4\frac{2}{3}$
11. $3\frac{5}{6}$
12. $1\frac{7}{8}$
13. $6\frac{1}{2}$
14. $3\frac{2}{5}$
15. $4\frac{2}{9}$
16. $3\frac{4}{5}$
17. $2\frac{3}{5}$
18. $2\frac{1}{3}$
19. $4\frac{3}{4}$
20. $2\frac{5}{8}$
21. $5\frac{3}{8}$
22. $1\frac{1}{4}$
23. $5\frac{1}{2}$
24. $2\frac{2}{3}$
25. $5\frac{3}{4}$
26. $4\frac{5}{6}$
27. $6\frac{5}{8}$
28. $2\frac{4}{5}$
29. $5\frac{2}{3}$
30. $3\frac{5}{8}$
31. $1\frac{3}{8}$
32. $5\frac{5}{6}$
33. $2\frac{3}{10}$
34. $1\frac{5}{8}$
35. $3\frac{1}{8}$
36. $6\frac{3}{4}$

## Set 36 Write in simplest form.

1. $\frac{4}{3}$ $1\frac{1}{3}$
2. $\frac{5}{4}$
3. $\frac{3}{6}$
4. $\frac{4}{2}$
5. $\frac{9}{8}$
6. $\frac{5}{3}$
7. $\frac{4}{6}$
8. $\frac{9}{2}$
9. $\frac{11}{6}$
10. $\frac{10}{2}$
11. $\frac{6}{16}$
12. $\frac{24}{4}$
13. $\frac{7}{6}$
14. $\frac{8}{5}$
15. $\frac{10}{4}$
16. $\frac{12}{16}$
17. $\frac{7}{4}$
18. $\frac{7}{3}$
19. $\frac{10}{3}$
20. $\frac{9}{15}$
21. $\frac{8}{4}$
22. $\frac{4}{12}$
23. $\frac{7}{5}$
24. $\frac{7}{2}$
25. $\frac{12}{6}$
26. $\frac{15}{20}$
27. $\frac{11}{8}$
28. $\frac{15}{5}$
29. $\frac{6}{5}$
30. $\frac{8}{3}$
31. $\frac{5}{2}$
32. $\frac{11}{4}$
33. $\frac{9}{5}$
34. $\frac{6}{4}$
35. $\frac{12}{18}$
36. $\frac{13}{6}$

**Set 37 Complete.**

1. $\frac{1}{2}$ of 8 = 4
2. $\frac{1}{3}$ of 18 = ?
3. $\frac{1}{5}$ of 20 = ?
4. $\frac{1}{2}$ of 12 = ?
5. $\frac{1}{4}$ of 16 = ?
6. $\frac{1}{6}$ of 18 = ?
7. $\frac{1}{8}$ of 24 = ?
8. $\frac{1}{4}$ of 24 = ?
9. $\frac{1}{3}$ of 9 = ?
10. $\frac{1}{3}$ of 15 = ?
11. $\frac{1}{2}$ of 6 = ?
12. $\frac{1}{6}$ of 12 = ?
13. $\frac{1}{8}$ of 32 = ?
14. $\frac{1}{5}$ of 30 = ?
15. $\frac{1}{6}$ of 24 = ?
16. $\frac{1}{4}$ of 20 = ?
17. $\frac{1}{4}$ of 12 = ?
18. $\frac{1}{2}$ of 20 = ?
19. $\frac{1}{8}$ of 8 = ?
20. $\frac{1}{6}$ of 30 = ?
21. $\frac{1}{3}$ of 24 = ?
22. $\frac{1}{8}$ of 40 = ?
23. $\frac{1}{3}$ of 21 = ?
24. $\frac{1}{6}$ of 42 = ?

**Set 38 Complete.**

1. $\frac{2}{3}$ of 15 = 10
2. $\frac{5}{8}$ of 40 = ?
3. $\frac{3}{5}$ of 15 = ?
4. $\frac{2}{3}$ of 30 = ?
5. $\frac{5}{3}$ of 21 = ?
6. $\frac{4}{3}$ of 12 = ?
7. $\frac{3}{4}$ of 36 = ?
8. $\frac{5}{6}$ of 36 = ?
9. $\frac{3}{4}$ of 12 = ?
10. $\frac{3}{2}$ of 16 = ?
11. $\frac{2}{3}$ of 24 = ?
12. $\frac{5}{9}$ of 18 = ?
13. $\frac{4}{5}$ of 20 = ?
14. $\frac{4}{3}$ of 18 = ?
15. $\frac{3}{8}$ of 24 = ?
16. $\frac{3}{4}$ of 24 = ?
17. $\frac{2}{3}$ of 36 = ?
18. $\frac{3}{2}$ of 12 = ?
19. $\frac{6}{5}$ of 30 = ?
20. $\frac{7}{8}$ of 24 = ?
21. $\frac{5}{2}$ of 18 = ?
22. $\frac{5}{3}$ of 15 = ?
23. $\frac{3}{4}$ of 40 = ?
24. $\frac{5}{6}$ of 30 = ?

**Set 39 Give each product in simplest form.**

1. $\frac{1}{4} \times \frac{1}{2} = \frac{1}{8}$
2. $\frac{1}{3} \times \frac{3}{4}$
3. $\frac{5}{8} \times \frac{4}{5}$
4. $\frac{2}{9} \times \frac{3}{4}$
5. $\frac{5}{8} \times \frac{3}{4}$
6. $\frac{2}{3} \times \frac{2}{3}$
7. $\frac{5}{6} \times \frac{3}{5}$
8. $\frac{7}{8} \times \frac{4}{7}$
9. $\frac{3}{10} \times \frac{5}{2}$
10. $3 \times \frac{1}{2}$
11. $\frac{4}{3} \times \frac{7}{10}$
12. $\frac{4}{3} \times \frac{4}{3}$
13. $\frac{4}{5} \times 3$
14. $\frac{3}{4} \times \frac{4}{3}$
15. $\frac{2}{3} \times \frac{3}{2}$
16. $2 \times \frac{3}{8}$
17. $\frac{5}{8} \times \frac{3}{5}$
18. $\frac{5}{7} \times \frac{5}{7}$
19. $\frac{5}{6} \times \frac{4}{5}$
20. $\frac{5}{9} \times \frac{3}{2}$
21. $6 \times \frac{2}{3}$
22. $\frac{6}{5} \times \frac{5}{6}$
23. $\frac{3}{10} \times \frac{5}{3}$
24. $\frac{3}{4} \times 3$

**Set 40 In 821.946, what digit is in the**

1. hundreds place? *8*
2. tens place?
3. ones place?
4. hundredths place?
5. thousandths place?
6. tenths place?

**In 375.691, what digit is in the**

7. tenths place?
8. hundreds place?
9. tens place?
10. ones place?
11. thousandths place?
12. hundredths place?

**Set 41 <, =, or >?**

| | | |
|---|---|---|
| 1. 7.28 (<) 7.36 | 2. 52.8 ● 51.9 | 3. 35.6 ● 3.52 |
| 4. 6.080 ● 6.08 | 5. 93.74 ● 93.7 | 6. 5.3 ● 5.30 |
| 7. 18.394 ● 18.4 | 8. 63.41 ● 6.341 | 9. 17.281 ● 17.29 |
| 10. 16.04 ● 16.040 | 11. 5.964 ● 5.963 | 12. 36.04 ● 36.40 |
| 13. 43.61 ● 4.4 | 14. 7.528 ● 7.582 | 15. 0.367 ● 1.0 |
| 16. 0.9 ● 1.0 | 17. 4.258 ● 41.32 | 18. 1.0 ● 0.999 |

**Set 42 Give each sum.**

| | | | | | |
|---|---|---|---|---|---|
| 1. 4.6 + 7.8 = *12.4* | 2. 0.49 + 0.76 | 3. 2.04 + 1.67 | 4. 2.35 + 9.93 | 5. 82.3 + 79.9 | 6. 0.237 + 0.994 |
| 7. 5.75 + 9.39 | 8. 6.78 + 4.56 | 9. 3.76 + 9.88 | 10. 83.8 + 51.9 | 11. 20.3 + 50.8 | 12. 99.9 + 99.9 |
| 13. 0.275 + 0.396 | 14. 0.582 + 0.491 | 15. 6.84 + 4.75 | 16. 8.94 + 3.67 | 17. 9.74 + 3.62 | 18. 0.358 + 0.297 |
| 19. 5.963 + 0.888 | 20. 4.396 + 7.519 | 21. 53.74 + 19.97 | 22. 38.07 + 55.11 | 23. 938.1 + 642.7 | 24. 6.543 + 2.944 |

**Set 43 Give each difference.**

1. $9.6 - 4.3$ = 5.3
2. $4.9 - 2.4$
3. $6.09 - 4.04$
4. $46.0 - 39.5$
5. $84.6 - 23.9$
6. $0.463 - 0.397$
7. $4.60 - 1.39$
8. $95.8 - 23.9$
9. $6.04 - 4.87$
10. $56.3 - 21.8$
11. $0.835 - 0.598$
12. $6.00 - 3.95$
13. $0.342 - 0.168$
14. $92.3 - 25.8$
15. $0.205 - 0.096$
16. $8.61 - 3.95$
17. $0.431 - 0.278$
18. $0.520 - 0.274$
19. $8.00 - 3.59$
20. $0.821 - 0.477$
21. $6.00 - 2.57$
22. $9.43 - 6.85$
23. $0.802 - 0.574$
24. $7.01 - 5.98$

**Set 44 Give each sum or difference.**

1. $3.64 + 2.5$

$$\begin{array}{r} 3.64 \\ +2.5 \\ \hline 6.14 \end{array}$$

2. $3.64 - 2.5$
3. $4.78 + 1.9$
4. $4.78 - 1.9$
5. $0.881 + 0.69$
6. $0.881 - 0.69$
7. $4.3 + 2.65$
8. $4.3 - 2.65$
9. $0.9 + 0.354$
10. $0.9 - 0.354$
11. $5 + 3.42$
12. $5 - 3.42$
13. $2.64 + 0.8$
14. $2.64 - 0.8$
15. $2.8 + 5.9$
16. $28 - 5.9$
17. $40 + 0.3$
18. $40 - 0.3$
19. $1.54 + 0.8$
20. $1.54 - 0.8$
21. $29.4 + 1.72$
22. $29.4 - 1.72$

**Set 45 Complete.**

1. 20 mm = 2 cm
2. 40 mm = ? cm
3. 70 mm = ? cm
4. 100 mm = ? cm
5. 24 mm = ? cm
6. 52 mm = ? cm
7. 83 mm = ? cm
8. 125 mm = ? cm
9. 6 cm = ? mm
10. 5 cm = ? mm
11. 8 cm = ? mm
12. 13 cm = ? mm
13. 200 mm = ? cm
14. 75 cm = ? mm
15. 135 mm = ? cm
16. 83 cm = ? mm
17. 270 mm = ? cm
18. 96 cm = ? mm

**Set 46** **Complete.**

1. 3 m = 300 cm
2. 8 m = ? cm
3. 15 m = ? cm
4. 28 m = ? cm
5. 1.5 m = ? cm
6. 2.4 m = ? cm
7. 13.6 m = ? cm
8. 34.3 m = ? cm
9. 500 cm = ? m
10. 350 cm = ? m
11. 442 cm = ? m
12. 625 cm = ? m
13. 4 km = ? m
14. 12 km = ? m
15. 1100 m = ? km
16. 2.85 km = ? m
17. 2500 m = ? km
18. 4.6 km = ? m

**Set 47** **Complete.**

1. 3000 mL = 3 L
2. 2500 mL = ? L
3. 3400 mL = ? L
4. 2560 mL = ? L
5. 2 L = ? mL
6. 5 L = ? mL
7. 5.2 L = ? mL
8. 8.4 L = ? mL
9. 4000 g = ? kg
10. 7000 g = ? kg
11. 6800 g = ? kg
12. 800 g = ? kg
13. 5 kg = ? g
14. 6.2 kg = ? g
15. 0.782 kg = ? g
16. 8500 g = ? kg
17. 5.34 kg = ? g
18. 4600 g = ? kg

**Set 48** **Complete.**

1. 8 c = 4 pt
2. 4 qt = ? gal
3. 6 qt = ? half-gal
4. 14 pt = ? qt
5. 5 qt = ? pt
6. 3 gal = ? qt
7. $5\frac{1}{2}$ gal = ? qt
8. 18 pt = ? c
9. $4\frac{1}{2}$ qt = ? pt
10. $5\frac{1}{2}$ pt = ? c
11. $2\frac{1}{2}$ gal = ? qt
12. $1\frac{3}{4}$ gal = ? qt
13. 5 lb = ? oz
14. 9 lb = ? oz
15. 150 oz = 9 lb ? oz
16. 108 oz = ? lb 12 oz
17. 3 tons = ? lb
18. 10,000 lb = ? tons

**Set 49** **Solve.**

1. 3 for 49¢ How much for 1? 17¢
2. 4 for 69¢ How much for 2?
3. 3 for \$2.00 How much for 2?
4. 5 for \$1.19 How much for 7?
5. 3 for 89¢ How much for 2?
6. 4 for 85¢ How much for 5?
7. 6 for \$1.39 How much for 5?
8. 5 for \$2.29 How much for 2?
9. 4 for \$3.25 How much for 6?

**Set 50** **Compute the sale price.**

1. \$1.50 $\frac{1}{3}$ off \$1.00
2. \$2.80 $\frac{1}{4}$ off
3. \$4.56 $\frac{1}{2}$ off
4. \$5.25 $\frac{1}{4}$ off
5. \$4.89 $\frac{1}{3}$ off
6. \$6.00 $\frac{1}{5}$ off
7. \$8.20 $\frac{1}{6}$ off
8. \$7.75 $\frac{2}{3}$ off
9. \$8.20 $\frac{3}{4}$ off
10. \$5.79 $\frac{2}{5}$ off
11. \$9.61 $\frac{3}{4}$ off
12. \$7.38 $\frac{3}{5}$ off

**Set 51** **Add or subtract. Give answers in simplest form.**

1. $3\frac{1}{3} + 2\frac{1}{3} = 5\frac{2}{3}$
2. $4\frac{1}{4} + 3\frac{1}{8}$
3. $5\frac{1}{2} + 2\frac{3}{8}$
4. $6\frac{2}{5} + 3\frac{1}{10}$
5. $6\frac{1}{3} + 6\frac{1}{2}$
6. $8\frac{2}{5} + 4\frac{1}{2}$
7. $8\frac{7}{8} - 2\frac{3}{8}$
8. $5\frac{3}{4} - 3\frac{1}{2}$
9. $9\frac{5}{8} - 2\frac{1}{4}$
10. $3\frac{5}{6} - 1\frac{1}{3}$
11. $6\frac{5}{8} - 3\frac{1}{4}$
12. $4\frac{1}{2} - 1\frac{1}{3}$
13. $7\frac{2}{3} + 2\frac{1}{4}$
14. $7\frac{2}{3} - 2\frac{1}{4}$
15. $8\frac{5}{8} + 6\frac{1}{3}$
16. $8\frac{5}{8} - 6\frac{1}{3}$
17. $9\frac{5}{8} + 3\frac{1}{6}$
18. $9\frac{5}{8} - 3\frac{1}{6}$

**Set 52** **Add. Give answers in simplest form.**

1. $5\frac{1}{2} = 5\frac{2}{4}$ / $+2\frac{3}{4} = 2\frac{3}{4}$ / $8\frac{1}{4}$
2. $5\frac{1}{4} + 3\frac{3}{4}$
3. $5\frac{1}{2} + 3\frac{1}{4}$
4. $8\frac{3}{8} + 9\frac{3}{4}$
5. $6\frac{1}{3} + 4\frac{7}{9}$
6. $7\frac{3}{8} + 9\frac{5}{8}$
7. $7\frac{3}{4} + 1\frac{5}{8}$
8. $8\frac{2}{3} + 6\frac{1}{2}$
9. $9\frac{1}{4} + 6\frac{1}{2}$
10. $7\frac{5}{9} + 4\frac{2}{3}$
11. $4\frac{3}{4} + 7\frac{2}{3}$
12. $7\frac{2}{3} + 3\frac{5}{6}$
13. $9\frac{5}{6} + 5\frac{3}{4}$
14. $7\frac{3}{5} + 8\frac{7}{10}$
15. $8\frac{4}{5} + 8\frac{1}{2}$
16. $5\frac{7}{8} + 6\frac{2}{3}$
17. $9\frac{5}{6} + 9\frac{5}{6}$
18. $7\frac{5}{6} + 4\frac{7}{8}$

**Set 53** **Subtract. Give answers in simplest form.**

1. $8\frac{1}{5} = 7\frac{6}{5}$ / $-2\frac{4}{5} = 2\frac{4}{5}$ / $5\frac{2}{5}$
2. $3\frac{1}{6} - 1\frac{5}{6}$
3. $7 - 3\frac{2}{3}$
4. $10\frac{1}{4} - 5\frac{3}{4}$
5. $9\frac{3}{8} - \frac{5}{8}$
6. $7\frac{1}{3} - \frac{2}{3}$
7. $5\frac{1}{4} - 2\frac{1}{2}$
8. $9\frac{2}{3} - 6\frac{3}{4}$
9. $8\frac{1}{3} - 4\frac{5}{9}$
10. $5\frac{2}{5} - 3\frac{9}{10}$
11. $7 - 6\frac{5}{9}$
12. $9\frac{1}{6} - 3\frac{3}{4}$
13. $8 - 3\frac{2}{3}$
14. $7\frac{1}{2} - 4\frac{5}{6}$
15. $9\frac{2}{3} - 5\frac{7}{8}$
16. $8\frac{2}{5} - 3\frac{7}{10}$
17. $6\frac{2}{3} - 4\frac{5}{6}$
18. $7 - 5\frac{5}{8}$

**Set 54** **Multiply.**

1. $5.29 \times 0.5$ = 2.645
2. $35.8 \times 0.4$
3. $71.5 \times 0.6$
4. $42.7 \times 0.3$
5. $0.826 \times 0.7$
6. $6.66 \times 0.2$
7. $81.1 \times 0.8$
8. $74.2 \times 0.6$
9. $8.56 \times 0.9$
10. $0.674 \times 0.3$
11. $8.56 \times 0.6$
12. $90.3 \times 0.8$
13. $5.74 \times 0.5$
14. $29.6 \times 0.7$
15. $0.359 \times 0.9$

**Set 55 Multiply.**

1. 0.3 × 0.2 = 0.06
2. 0.4 × 0.2
3. 0.3 × 0.3
4. 0.03 × 0.2
5. 0.04 × 0.2
6. 1.04 × 2.5
7. 34.6 × 3.7
8. 7.81 × 5.8
9. 50.3 × 63
10. 605 × 4.3
11. 71.7 × 1.58
12. 82.9 × 20.9
13. 6.48 × 534
14. 83.5 × 3.21
15. 96.3 × 2.76
16. 374 × 2.96
17. 52.1 × 250
18. 3.06 × 2.29
19. 48.3 × 60.5
20. 7.42 × 39.1

**Set 56 Divide.**

1. 3)52.5 = 17.5  
   -3  
   22  
   -21  
   15  
   -15  
   0
2. 5)3.045
3. 4)27.36
4. 8)4.968
5. 6)94.44
6. 9)7.533
7. 7)25.34
8. 2)4.798
9. 8)579.2
10. 9)3.951
11. 23)14.72
12. 15)1.875
13. 42)348.6
14. 68)29.24
15. 74)466.2
16. 37)34.78
17. 52)3.276
18. 64)65.92
19. 85)2.635
20. 35)29.19
21. 49)313.6
22. 43)37.41
23. 38)307.8
24. 81)44.55
25. 92)1.472
26. 67)5.092
27. 73)62.05
28. 83)7.802

**Set 57 Divide.**

1. 8)3.000 = 0.375  
   -24  
   60  
   -56  
   40  
   -40  
   0
2. 4)3
3. 8)9
4. 4)5
5. 8)11
6. 2)1.7
7. 4)1.5
8. 4)29
9. 8)47
10. 2)4.5
11. 4)2.06
12. 5)2.19
13. 5)273
14. 78)2.652
15. 54)1.998
16. 34)7.82
17. 43)98.9
18. 65)0.78
19. 43)98.9
20. 27)1.62
21. 78)0.936
22. 78)2.652
23. 54)1.998
24. 10)68.3
25. 10)7.83
26. 10)10.6
27. 10)254
28. 10)1.086

# EXTRA PROBLEM SOLVING

## Set 1

**Use the chart to find the total number of calories in each meal.**

| *Food* | *Calories* |
|---|---|
| beans, green, 1 cup | 30 |
| blueberry muffin | 150 |
| bun, hamburger | 100 |
| butter, pat | 50 |
| coleslaw, 1 cup | 110 |
| cornflakes, 1 cup | 50 |
| egg, boiled | 100 |
| hamburger patty | 200 |
| ice cream, 1 cup | 300 |
| lettuce | 15 |
| milk, 1 cup | 150 |
| noodle soup, 1 cup | 150 |
| oatmeal, 1 cup | 100 |
| orange juice, 1 cup | 100 |
| toast, 1 slice | 75 |
| tomato, 1 medium | 40 |

**1.** 1 boiled egg
1 slice of toast
1 pat of butter
1 cup of milk

**2.** 2 boiled eggs
1 cup of orange juice
1 blueberry muffin

**3.** 1 cup of oatmeal
1 cup of milk
2 slices of toast
2 pats of butter

**4.** 1 cup of cornflakes
1 cup of milk
1 blueberry muffin
2 pats of butter
1 cup of orange juice

## Set 2

**Use the chart to find the total number of calories in each meal.**

**1.** 2 cups of noodle soup
1 cup of milk
1 medium tomato
1 cup of coleslaw

**2.** 2 hamburger patties
2 hamburger buns
1 cup of milk
1 cup of ice cream

**Tell which meal has fewer calories. How many fewer?**

**3. a.** 2 boiled eggs
1 blueberry muffin
2 pats of butter
1 cup of orange juice

**b.** 1 cup of noodle soup
1 hamburger patty
1 cup of green beans
1 cup of milk

## Set 3

1. The Wilsons are planning a trip to Rapid River, which is 412 kilometers from their home. How far will the round trip be?

2. The Wilsons live 412 kilometers from Rapid River. During the first morning they drove 186 kilometers. How much farther did they have to drive to get to the river?

3. Vince took $7.65 and Paige took $9.00. How much more did Paige take?

4. Paige could take 36 pictures on a roll of film she had brought with her. During the first day she took 29 pictures. How many more pictures could she take?

5. Vince bought some lemonade for $.75, a sandwich for $1.65, and some insect repellent for $1.79. How much did he spend?

6. Paige bought 2 postcards for $.22 each and a *Tales of Rapid River* booklet for $3.79. How much did she spend in all?

## Set 4

1. For a 3-day float trip, the adult tickets were $32.50 and the children's tickets were $18.75. How much did 2 adult tickets and 2 children's tickets cost?

2. The total number of people that the 5 rafts could take was 205. 138 tickets had been sold. How many more could be sold?

3. The Wilsons arrived at 8:05 A.M. The rafts were to leave at 9:15 A.M. How many minutes did they have to wait?

4. Vince helped serve the hot dogs at the first night's camp. He served 2 each to 118 people and 1 each to 87 people. How many hot dogs did he serve?

5. The 2-day float trip covered 121 kilometers. On the first day they floated 78 kilometers. How far did they float on the second day?

6. All 205 people returned to the starting point in 2 large buses and 3 small buses. The large buses took 112 people. How many people did the small buses take?

| Sample rates from Champaign, Illinois, to | WEEKDAY Full rate | | EVENING 35% discount | | NIGHT & WEEKEND 60% discount | |
|---|---|---|---|---|---|---|
| | First Minute | Each Additional Minute | First Minute | Each Additional Minute | First Minute | Each Additional Minute |
| Atlanta, Ga. | .50 | .34 | .32 | .23 | .20 | .14 |
| Boston, Mass. | .50 | .34 | .32 | .23 | .20 | .14 |
| Denver, Colo. | .50 | .34 | .32 | .23 | .20 | .14 |
| Detroit, Mich. | .48 | .34 | .31 | .23 | .19 | .14 |
| Los Angeles, Cal. | .52 | .36 | .33 | .24 | .20 | .15 |
| Miami, Fla. | .52 | .36 | .33 | .24 | .20 | .15 |
| Milwaukee, Wis. | .46 | .32 | .29 | .21 | .18 | .13 |
| Minneapolis, Minn. | .48 | .34 | .31 | .23 | .19 | .14 |
| New Orleans, La. | .50 | .34 | .32 | .23 | .20 | .14 |
| New York, N.Y. | .50 | .34 | .32 | .23 | .20 | .14 |
| Seattle, Wash. | .52 | .36 | .33 | .24 | .20 | .15 |
| Washington, D.C. | .50 | .34 | .32 | .23 | .20 | .14 |

Table title: **DIAL-DIRECT**

## Set 5

**Use the table.**

1. For an evening call to Los Angeles, what is the charge for the first minute? The second minute?
2. What is the charge for a 2-minute evening call to Los Angeles? *Hint:* See exercise 1.

**Use the table to find the cost of a**

3. 3-minute weekday call to Atlanta.
4. 5-minute weekday call to New Orleans.
5. 15-minute evening call to Seattle.
6. 28-minute night call to Detroit.
7. 2-hour evening call to New York.

## Set 6

**Use the table.**

**How much cheaper is it to call at night than to call during a weekday for a**

1. 9-minute call to Boston?
2. 17-minute call to Miami?
3. 25-minute call to Washington, D.C.?
4. 38-minute call to Denver?
5. 1-hour call to Minneapolis?

**Use the table.**

6. Which costs more, one 4-minute call or four 1-minute calls to New York at night? How much more?
7. ★ One week Carl called his friend in Milwaukee each evening except Sunday. Each call was 10 minutes long. What was the total cost of the 6 calls?

## Set 7

1. On bat night 256 bats were given away. Each bat cost \$8. What was the total cost of the bats?

2. There are 162 games played each season. The number of games played away is 79. How many games are played at home?

3. There are 28 seats in each row of a section. There are 12 rows in the section. How many seats are in the section?

4. An adult ticket costs \$7.25. How much do 5 adult tickets cost?

5. There are 32,508 seats in the stadium. The number of people that came to the game was 19,759. How many empty seats were there?

6. A box of popcorn costs \$.75. A bag of peanuts costs \$.60. A carton of lemonade costs \$.55. How much do a bag of peanuts and a carton of lemonade cost?

## Set 8

1. A program costs \$1.25. Paula bought a program and gave the seller \$5. How much change did she get?

2. There are 3252 seats in the bleachers. The 6 sections in the bleachers are the same size. How many seats are in each section?

3. There are 162 games played each season. The number of games already played is 95. How many games are left to be played?

4. The number of games already played is 95. The number of hits made by the 9 players on one team is 570. What is the average number of hits per game?

5. There are 36 teams with 9 players on each team. How many players are there?

6. The total attendance for 9 days was 168,840. What was the average attendance each day?

## Set 9

1. Leon wants to buy a $139.95 camera. He can trade his old camera in for $25. How much will the difference be?

2. Leon has $146.18. If he trades his old camera in on the new one, how much money will be left? (See exercise 1.)

3. Later Leon bought a roll of black-and-white film for $1.97, a roll of colored film for $2.85, and some flashbulbs for $2.59. What was the total cost?

4. Thirty-six pictures can be taken with one large roll of film. How many pictures can be taken with 4 rolls?

5. The Shutter Store charges $3.79 for a roll of colored film with 24 exposures and $8.69 for developing the film. What is the average cost of each picture?

6. Leon wants to buy $45 worth of darkroom supplies. He has $31.19 and can earn $6.50 for mowing a lawn. How much more will he need then?

## Set 10

1. Before going on his summer vacation, Leon bought 3 rolls of 24-exposure film and 2 rolls of 36-exposure film. How many pictures can he take?

2. One photo company charges 8¢ a picture for developing black-and-white film. How much would they charge for 3 rolls of 24-exposure film?

3. One store charges $1.59 for developing 24 black-and-white pictures. Another charges 8¢ for each picture. How much cheaper is the first store?

4. Leon took a total of 144 pictures during one week. He used rolls with 24 exposures. How many rolls did he use?

5. Leon bought two 48-page albums that hold 8 pictures per page. How many pictures will the albums hold?

6. The Shutter Store charged $3.85 for colored film and $8.69 for developing. During a special, they sold and developed the film for $9.95. How much would the savings be on 2 rolls?

## Set 11

1. Mrs. Rogers' class could make 9 small valentines from 1 piece of construction paper. How many could be made from 36 pieces of construction paper?

2. The class had \$13.50 to buy materials for making valentines. They spent \$4.75 for paper, \$3.15 for paste, \$2.50 for glitter, and \$2.65 for ribbon. How much money did they have left?

3. Mrs. Rogers bought 14 yards of ribbon for \$4.48. What was the price per yard?

4. They got 9 jars of paste for \$3.15. What was the price of 1 jar?

5. The class had 3 packages of glitter. Each package could be used on 12 valentines. How many glitter valentines could the class make?

6. The class had $\frac{3}{4}$ yard of red ribbon and used $\frac{2}{3}$ yard. How much red ribbon was left?

## Set 12

1. Allison made 18 valentines. She kept 6 and gave the rest to her classmates. How many did she give to her classmates?

2. One kind of valentine required 8 inches of lace. How much lace was needed for 4 of these valentines?

3. Allison used $\frac{2}{9}$ yard of red ribbon and $\frac{1}{3}$ yard of white ribbon. How many yards was that?

4. Alfredo used $\frac{1}{6}$ yard of pink ribbon for one valentine and $\frac{1}{4}$ yard of pink ribbon for another valentine. How much pink ribbon did he use?

5. Eight students each brought 2 dozen cookies to a valentine party. How many cookies were brought to the party?

6. Three students made 9 quarts of punch. Since there are 32 ounces in a quart, how many ounces should each of the 24 children get?

## Set 13

1. Joan bought $\frac{3}{8}$ pound of pecans and $\frac{3}{4}$ pound of peanuts. How many pounds of nuts did she buy?

2. Laura bought $\frac{1}{2}$ pound of cashews, $\frac{1}{4}$ pound of almonds, and $\frac{2}{3}$ pound of peanuts. How many more pounds of peanuts than almonds did she buy?

3. Kevin bought 12 ounces of peanuts. What fraction of a pound is that?

4. Mike bought $\frac{2}{3}$ pound of pecans for \$1.10 and $\frac{3}{4}$ pound of filberts for \$1.35. How much change did he get from a \$5 bill?

5. Bill bought $\frac{3}{4}$ pound of peanuts. Jennifer bought $\frac{1}{2}$ pound more of peanuts than Bill bought. How many pounds of peanuts did Jennifer buy?

6. Maria bought 6 boxes of mixed nuts with 8 ounces in each box. How many pounds of nuts did she buy?

## Set 14

1. Larry bought $\frac{3}{4}$ pound of filberts at \$1.84 a pound. How much did he spend?

2. Jake bought $\frac{2}{3}$ pound of pecans at \$1.65 a pound. He gave the clerk \$1.50. How much change did he get?

3. Jill bought $\frac{1}{2}$ pound more of peanuts than Jack did. Jack bought $\frac{3}{8}$ pound. How many pounds did Jill buy?

4. Carol bought $1\frac{3}{4}$ pounds of pecans. How many ounces is that?

5. Lacey bought $\frac{3}{4}$ pound of cashews. She ate $\frac{1}{3}$ of what she bought. What fraction of a pound of cashews did she have left?

6. Sue bought $\frac{2}{3}$ pound of pecans at \$1.65 a pound and $\frac{3}{4}$ pound of filberts at \$1.80 a pound. How much change did she get from a \$5 bill?

## Set 15

1. This year 834 people watched the all-city track meet. This was an increase of 106 people over last year. How many people watched the meet last year?

2. The bleachers had 16 rows with 64 seats in a row. How many people would they hold?

3. If 834 people watched the track meet, how many empty seats were there? (See exercise 2)

4. Of the 75 students who competed in the track meet, $\frac{2}{5}$ were girls. How many girls is that?

5. There were 352 adult tickets and 482 student tickets sold for the track meet. Adult tickets sold for $.75 and student tickets sold for $.35. What was the total amount spent on tickets?

6. Sales of drinks totaled $82.65 and sales of popcorn totaled $72.50. Expenses totaled $37.52. How much profit was made on refreshments?

## Set 16

1. Members of the swimming team sold lemonade at the track meet. They sold 92 lemonades for $.45 each. How much money did they get?

2. The times for the 100-meter dash were: Jones, 12.24 seconds; Davis, 12.42 seconds; Weaver, 11.94 seconds; Montez, 11.09 seconds. Who had the best (least) time?

3. In the 100-meter dash, Montez (11.09 seconds) set a new all-city record by 0.18 second. What was the old record?

4. The relay team ran the 4 laps in the 1600-meter relay in 59.3, 60.1, 59.5, and 58.9 seconds. What was the total time?

5. The first- and last-place times in the 200-meter dash were 26.03 seconds and 29.58 seconds. What was the difference in the times?

6. A year ago, Grace ran the 400-meter race in 62.23 seconds. This year she bettered her time by 2.47 seconds. What was her time this year?

## Set 17

**Use the table. Assume each person is an average adult.**

| *Some Weekly Food Requirements For an Average Adult* | |
|---|---|
| cereals and bread | 0.9 to 3.2 kg |
| milk | 4.7 to 5.7 L |
| meat | 1.1 to 2 kg |
| citrus fruit | 1.1 to 2.3 kg |
| leafy green and yellow vegetables | 1.6 to 1.8 kg |
| other fruits and vegetables | 1.4 to 2.7 kg |
| dried beans, peas, and nuts | 57 to 110 gm |

1. Dan drinks 0.7 liter of milk each day. Does he satisfy the weekly requirement of milk?

2. Ellen says she eats 0.06 kilogram of meat on each school day and 0.35 kilogram of meat on each Saturday and Sunday. Does she satisfy the weekly requirement of meat?

3. Barbara eats about 2.1 kilograms of cereals and bread each week. How many kilograms of cereals and bread does she average per day?

4. Patrick eats about 0.4 kilogram of citrus fruit each day. How many more kilograms of citrus fruit than is required does he eat each week?

## Set 18

1. One grapefruit has 1.8 grams of protein. How many grams of protein are in half of a grapefruit?

2. One glass of milk has 17.1 grams of protein. Evelyn drinks four glasses of milk each day. How many grams of protein is that?

3. One slice of enriched white bread has 1.9 grams of protein. One slice of whole wheat bread has 2.1 grams of protein. How many more grams of protein does the whole wheat bread have?

4. A lamb chop has 15.8 grams of protein, a potato has 2.5 grams of protein, and a serving of peas has 3.9 grams of protein. If these foods were a meal, how many grams of protein would be in the meal?

5. One serving of ice cream has 3 grams of protein. One hamburger patty has 24 grams of protein. How many servings of ice cream have as many grams of protein as the hamburger?

6. An apple has 0.4 gram of protein. An orange has 1.4 grams of protein. One day Mike ate 3 apples and 2 oranges. How many grams of protein was that?

## Set 19

1. One car averages 22 miles per gallon at a speed of 35 miles per hour. How far could the car travel at that rate on 15 gallons?

2. A small car averaged 34 miles per gallon at 55 miles per hour. How far could it go at that rate on 12 gallons?

3. A small car traveled 420 miles on 15 gallons of gas at a steady rate of 50 miles per hour. How many miles did it average per gallon?

4. The gas tank of a mid-sized car held 18 gallons and was $\frac{2}{3}$ full. The car could average $22\frac{1}{2}$ miles per gallon in city driving. How far could it travel in the city before it ran out of gas?

5. A small car averaged 25 miles per gallon at 35 miles per hour, and 27 miles per gallon at 45 miles per hour. How much farther could it travel on 12 gallons of gas at 45 miles per hour than at 35 miles per hour?

6. A car averaged 27.4 miles per gallon at 45 miles per hour, and 2.7 miles per gallon more at 55 miles per hour. How many miles per gallon did the car average at 55 miles per hour?

## Set 20

1. How much would 15 gallons of gas cost if the price of gas was $1.26 a gallon?

2. There were 7 gallons of gas in a tank that held $2\frac{1}{2}$ times that amount. How many gallons did the tank hold?

3. There were 6 gallons of gas in a tank that held $3\frac{1}{4}$ times that amount. How many more gallons of gas could be put into the tank?

4. There were $9\frac{1}{2}$ gallons of gas in a tank. $7\frac{3}{4}$ more gallons of gas were put into the tank. How much gas was in the tank then?

5. There were 7.5 gallons of gas in a tank that held 18.5 gallons. 8.7 more gallons were put into the tank. How far from being full was it?

6. The gas tank of a large car held 19 gallons and was $\frac{1}{2}$ full. The car averaged $18\frac{1}{2}$ miles per gallon on the highway. How far could it go on the gas that was in the tank?

## Set 21

1. It is 683 miles from Chicago to Tulsa and 1452 miles from Tulsa to Los Angeles. How far is the entire route?

2. It is 1067 miles from Milwaukee to Jacksonville. A trucker had driven 679 miles. How many miles did he have left to go?

3. A certain truck could haul 15 tons (30,000 pounds). How many 75-pound boxes of canned milk could it haul?

4. The tanks of a certain truck hold 82.5 gallons. One day it took 57.6 gallons of fuel to fill the tanks. How much fuel was in the tanks before they were filled?

5. The license plates cost $13.84 per year, and the insurance costs $87.35 a month. How much does it cost for a license and insurance each year?

6. One year a trucker drove his truck 117,408 miles. How many miles did he average each month?

## Set 22

1. The Kellys have a truck payment of $683.45 a month. How much is that per year?

2. During 12 hours, Mr. and Mrs. Kelly drove their truck 625.2 miles. How many miles per hour did they average?

3. A certain toll road charged them 6.74¢ per mile. How much did they pay for 87.5 miles? Round your answer to the nearest cent.

4. On the return trip they hope to average 6.3 miles per gallon. If so, could they travel 500 miles after filling their tanks, which hold 82.5 gallons of fuel?

5. A trucker paid $1.15 per gallon for 57.6 gallons of fuel. What was the total cost?

6. During the next 235.8 miles the truck used 52 gallons of fuel. How many miles per gallon did it average? Round your answer to the nearest tenth.

# GLOSSARY

**acute angle** An angle whose measure is less than 90°.

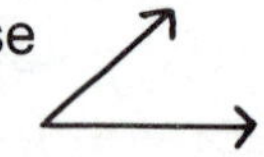

**addend** A number used in an addition problem.

addends 9 + 4 = 13 ← sum

**adding 0 property** The sum of any number and 0 is that same number.

**A.M.** A symbol used for times after 12:00 midnight and before 12:00 noon.

**angle** A figure formed by two rays with the same endpoint.

**average** The average of 4, 5, 5, 7, and 9 is 6. To find the average, add the numbers and divide by the number of numbers.

**BASIC** A computer language used with many microcomputers.

**bisect** To cut into halves.

**Celsius temperature (°C)** The metric temperature scale in which 0°C is the freezing point of water and 100°C is the boiling point of water.

**centimeter (cm)** A unit of length in the metric system. 100 centimeters is 1 meter.

**character** A letter, number, or symbol.

**chord** A line segment with endpoints on a circle.

**circle** A curved figure with all points a given distance from the center.

**circumference** The distance around a circle.

**command** One of the directions used in a computer program.

**common denominator** A common denominator for $\frac{1}{2}$ and $\frac{1}{3}$ is 6, because $\frac{1}{2} = \frac{3}{6}$ and $\frac{1}{3} = \frac{2}{6}$.

**common factor** 2 is a common factor of 4 and 6, because 2 is a factor of 4 and a factor of 6.

**common multiple** 30 is a common multiple of 5 and 6, because it is a multiple of 5 and a multiple of 6.

**compass** An instrument used to draw circles.

**computer** A machine that can count, store information, and process information.

**cone** A solid figure shaped like this:

**congruent figures** Figures that have the same size and shape.

**corresponding parts** In congruent figures, the parts that fit are called corresponding parts.

**cube** A rectangular solid whose flat surfaces are all squares.

**cylinder** A solid figure shaped like this:

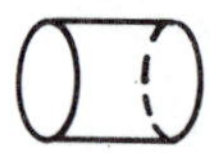

**decimal** or **decimal fraction** A number such as 0.4 (four tenths) and 3.2 (three and two tenths). A decimal is sometimes called a decimal fraction.

**degree** A unit for measuring angles. This is a 1° (1 degree) angle.

**denominator** In $\frac{2}{3}$, the denominator is 3.

**diameter** The distance across a circle through its center.

**difference** The answer to a subtraction problem. $7 - 3 = 4$ ← difference

**digit** Any one of the symbols 0, 1, 2, 3, 4, 5, 6, 7, 8, and 9.

**discount** An amount subtracted from the regular price of an item.

**diskette** A round, magnetic object, used to store computer information.

**display** Any information shown on a computer screen.

**divisor** The number that is divided by. A divisor is also another name for a factor. divisor → $4\overline{)36}$

**equation** A sentence with an equal sign, such as $3 \times 9 = 27$.

**equilateral triangle** A triangle with all sides congruent.

**equivalent fractions** Fractions for the same number. $\frac{1}{2}$, $\frac{2}{4}$, and $\frac{3}{6}$ are equivalent fractions.

**estimate** To use rounded numbers to check whether an answer is correct.

**even numbers** 0 and the numbers that are multiples of 2.

**expanded numeral** A numeral used to show the place value at each digit in a standard numeral.

$$248 = 200 + 40 + 8$$

**factors** Numbers that are used in a multiplication problem.

factors → $8 \times 6 = 48$ ← product

**Fahrenheit temperature scale (°F)** The temperature scale in which 32°F is the freezing point of water and 212°F is the boiling point of water.

**fraction** A number such as $\frac{1}{2}$, $\frac{3}{4}$, and $\frac{4}{6}$.

**gram (g)** A unit of mass (weight) in the metric system. 1000 grams is 1 kilogram.

**graph** A picture used to show some information. A graph can be a bar graph, a picture graph, a circle graph, or a line graph.

**graphics** Information shown in the form of pictures.

**greater than** A comparison of two numbers that are not the same. The symbol is >. For example, $7 > 2$.

**greatest common factor** The largest number that is a common factor of two numbers. 1, 3, and 9 are all common factors of 9 and 18, but 9 is the greatest common factor of 9 and 18.

**grouping property of addition** Changing the grouping of the addends does not change the sum.

$$(4 + 1) + 5 = 4 + (1 + 5)$$

**grouping property of multiplication** Changing the grouping of the factors does not change the product.

$$(4 \times 2) \times 3 = 4 \times (2 \times 3)$$

**hexagon** A plane figure with six sides.

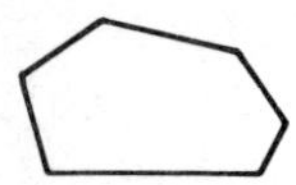

**interest** A payment for the use of money.

**intersecting lines** Lines in a plane that cross at a point.

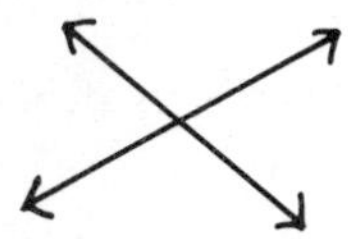

**isosceles triangle** A triangle with two congruent sides.

**kilogram (kg)** A unit of mass (weight) in the metric system. 1 kilogram is 1000 grams.

**kilometer (km)** A unit of length in the metric system. 1 kilometer is 1000 meters.

**least common multiple** The smallest number (not 0) that is a common multiple of two numbers. 6 is the least common multiple of 2 and 3.

**less than** A comparison of two numbers that are not the same. The symbol is $<$. For example, $3 < 8$. (Another comparison is *greater than*.)

**line of symmetry** If a figure can be folded along a line so the two parts of the figure match, the fold line is a line of symmetry.

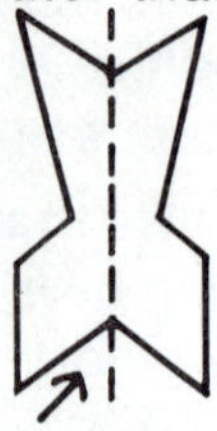

line of symmetry

**liter (L)** A unit of liquid measure in the metric system. 1 liter is 1000 milliliters.

**lowest terms** A fraction is in lowest terms if the only common factor of its numerator and denominator is 1.

**meter (m)** A unit of length in the metric system. 1 meter is 100 centimeters.

**microcomputer** A small, complete computer system that can do one job at a time.

**milliliter (mL)** A unit of liquid measure in the metric system. 1000 milliliters is 1 liter.

**millimeter** A unit of length in the metric system. 1000 millimeters is 1 meter.

**mixed number** A number that has a whole-number part and a fraction part. $2\frac{3}{4}$ is a mixed number.

**multiple** A product. 4, 8, 12, 16, 20, and so on, are multiples of 4.

**multiplying by 1 property** The product of a number and 1 is that same number. $4 \times 1 = 4$

**multiplying by 0 property** The product of a number and 0 is 0. $4 \times 0 = 0$

**numeral** A way of writing a number. These are some numerals for the number 4:

four 4 IV $6 - 2$

**numerator** In $\frac{2}{3}$, the numerator is 2.

**obtuse angle** An angle whose measure is greater than 90°.

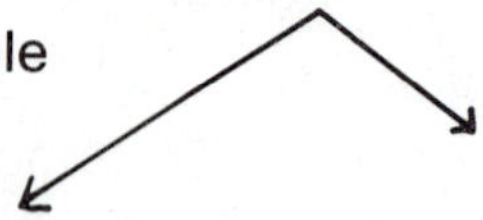

**odd numbers** Numbers that are not multiples of 2. The numbers 1, 3, 5, 7, 9, and 11 are odd.

**order property of addition** Changing the order of the addends does not change the sum.

$$9 + 8 = 8 + 9$$

**order property of multiplication** Changing the order of the factors does not change the product.

$$5 \times 2 = 2 \times 5$$

**outcome** A possible result of an experiment.

**parallel lines** Lines in a plane that never meet and are always the same distance apart.

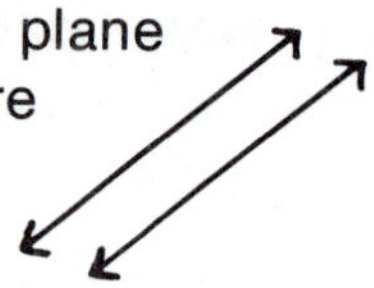

**parallelogram** A four-sided plane figure with opposite sides parallel.

**pentagon** A plane figure with five sides.

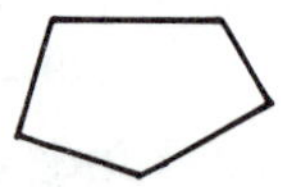

**perimeter** The distance around a figure.

**perpendicular lines** Two lines that intersect to form right angles.

**place value** A system for writing numbers in which the value of a digit is determined by its position.

**P.M.** A symbol used for times after 12:00 noon and before 12:00 midnight.

**prime number** 2, 3, 5, 7, 11, 13, and so on, are prime numbers. They have exactly two factors, 1 and the number itself.

**procedure** A list of commands that is used often. You can use the name of the procedure as a command in another program.

**product** The answer to a multiplication problem. 7 × 8 = 56 ← product

**protractor** A "ruler" for measuring angles.

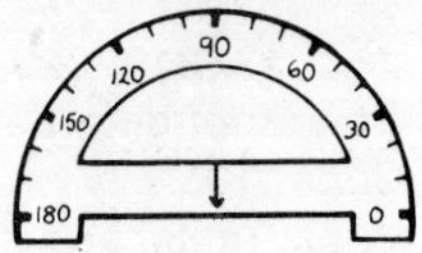


**quadrilateral** A plane figure with four sides.

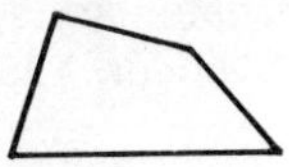

**quotient** The answer to a division problem.

$$8\overline{)56} = 7 \leftarrow \text{quotient}$$

**radius** The distance from the center of a circle to the circle.

**ratio** A comparison of two quantities by division.

**ray** A part of a line that has one endpoint. This is ray *AB*.

**rectangle** A plane figure with four sides and four right angles.

**rectangular solid** A box-shaped solid figure whose flat surfaces are all rectangles.

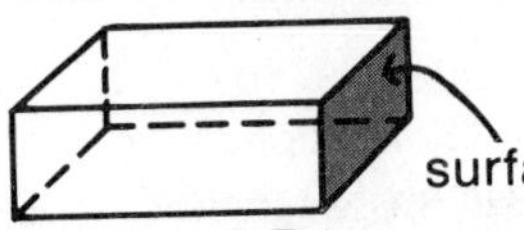


**remainder** The number "left over" after a division.

```
   5 R1
3)16
 -15
   1 ← remainder
```

**rhombus** A parallelogram with all sides congruent.

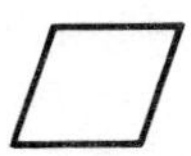

**right angle** An angle whose measure is 90°.

**Roman numerals** A system of writing numbers using letters. For example,

MCDLXIV = 1464

**round** To replace an exact number by another one that is easier to use.

**sales tax** A tax paid when you make a purchase.

**scalene triangle** A triangle with no congruent sides.

**side of an angle** One of the rays that make up the angle.

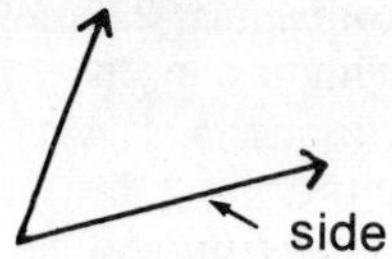


**side of a plane figure** One of the segments that make up a figure.

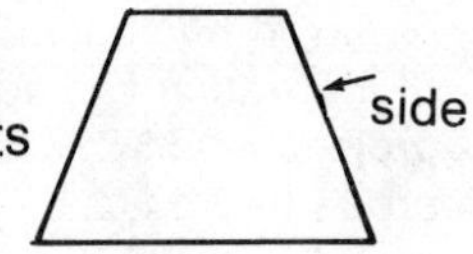


**similar figures** Figures which are the same shape but not necessarily the same size.

**sphere** A solid figure that is the shape of a ball.

**square** A rectangle with all four sides congruent.

**sum** The answer to an addition problem. 23 + 58 = 81 ← sum

**terms** (of a fraction) The numerator and denominator of a fraction.

**total** Another name for *sum*.

**trapezoid** A four-sided plane figure with only two sides parallel.

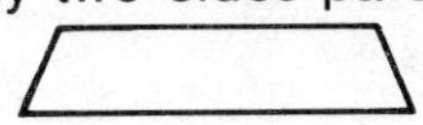

**triangle** A figure with three sides.

**vertex** The point at the "corner" of an angle, plane figure, or solid figure.

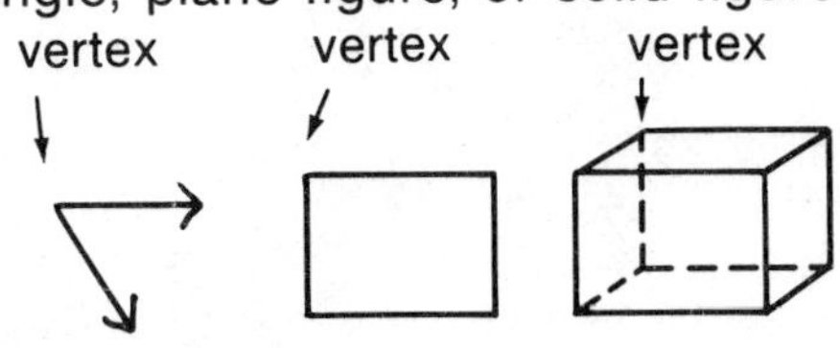


**whole number** Any of the numbers 0, 1, 2, 3, 4, and so on.

# Index


Acute angle, 140, 141
Addend, 24–29, 40, 237
Adding zero property, 24
Addition
  basic facts, 24–27
  of decimals, 236, 237, 240, 241
  estimation in, 32, 33
  of fractions, 174–179, 218
  games, 49, 315
  of larger numbers, 34, 35
  with a missing addend, 25, 29, 39, 40, 343
  of mixed numbers, 308–311, 314, 315, 318
  of more than two addends, 28, 29, 36, 48
  project for, 55
  properties, 24, 28
  of 3-digit numbers, 32, 33
  of 2-digit numbers, 30, 31
  zero in, 24, 240
A.M., 252–255
Angles, 136–141, 145, 148, 149, 153, 159
Area, 13, 262, 263, 274, 275, 283
Average, 114, 115

Bar graph, 27, 39, 55, 74, 101, 115, 197, 247, 321
Base, 262
Bisect, 145

Calculators, 15, 57, 93, 277, 281, 336
Center of circle, 144
Centimeter, 256–265, 267, 331
Checking accounts, 47
Chord, 144
Circle, 144, 145, 306
Circle graph, 179, 198, 223
Circumference, 261
Common denominator, 172–185, 308, 310, 312
Common factor, 102
Common multiples, 64, 65
Compass, 145
Computers
  byte, 89
  command, 346, 348, 350
  decision, 354, 355
  input/output, 43, 107, 318
  history of, 17
  memory, 89
  procedure, 350–353
  program, 346–355
  uses of, 47, 73, 242, 297
Cone, 135
Congruent figures, 146–149
Consumer mathematics, 31, 41, 43, 45, 47, 50, 51, 53, 71, 75, 81, 87, 91, 107, 111, 113, 119, 121, 124–127, 189, 210, 211, 219, 235, 239, 241, 244, 245, 279, 285–304, 313, 316, 317, 319, 329, 333, 338, 339
Corresponding parts, 148, 149
Cube, 135
Cubic units, 264, 265, 274, 275
Customary system, 270–279
Cylinder, 135

Decimals
  addition, 236, 237, 240, 241, 249, 343
  comparing and ordering, 229, 232, 233
  division, 332–335, 343
  estimating, 336, 337
  game, 243, 337
  multiplication, 326–331, 343
  place value of, 228–231
  regrouping, 236–239, 332–335
  rounding, 234, 235
  subtraction, 238–241, 249, 343
Degrees, 138, 140, 141, 223
Denominator, 164, 166, 168, 170, 172, 174, 176, 178, 180, 182–184, 192, 208, 214, 306, 308, 310, 312
Diagonal, 161
Diameter, 144, 145, 271, 316
Difference, 38–41, 43, 46, 180, 182, 184, 193, 238, 239, 249, 308, 309, 312, 314
Discounts, 292, 293
Distributive property of multiplication, 80
Divisibility, 131
Division
  basic facts, 98–101
  check, by multiplication, 120
  of decimals, 332–335
  estimation in, 112, 113, 117, 118
  as finding a missing factor, 98, 99
  of fractions, 216, 217
  game, 123
  of greater numbers, 110–113, 120, 121
  with regrouping, 108–113, 116–121
  with remainder, 98, 99, 108–113, 116–121, 123
  by 2, 3, 4, or 5, 131
  by a 2-digit number, 116–119
  of a 2-digit number, 106, 107
  zero in, 103, 112, 113
Divisor, 98, 102, 118

Egyptian numerals, 21
Endpoint, 135–137, 144

Equilateral triangle, 153
Equivalent fractions, 166–172, 184
Estimating, 32, 33, 42–44, 68, 69, 84–87, 112, 113, 116, 117, 336, 337
as a check, 42, 43, 68, 69, 87
in measurement, 247, 259, 261, 268, 269, 281, 321, 341
Even numbers, 65, 103
Expanded numerals, 2, 3, 5

Factors, 60, 83, 98, 99, 102, 103, 129, 168, 170, 328
Fractions
addition of, 174–179, 188, 193, 218, 308, 310
comparing, 172, 173, 187
division of, 216, 217
equivalent, 166–172, 184
game, 171
in lowest terms, 168–171
and mixed numbers, 190–193, 210, 211, 306, 307, 323
multiplication of, 212–215, 218
of a number, 206–209, 213, 214, 223, 292, 293
of outcomes, 196, 197, 201
projects for, 201, 223
and regions, 164, 166, 167, 172, 174, 180, 192, 193, 206, 212–214, 228
and sets, 164, 165, 206, 208, 216
subtraction of, 180–185, 188, 193, 218
terms of, 164

Geometry, 134–162
Gram, 269, 331, 341
Graphing, 27, 39, 55, 63, 74, 101, 115, 154–157, 179, 197, 198, 223, 247
projects for, 55, 101, 129, 223, 247, 321
Greater than, 6, 7, 68, 172, 173, 232, 235
Greatest common factor, 102, 103, 168, 170
Grid, 154, 155
Grouping properties, 28, 60
Grouping symbols, 48, 122, 336

Hexagon, 135
Hundredths, 228, 230, 231, 236, 242, 258, 334

Inch, 270–275, 278, 279, 316
Interest, 303
Intersecting lines, 143
Inverse operations, 40, 98, 99
Isosceles triangle, 153

Kilogram, 269, 331, 333
Kilometer, 258, 259, 267, 331, 333

Least common multiple, 64, 65, 178, 184
Less than, 6, 7, 68, 172, 173, 232, 235
Line, 136, 142, 143
of symmetry, 150, 151, 153
Linear measurement, 256-261, 270–273
Line graph, 63, 156, 157
Liquid measurement, 268, 269, 276, 281
Liter, 268, 281, 331
Lower and lowest terms, 168–170, 176–178, 181–186, 192

Mean, 114, 115
Measurement, 251–284
projects for, 268, 269, 281, 321, 341
Measuring angles, 138–141, 223
Meter, 258, 259, 261, 263, 265, 267, 331
Metric system, 256–269, 281, 331
Milliliter, 268, 331
Millimeter, 256–258, 261, 263, 265
Mixed numbers, 190–193, 210–211, 223, 306, 307
addition and subtraction, 308-315, 318
multiplication, 323
Multiples, 64, 65, 176, 182
common, 64, 65, 178, 184
Multiplication
basic facts, 60–63
as check for division, 120
estimation, 68, 69, 84–87
of decimals, 326–331
of fractions, 212–218
game, 83
of larger numbers, 70–72, 88
of mixed numbers, 323
by multiples of 10 and 100, 76–79
projects for, 93, 223
properties of, 60, 80
with regrouping, 66–72
of 2- and 3-digit numbers, 66–69, 72, 80–88
zeros in, 60, 86–88
Multiplying by 1 property, 60
Multiplying by 0 property, 60

Number line, 232, 234
Number pair, 154–156
Numbers
comparing, 6, 7, 104, 172, 173, 232, 233
games, 243, 315
patterns, 57, 95, 167
Numeral, 16, 17
Numerator, 164, 166, 168–170, 174, 180, 192, 208, 214, 306

Obtuse angle, 140, 141
Odd number, 65, 103

Order properties, 24, 36
Outcome, 196, 197, 201

Parallel lines, 142, 143, 153
Parallelogram, 152
Pentagon, 135
Percent, 294–296
Perimeter, 260, 261, 271, 274, 275, 311
Perpendicular lines, 142, 143
Place value, 1–21, 228–231
Plane, 134
Plane figures, 134, 135
P.M., 252–255
Point, 134
Prime number, 129
Probability, 196, 197, 201, 225
Problem solving
  choosing the operation, 104, 105
  drawing pictures, 266, 267
  estimating in, 77, 89
  extra, 379–389
  using a graph or table, 7, 11, 13, 27, 33, 39, 48, 55, 74, 75, 77, 89, 114, 115, 125, 129, 156, 157, 179, 185, 195–198, 219, 220, 223, 233, 242, 259, 294–296, 298, 299, 301, 303, 317, 323
  using maps, 37, 69, 122, 199, 244, 314
  making up a problem, 245
  more than one way to solve, 338, 339
  not enough information, 126, 127
  organizing and interpreting data, 55, 129, 156, 157, 223, 247, 321
  finding patterns or rules, 19, 57, 65, 71, 77, 95, 135, 139, 161, 167, 283, 329–331, 335
  planning what to do, 50, 104, 157
  reading a floor or building plan, 278, 279, 316, 317
  steps for, 50, 52, 104, 126, 266, 328
  too much information, 52, 53
  two or more steps, 90, 91, 105, 338, 339
Product, 60–62, 68–71, 76, 81–88, 178, 212, 214, 215, 327, 329, 330
Protractor, 138–140, 223

Quadrilateral, 152, 153
Quotient, 98, 99, 106, 112, 113, 117, 120, 335

Radius, 144, 145, 223
Ratios, 194, 195
Ray, 136, 137, 140
Rectangle, 135, 141, 152, 212, 261–263, 274, 275
Rectangular solid, 135, 264, 265, 274, 275
Remainder, 98, 99, 108, 118, 120, 123
Rhombus, 152
Right angle, 140–142
Roman numerals, 16, 17
Rounding, 8–11, 15, 32, 33, 42, 43, 68, 69, 118, 119, 234, 235, 247, 288, 290, 292

Sale price, 298, 299
Sales tax, 294–296, 298, 299
Scalene triangle, 153
Segment, 134, 135, 137, 142, 143, 145, 148, 149, 155, 159, 161, 270
Side of a plane figure, 134–137, 148, 161, 260, 261
Similar figures, 146
Simplest form, 192, 193, 202, 214, 215, 307, 312, 314, 318, 323
Sphere, 135
Square, 134, 152, 212, 261
Square unit, 262, 263, 274, 275
Subtraction
  basic facts, 38, 39
  checking, 40
  of decimals, 238–241, 243
  estimation in, 42, 43
  of fractions, 180–185, 188, 218
  game, 49, 315
  of larger numbers, 46
  of mixed numbers, 308, 309, 312–315, 318
  of 2- and 3-digit numbers, 40–43
  zeros in, 44, 45
Sum, 24–27, 32, 33, 174, 176, 179, 193, 203, 236, 249, 309–311, 314
Surface area, 283
Symmetry, 150, 151, 153

Tenths, 228–231, 235, 236, 249, 332
Terms, 168–170
Thousandths, 230, 231, 249, 334
Time, 67, 93, 207, 209, 210, 242, 247, 252–255, 319
Total, 31, 239, 289, 291, 294, 295, 299
Trapezoid, 152
Triangle, 135, 137, 141, 148–150, 152, 153, 260–263

Vertex, 135, 136
Volume, 264, 265, 268, 274–276, 329

Weight, 269, 277, 323, 341
Whole number, 190–193, 234, 235, 308

Zero, 44, 45, 86, 103, 112, 113, 228, 240, 330, 334
  properties, 24, 60




2 3 4 5 6 7 8 9 0